QUARKS,

QUASARS,
and
QUANDARIES

Proceedings of the Conference on the Teaching of Modern Physics
held at Fermi National Accelerator Laboratory

April 24th through 27th, 1986

The Conference on the Teaching of Modern Physics
was partially supported by the
National Science Foundation
under grant number TEI-8550610,
Leon Lederman and Jack M. Wilson,
Principal Investigators.

Edited by
Gordon J. Aubrecht II

American Association of Physics Teachers
College Park, MD

Published by
American Association of Physics Teachers
5112 Berwyn Road
College Park, Maryland 20740

© 1987, American Association of Physics Teachers
Beth Harder, Managing Editor

Production Staff
 Karen S. Garvin, Managing Editor; Design and Typesetting
 Bennett S. Garvin, Production Assistant

Cover Design: Angela Gonzales, Fermilab

ISBN #0-917853-26-1

Contents

Preface

This book is the Proceedings of Quarks, Quasars, and Quandaries: Conference on the Teaching of Modern Physics, which was held at Fermilab from 24 to 27 April 1986. A total of 135 physicists were in attendance: 52 high school teachers; 58 college and university teachers, including the AAPT delegation; 10 university teachers from the Latin American countries Argentina, Brasil, Chile, Mexico, and Venezuela; and 15 Fermilab physicists.

The aim of the Conference was to develop materials and strategies for the introduction of topics from modern-day physics into high school and introductory college classes. The lectures printed here represent attempts to describe the development of elementary particle physics, both in its experimental and theoretical aspects, and to give participants some appreciation of the impact of general relativity and cosmology on current scientific work. The participants spent four days working: Listening to the plenary lectures was followed by crafting of materials which we hope will be useful in the classroom.

The spirit of the late Roman Ulli Sexl, Professor at the University of Vienna and chairman of the International Commission on Physics Education, who fathered the first Conference on the Teaching of Modern Physics at CERN, lives on through this later evocation at Fermilab. We will miss his advice and enthusiasm for physics.

The Conference could not have succeeded without the hospitality of Fermilab, and the dedicated work by Leon Lederman, its Director. The work of Jack M. Wilson, Executive Officer of the American Association of Physics Teachers, was essential for the success of the preparation of the Conference.

Drasko Jovanovic, Director of Program Planning at Fermilab, and Judy Zielinski, who was coordinating the conference for Fermilab and acted as a general factotum for Drasko and me, exerted amazing efforts on behalf of the Conference. They deserve special thanks. The effort would not have succeeded without the Fermilab staff who labored so mightily on our behalf: Stanka Jovanovic (president of Friends of Fermilab), Marge Bardeen (who helped coordinate the materials production), and Marilyn Smith, Pat Hatcher, Lisa Lopez, Gladys Pool, and Eva Williams, who typed, collated, and assisted cheerfully in any way they could. Fermilab physicists Charles Brown, David Brown (visiting from Harvard), Richard Carrigan, Carlos Hojvat, Richard Holman, Rolland Johnson, Drasko Jovanovic, Leo Michelotti, Stephen Parke, Stephen Pordes, and Terry Walker gave generously of their time to assist the groups in their attempts to come to grips with the content of modern physics. They also acted as resources during the subgroup meetings and assisted greatly in interpreting particle physics and cosmology and translating some ideas from lecture into prototype materials. Many were scheduled to meet only one or two times with the groups, but became habituees. Thank you for all your help.

The conference was conducted by the American Association of Physics Teachers, with major grant support from the National Science Foundation (#TEI-8550610). The American

Physical Society and the National Science Foundation gave joint grant support. The International Commission on Physics Education (ICPE) of the International Union of Pure and Applied Physics, sponsored the Conference and cooperated with the AAPT and Fermilab in its planning. Fermilab itself (supported by the Department of Energy) contributed substantial monetary support. Through the hard work of Stanka Jovanovic (president of the Friends of Fermilab) and the Friends of Fermilab, several other organizations contributed to the conference as well. This support was particularly essential in helping the physicists from Latin America to attend. These other contributors are the Brooks and Hope B. McCormick Foundation, AT&T, George J. Ball, Inc., W. W. Grainger, Inc., and Albert H. Ramp Architects/Engineers Inc.

I would like to thank everyone who participated or contributed to this volume. I most especially want to thank Jack Wilson for giving this job to me and for his essential support during the time it was happening (even when I was at O'Hare at a time I'd said I'd be at Midway). The views presented here represent those of the authors and are not official positions of the National Science Foundation, the Department of Energy, or any other organization connected with this Conference.

Gordon J. Aubrecht II

In memoriam

Roman U. Sexl,
Chairman, International Commission on Physics Education
Originator of the Conferences on the Teaching of Modern Physics

Opening Session

Welcome

Leon Lederman
Director, Fermi National Accelerator Laboratory
Batavia, Illinois 60510

It is traditional for the Director to "welcome" each assembly that takes place at Fermilab. Since we have two or three million assemblies every year, it is not too easy to say anything unique, original, fresh or memorable. Most of my welcomes are duplicative, stale and forgettable.

However, this is not a routine assembly aimed at illuminating some arcane aspect of mankind's attempt to understand and control nature. This is an assembly of some of the most outstanding teachers in the nation, teaching the most fundamental of all disciplines—physics. What is more, we are gathered here to address the issue of waning enrollments in hard science at a time when the need for scientists and for public understanding of science has never been more urgent.

This particular assembly is even more sharply focused—we are not here to debate the philosophical, socioeconomic issues in science education but we are here to consider one very well defined approach—that is to rejuventate the traditional first year courses in physics by embedding somehow, by hook or crook, aspects of modern physics which will quicken the pulse, make the book race, shorten the breath and in general, present to our innocent students, a subject of irresistible interest, an unforgettable experience which, if it doesn't convert them all to physics majors, will at least leave them with a sense of the power and the excitement of a subject, alive with profound and interesting things relevant to our culture and to our comfort.

Well! with this as our charge, it is clear why this conference is different from all other conferences and why I want to give you a very special welcome, way beyond the call of duty. I really do want this to be a spectacularly successful meeting and if there is anything the laboratory can do to insure this, and if we haven't already thought about it, you have only to ask. Welcome!

Welcome To The Conference On The Teaching of Modern Physics

Jack M. Wilson
American Association of Physics Teachers
Department of Physics
University of Maryland
College Park, MD 20742

On behalf of the American Association of Physics Teachers I would like to welcome you to the Conference on the Teaching of Modern Physics. This Conference is sponsored jointly by the American Association of Physics Teachers and Fermi National Laboratory. Funding for the Conference comes from AAPT, Fermilab, and the National Science Foundation. The Conference is the second in the series of such conferences that originated with the International Commission on Physics Education, a commission of the International Union of Pure and Applied Physics.

The Chairman of that commission, Professor Roman Sexl of the University of Vienna, began the series with a similar conference held at CERN in Geneva, Switzerland in 1985. ICPE hoped to stimulate similar conferences in other countries. A.P. French, President of AAPT, and I were invited to the CERN Conference to explore the possibility of a follow-up conference in the United States. Upon our return we met with Dr. Leon Lederman, the Director of Fermilab, who embraced the idea enthusiastically.

Many of us at the American Association of Physics Teachers feel that it is time to reexamine *what* we teach in physics and the *way* we teach physics in the light of research into the learning process, in the light of the new educational technologies available, and in the light of the changes in physics itself.

If one were to resurrect a 19th Century physicist and transport him into a introductory physics classroom today he would be likely to feel quite comfortable with most of the material being presented in the classroom. If, on the other hand, you put that same physicist in the middle of a typical research conference, he would recognize little of what was being discussed. There is an enormous disparity between the way physics is done and the way physics is taught. The introductory physics course has remained static over several decades and has really changed little over the last fifty years. It is this issue that is of primary concern to us at this conference.

We have brought together approximately 100 teachers of introductory physics drawn from both the high schools and the universities. We will have some of the leading research physicists in the world working with the group. The teachers of physics, through interactions with research scientists, may be able to develop some ideas for how the teaching of introductory physics might be improved—through integration of modern physics topics, or through a radical restructuring of the physics course.

Because of the breadth of modern physics and the short time available to us at this meeting, we felt the need to focus on some small subset of topics. It seemed particularly appropriate, given our location, to focus on particle physics and cosmology. Plans are forming for later meetings to consider topics in solid state physics or materials science. We have also invited a number of physicists from Latin America to attend in the hope that this conference could stimulate a similar conference in Latin America, much as the CERN Conference led to this conference.

From this conference we will develop a collection of proceedings and a collection of materials developed by the teachers and professors in attendance. Material developed at this meeting will be revised, class tested, and expanded during the coming months and discussed at a follow-up meeting in San Francisco to be held in January of 1987.

We recognize that this is not a simple task we have undertaken. Many have tried to restructure the introductory physics curriculum without success. Various techniques have been the subject of experimentation, with varying levels of success. We are unlikely to solve the problems of introductory physics courses in a short conference such as this, but we hope that the conference will stimulate some of the activities that will be required to work towards a long term revision of introductory physics courses.

It is always difficult to follow Leon Lederman because he always knows twice as many physics jokes as anyone else, but let me leave you with a short description of the character of physicists. There once were three individuals who faced execution by guillotine: a physicist; a lawyer; and a businessman.

The businessman was the first one brought to the guillotine. He was asked if he had any last wishes, and whether he wanted to go to the guillotine face up or face down. The businessman replied "face down," and took his place at the guillotine. The executioner hauled the blade into the air and pulled the tripwire. The blade fell a short distance and jammed. Since tradition required that the criminal go free in such cases, the businessman was spared execution and released.

The lawyer was brought forward next and was asked the same question. He replied he wished to go to the guillotine face down, and he took his place as well. The executioner looked over the guillotine to determine that it was in proper working order and hauled the blade to the top once again. When he pulled the tripwire the blade once again jammed, and they were forced to release the lawyer.

After a flurry of activity to repair the guillotine, the physicist was brought forward. When they asked the physicist how he wanted to go to the guillotine, he replied "face up." He was led to the guillotine and took his place. The executioner hauled the blade to the top of the guillotine. As he prepared to pull the tripwire, the physicist blurted out "Wait! Wait! I think I see the problem."

I hope that you will leave this conference looking at the problems of physics education face up.

Welcome from the International Commission on Physics Education

E. Leonard Jossem
Secretary, International Commission on Physics Education
Department of Physics
Ohio State University
Columbus, OH 43210

I am very pleased, as Secretary of the International Commission of Physics Education of the International Union of Pure and Applied Physics to have the opportunity to welcome you to this conference. I am also very sorry that the Chairman of the Commission, Professor Roman Sexl, is not here to welcome you himself. Unfortunately, he is gravely ill and is unable to attend.

I would like to follow the example set by Leon Lederman and by Jack Wilson and also tell a story. This one concerns a physics teacher who came home one evening and was asked by his wife, "How did things go today?" "Well," he replied, "It was very interesting. I explained a very important part of physics to the students today, and when I was finished, they said they didn't understand it, and would I explain it again. So I did, but they said they still didn't understand it. So I explained it a third time, and then I understood it." This just tells us what we all know, that the best way to learn a subject is to have to teach it, to have to explain it clearly to someone else.

We hope that this conference will be useful to you in this continuing process of learning and teaching, and that it will stimulate the invention of new ideas and strategies of presenting the content of physics to our students. On behalf of the I.C.P.E., let me wish you all success in your work. Thank you.

Orientation for the Conference on the Teaching of Modern Physics

Gordon J. Aubrecht II

American Association of Physics Teachers
Department of Physics
University of Maryland
College Park, Maryland 20742
(On leave from the Ohio State University)

I am happy to be able today to extend a warm welcome to you all, especially those of you from Latin America who have had so far to travel. We are indebted to Fermilab for the use of these fine facilities, to Professor Lederman for his hospitality, and to the fine staff for all the work they have done so far and for all that you will cause them to do. I would especially like to mention Professor Drasko Jovanovic of Fermilab, who has been working with me deciding on the organizational aspects of this Conference, and Judy Zielinski, one of the beautiful women at the registration desk, who has done a fine job watching out for the details in preparing for the Conference with me. I hope you will have a chance to see some of this lab, one of America's premier high energy physics research installations, while you are here.

To begin this Conference, it is appropriate to review what has come before. You have heard greetings from the distinguished Director of Fermilab, Professor Leon Lederman, from the Executive Officer of the American Association of Physics Teachers, Professor Jack M. Wilson, and from the Secretary and acting Chairman of the International Commission on Physics Education, Professor E. Leonard Jossem. Now you will be getting from me "greetings" more of the ilk of that which comes unbidden in the mail. After a bit of history, I will describe the organization of the next four days.

THE HISTORY OF THE CONFERENCE

This conference had its roots in an earlier one held at CERN in September 1984. The CERN conference was an experiment conceived by the late Professor Roman Sexl, chairman of the International Commission on Physics Education (ICPE) of the International Union of Pure and Applied Physics. Professor Sexl and colleagues on the ICPE and at CERN organized the conference to explore possibilities that might follow from bringing together experts in a research area and physics teachers from high schools and colleges. The organizers hoped that interaction at the conference would promote mutual understanding and produce materials suitable for inclusion in introductory physics classes. One of the lessons learned from the experiment was the importance of allowing enough time for interactions to occur, for ideas to be absorbed, and for new ideas to arise.

At the conclusion of the CERN conference, there was a discussion between members of the Commission and A. P. French, W. C. Kelly, and J. M. Wilson of the AAPT on the possibility of having the next such conference in the U.S. Fermilab was suggested as a potential site. After this delegation returned, Leon M. Lederman, the Director of Fermilab, was approached about the idea. Professor Lederman, who has involved Fermilab heavily in science education, was enthusiastic and offered the use of Fermilab facilities for the conference. A steering committee was set up to consider plans for the conference and a proposal to provide support for fifty high school teachers to attend the conference was submitted to the National Science Foundation on behalf of AAPT.

The first steering committee meeting was in March of 1985, and the committee met again in October of 1985 and in January of 1986. During this period, Drasko Jovanovic, Director of Program Planning at Fermilab, and Gordon Aubrecht, Visiting Fellow at AAPT, were designated as conference coordinators/directors on behalf of the respective organizations.

The steering committee decided that this Conference would concentrate on the teaching of particle physics and cosmology, as the CERN Conference before it did. The committee chose the topics on which the lectures would be given and the lecturers you will soon begin meeting. It was also decided to hold the conference partly over a weekend. This would minimize the time that teachers needed to be away from their classes and make it easier for them to obtain the necessary administrative approval to attend. We decided that we wished to make a great deal of time available during the Conference for discussion. The steering committee also decided to have a followup to this Conference take place sometime after our meeting here. Production and testing of teacher-generated materials will take place between now and the followup at the AAPT/APS joint meeting in San Francisco in January, 1987. This followup will focus on coordination of the materials produced and should stimulate further work.

HOW WE ANTICIPATE THIS CONFERENCE WILL WORK

The meeting is organized around the five plenary lectures, on cosmology, on accelerators and detectors, on grand unified theories and elementary particles, on symmetry, and on observational tests of general relativity, as well as the plenary roundtable on innovative ideas in teaching modern physics, the international discussions session, and the poster session. You already have been assigned to one of the interest groups on the basis of your choice of topic. The breakup into the interest groups represents our attempt to assist the production of varied classroom materials on the five topics represented at this conference.

The interest groups will be chaired by the following people: Cosmology, James Ruebush; Accelerators and Detectors, William Conway; GUTs and Elementary Particles, Walter Schearer; Symmetry, Ward Haselhorst; and Observational Tests of General Relativity, JoAnn Johnson. You will be meeting with your interest group in the same room during the entire conference immediately after the plenary lectures. Each interest group will discuss each plenary lecture after it has taken place. The purpose of breaking you up into interest groups is to facilitate the production of varied materials on the five topics represented at this conference; we expect you to keep your interest topic at the back of your minds as you discuss the plenary lectures. During your group meeting, your chairman will give you any necessary further instructions.

The interest groups have been subdivided into six subtopic subgroups within each topic interest group: lecture presentations; demonstrations; software and audiovisuals; evaluation instruments; experiment and laboratory activities; and homework problems. The subgroups will be set up during the first discussion session and will remain as a subgroup during all subsequent discussions.

During lunch, you will probably want to continue your pre-lunch discussions within your interest groups. Simply choose your tablemates accordingly. During supper, we will try to have you sitting with other members of subgroups like yours from other topic interest groups. We have put time in after supper for these discussions to continue on an informal basis Thursday and Friday nights. On Saturday night, these subgroups will be working together after the dinner to prepare for the reports to be made Sunday

Each subgroup has been assigned a color so we can color-code the reports of each subgroup meeting. The colors are, respectively, sand, blue, pink, gold, green, and red. EACH SUB-GROUP WILL BE EXPECTED TO PRODUCE A REPORT FROM EACH TWO-HOUR MEETING. These reports will be delivered to the chairmen of your topic interest group. The chairs will then deliver these to the typists, who will be typing the reports for inclusion into the summary report, which will be distributed Sunday.

The subgroups will discuss the choice of topics as well as effective strategies to bring the materials into high school and beginning undergraduate courses. We expect that the ideas in this report will serve as the basis of a concerted effort by the conference participants to develop materials for use in the classroom.

Each subgroup will select one person to represent the subgroup during the summary report session. The five selected for, say, demonstrations will prepare their report Saturday evening with the help of their colleagues from their subgroup. Each group of five persons will provide an overview of the recommendations made by all subgroups working on that topic (e.g., demonstrations) for the rest of the participants. Each group of five people will have about 20 minutes to give the flavor of the recommendations to all the participants, for a total summary time of two hours.

HOW YOU WERE SELECTED

Approximately one hundred physics teachers, half from high schools and half from colleges and universities, were selected to take part in the conference. We have also invited a dozen Latin American physicists to Fermilab. They will, we hope, be responsible for organizing a further conference in this series. Another ten of us are here representing AAPT and helping run the Conference. In addition, a group of Fermilab physicists has volunteered to assist you during the group meetings. You will be meeting them soon.

Support was available for the high school teacher participants under a special NSF grant to AAPT. The support covered the cost of travel both to Fermilab and will cover the cost of the trip to San Francisco, local transportation, meals, lodging, and the conference fee. College and university participants were expected to obtain funding from their home institutions.

To inform possible applicants, we distributed some 25,000 letters outlining the Conference goals and containing an application form. We believe we reached most secondary science teachers in the country. In addition to the letters, the announcement of the Conference and an application form were printed in the *AAPT Announcer*. Posters were sent to all colleges and universities with a program in physics. We sent letters of announcement to all Latin American physicists on our list, which contains the names of members and of other physicists whose names have been added on the recommendation of local physicists knowledgable about the state of physics teaching there.

In the end, we had about 460 applications, from which we could accept only about 110 (including the Latin American invitees). We chose participants on the basis of their essays, in which they told us how they could contribute to the goals of the conference and how the conference could help them in their teaching. The high school participants also submitted letters from a colleague and a supervisor, which were used as well. It has turned out that about ten percent of you had some prior knowledge of one of the topics to be discussed here.

In the acceptance letter, I asked you to "prepare a preliminary version of a lesson, test, short textbook chapter, class activity, or other document or activity of your choice on one of the five topics to be covered at the conference." There were two main reasons I asked you to create whatever material you wanted to submit. The first was to help get you thinking about this material before the conference, since very few of you are currently teaching any of the topics. The second was to allow the exchange of ideas before the conference, to set the stage for further discussion at the conference. The material you submitted was gathered and bound in a book[1], which we sent out several weeks ago. These papers may prove to be useful to you in the course of the discussion group meetings.

WHERE YOU ARE GOING

We do not know how to teach modern physics topics effectively in introductory classes. If we did, you would not have to be here pondering how to do it. We have no magic formulas, no nostrums, no specific ways to recommend. We are confident that you are here to work hard to try to develop materials for the classroom. We do not know if you will succeed. In fact, we will not know about the overall success in producing materials until quite some time—measured in years—has elapsed. Testing and review will have to be done.

Short term, there will be some results we can count on. We are videotaping this conference. The videotapes will be edited, and versions of the plenary sessions prepared, with additional material from Fermilab. AAPT plans to make these tapes available to interested parties essentially at cost.

We are going to publish the plenary lectures and additional material in a volume of Proceedings which you will all be getting at some time in the future. Every effort will be made to get you the transparencies from each talk soon after the talk.

We hope that involvement of other teachers in curricular change will result from the materials the conferees generate, from the publication of the Proceedings itself, and from the production of videotapes of the plenary lectures.

We have a full four days set for you. Enjoy the physics. Think carefully and concretely about how to move this material into your classroom as you listen and, more importantly, discuss the content of the lectures. Learn from each other, so you can teach us all.

Reference

1. G.J. Aubrecht, ed., *Papers for the Conference on the Teaching of Modern Physics,* AAPT, 1986.

Plenary Lectures

Victor F. Weisskopf, Institute Professor and former head of the Department of Physics at the Massachusetts Institute of Technology, is widely known for his theoretical work in quantum electrodynamics, the structure of the atomic nucleus, and elementary particle physics.

A naturalized United States citizen since 1943, Dr. Weisskopf was born in Vienna, Austria, in 1908. He came to the United States in 1937 to join the faculty at the University of Rochester where he was instructor (1937–39) and assistant professor (1939–45). In 1943, he joined the Manhattan Project at Los Alamos, New Mexico, where he worked as associate head of the theory division.

In 1945, he was appointed professor of physics at the Massachusetts Institute of Technology.

Weisskopf was actively engaged in the rehabilitation of natural sciences in Europe after World War II and aided in the planning of the European international laboratory CERN in Geneva, Switzerland. In 1961, Dr. Weisskopf became Director-General of the European Center of Nuclear Research (CERN), in Geneva for 5 years, heading an international research establishment that operated the world's second most powerful large particle accelerator. Under his leadership CERN developed into one of the most successful research institutions.

Upon his return to M.I.T. in 1966, he was given the rank of Institute Professor, an honor bestowed sparingly by M.I.T. in recognition of faculty members of great distinction. In 1967 Weisskopf was appointed head of the Department of Physics, a position he held until his retirement in 1973.

Professor Weisskopf is the author of more than 200 papers on nuclear physics, quantum theory, radiation theory, science policy and nuclear disarmament. A collection of his essays appears under the title "Physics in the XX Century." He wrote *Theoretical Nuclear Physics* together with John M. Blatt (1952) and his book, *Knowledge and Wonder: The Natural World as Man Knows It* (Doubleday & Co., 1962; 2nd edition, M.I.T. Press, 1979), written for the intelligent layman, was selected by the Thomas Alva Edison Foundation as the best science book for the year for youth. The first volume of *Concepts of Particle Physics* with K. Gottfried appeared in 1984 (Oxford University Press); the second volume in 1986.

Professor Weisskopf is a member of the National Academy of Sciences and as associate of many foreign academies. He was elected as a member of the Pontifical Academy in 1978.

Qualitative Physics

Victor F. Weisskopf
Department of Physics
Massachusetts Institute of Technology
Cambridge, MA 02139

Prelude—The Joy of Insight

I would like to tell you how glad I am that you asked me to speak here to open this meeting. I am very much aware how important this is.

Before I talk about the subject, namely qualitative physics, subtitled "In Search of Simplicity" (some of you may have perhaps seen short sketches last year in the *American Journal of Physics*), I would like to make a few remarks about the problems which we face. I do not mean the scientific ones, but rather the cultural ones, you may call it political ones.

It is really true that, especially for our high school youngsters, education sorely needs improvements. This is one of the reasons why we are here today. I would like to call this the real window of vulnerability of our country. It is where we are most vulnerable but, unfortunately, much less is done about that window of vulnerability than about the imaginary one in weapons.

You read arguments for improvement in education in many, many journals so as to keep our competitive edge in the world markets. We have lost the competitive edge because much of our effort was directed toward weapons construction. Nevertheless the re-establishment of our competitive position may be a useful by-product of good education.

Another argument is that education is necessary to make our people able to understand the political, social, or technical situation in this world, in which so much science and technology is involved. That is a better argument, but still it should be a by-product of our efforts and not the main aim.

Perhaps the best argument I have found said "to contribute to a more interesting and cultural life by instilling a deeper awareness of what we see around us in nature." That is the true purpose of a scientific education. After all, science is part of our culture. Science is the pleasure to see, to understand, and to admire the world around us. I like to call this the joy of insight. It is the sense of wonder about nature.

Indeed, the joy of insight is something very important. I myself must say, if I look back at my life as a scientist and a teacher, I think the most important and beautiful moments were when I say, "ah-hah, now I see it a little better," and it is not necessarily when I myself have done something. When I hear a seminar or when I hear a good speaker, then I say, "ah, now I see," this is this joy of insight which pays for all the trouble one has had in this career.

To me the important part in science, especially for us science teachers, is to emphasize that science does not mean giving answers to definite questions. There is an interesting anecdote about Niels Bohr. Bohr was once approached by a journalist, and the journalist asked him, "Aren't you the man who knows the answers to most of the questions in science," and Bohr said "Oh, no, but perhaps I know a few more questions than the others."

And that is the point: Science is not rote knowledge, formulas, names. Science is curiosity, discovering things and asking "Why, why is it so?" Indeed, I would say that science is the opposite to knowledge. The aim of science is to question, to ask why and how. It is the process of questioning, not the acquisition of knowledge (as it is unfortunately mostly regarded and taught).

Try to have the kids wonder: "Oh, I was not aware of this," "What is it?" and "Why is it so?" You must always begin by asking questions, not by giving answers. You must create an interest. I like to say that we must create a vacuum in our head—not in the sense that there is nothing in there—a vacuum which sucks in the insight which you would like to have them discover.

Students don't need to know so much. "Oh, the guy doesn't even know the formula for alcohol." Well, so what? The essential part is, that there are molecules, and alcohol is made of molecules, and molecules are made of atoms, and there is such a thing as a chemical formula. Then he has to know where to find it. There are books. People forget that. (laughter) Look it up in the chemistry book. Then they will get the idea of what science is and not what it says. And then they will become eager to know more and not less. Unfortunately, today things are really very bad.

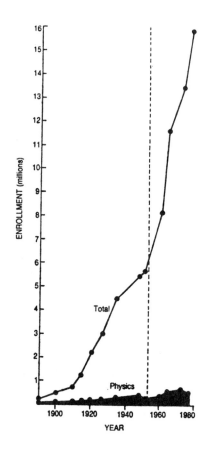

Fig. 1. Physics enrollment and total enrollments, 1890–1980.

For example, student interest in science is falling. The percentage who take different science subjects were 61 percent in 1969 and it was 37 percent in 1981. Look at Figure 1. Here is the enrollment in senior high schools from 1900 until today. As you can see, it is no joke. It is sad, to see the decreasing ratio of enrollment in physics to the total enrollment. It isn't much better in other sciences.

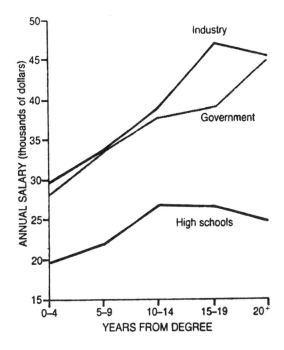

Fig. 2. Master's degree salaries for physicists employed in high schools, industry and government, plotted as a function of time from receipt of the degree. Data are from a 1982 survey.

Why is it so hard to get a sensible science education in our high schools? First of all, teachers' salaries are terrible. I don't need to tell you that.

Obviously, only the most idealistic people take such badly paid jobs, and that is the group we have here. I am so glad to see so many who sacrifice their own well-being for a higher purpose. You know the facts better than anybody else. Still, I would like to show a graph. Figure 2 shows annual salaries in thousands of dollars, for high schools, government, industry. High school teachers salaries are about a factor of .3 times those in industry, and that just should not be so. It has to change, but I don't know how to initiate such a change. It would cost an enormous amount of money, but it would really contribute to the United States in a far better way than many other expenditures. By the way, if you were to spend the money for SDI on teachers' salaries, it would really make a big difference. I think that would really do something for the safety of this country.

Now, speaking of money, I would like to show one other figure. It is the fractional part of the budget that NSF spends on education—science was actually pretty high in the beginning. It was almost 50 percent in 1950, then it went down and then came Sputnik—we need another Sputnik, I guess—and we had a jump and in the 1980s it's down to less than 10%. The budget has now increased, but not that much (in real dollars), and so again this shows the decrease of interest in this window of vulnerability, instead of an increase.

Before I start the physics, because I will speak about physics, I would like to just read you a few quotes out of a new book by my colleague, Lester Thurow, of MIT, *The Zero Sum Solution,* which just came out. Now, in it he says—and I think he has good sources:

"Eight percent of the people in New York are functionally illiterate; in Japan, one percent."

"Six percent of American undergraduate degrees are awarded in engineering; the comparable figures are 35 percent in the Soviet Union, 37 percent in West Germany. In the next few decades, there will be a time of tough international competition in high-tech products. Clearly America cannot compete in math and science personnel."

"Then, another comparison, in 1982–83, thirteen times as many math and science teachers left teaching as entered teaching. In '81–'82, half of the new math and science teachers hired were unqualified, had not taken enough math and science to teach those subjects."

Fig. 3. NSF commitment to science education, 1950–1984.

You and I know of stories of physical education teachers who have taught physics or attempted to teach physics. In the last ten years, the number of people studying to be math teachers fell 79 percent, and those studying to be science teachers fell 64 percent, and only half of those students actually end up to be teachers. The reasons, of course, are not hard to find. We have seen the salaries.

Over 40 percent of all science PhD candidates are foreign students, so at this rate, there will be no next generation of science and math teachers in high schools and colleges. Of course, it is clear what happens without good science instructors—it is not surprising, and look at these numbers—out of every 10,000 young people, 20 choose to be lawyers, 40 accountants, and 70 engineers in the United States. In Japan, there is only one lawyer among 10,000, 3 accountants, and 400 engineers. There are twice as many engineers on the payroll of the average Japanese firm, so that it should come as no surprise to learn that the Japanese products are better and cheaper than those here.

Of course, the high school teachers are in a very bad position, and some of the college teachers, because you have something that is called a curriculum, telling you that you have to teach that, and that. If you then leave something out the poor student may not pass the prescribed exam. For example, the student may not know the formula for alcohol.

Another terrible thing occurs in our schools of education, where you are taught education. My daughter is an elementary school teacher. Of course, she went through a school of education. She said, they are an effective means to make a naturally born teacher into a bad teacher. (laughter) What always worries me about these schools of education is that they forget about content, you learn *how* to teach and not *what* to teach, which is a contradiction.

In my time, when I was in high school, we called our teachers professors. That is another important point: the social status of the teacher. Ours had to have a PhD. A PhD doesn't necessarily make a good teacher, but it elevates his or her social position.

So here we are. There are lots of things that can be done. I am not a politician. I don't know how to do it, but it is absolutely clear that more money has to get into it, a lot more money. In order to get the money, the whole attitude towards education must be changed, and that is not an easy thing. The parents of the present students went to the same kind of high schools and mostly do not know how badly they were educated. It will be a long process but it must be done.

The Search for Simplicity

Now let me talk about something which I like very much. "In Search of Simplicity" is the actual title of the topic, qualitative physics. What I say now is perhaps less useful for high school students, although I think high school seniors may understand it, may profit from it. It is probably more at the adult level, because it is very important to say something simple about modern science.

I am writing a book, but it is not yet finished. Some of the material was in monthly columns in the *American Journal of Physics* last year. It is commonsense physics, qualitative physics, approximate physics, simple physics without details, aiming at orders of magnitude.

I always refer to a remark by Paul Ehrenfest, one of my teachers, who really brought out whatever pedagogic talent I may have; he said "Physics is simple, but subtle." And then I found a very nice quotation in Alfred Whitehead's writings, "The guiding motto of every natural philosopher should be: Seek simplicity and distrust it!"

First, I would like to talk about an interesting question: How long a time can a bee fly on one cubic millimeter of honey? This will turn out to be, essentially, a problem of molecular physics. By the way, that is one point which is always important in the case of teaching: that you start out with some interesting question from our environment and then go back to its causes. In order to get answer to our question, we have to know something about molecular binding, since it is a question concerning the chemistry of metabolism.

The second topic will be the question of how high mountains can be before they sink into the earth. This height has something to do with drops on leaky ceilings and with the waves on the surface of a lake.

The third is the question of thermal expansion (of solids). I doubt that I will get much more into my lecture. There is a principle I always adhere to: When I give a course, I say I do not want to cover the subject but to *uncover* part of it.

So, coming back to this, I may cover part of it, but I don't know if I will cover the whole thing.

It is interesting that all these things which we see around us are based on non-relativistic quantum mechanics. Only a few constants enter into these considerations: The Rydberg $Ry = \frac{1}{2} \frac{me^4}{\hbar^2} = 13.6$ eV as an energy measure; and the Bohr radius $a_B = h^2/me^2 = 5.3 \times 10^{-11}$m as our fundamental length. In chemical terms, one electron-volt per atom is 23 kilocalories per mole and one Rydberg is much more, 313 kilocalories per mole.

How Far a Bee Can Fly on 1 mm³ of Honey ("The Flight of the Bumblebee")

Back to our question: How long can a bee fly on $1mm^3$ of honey? This is mainly a chemical problem. How much energy can the digestion of $1mm^3$ honey produce? Digestion is combustion. The chemical reaction is the transformation of glucose into carbon dioxide and water with the use of the oxygen in the air: $C_6H_{12}O_6 + 6O_2 = 6CO_2 + 6H_2O$.

We have to know something about molecular binding. When two hydrogen atoms are far apart, their total energy is the sum of the individual energies, the electron energy is negative because it is bound. So hydrogen atoms get nearer, they begin to touch (at a separation of ~0.2 nm). When the electrons begin to interpenetrate they are attracted by the other protons, each electron feels two protons, so there is attraction to the protons. There is repulsion also, but the attraction wins out. If they get too close, the protons repel one another electrically. As a result, the interaction potential has a minimum at a separation of 1.22 Bohr radii, at which distance the bond strength is 4.6 eV, or 106 kcal/mol. We call electron pair bonds, such as in hydrogen, "good bonds."

Let us see how the oxygen is found in water and carbon dioxide. I call it a "plug and hole" bond: The oxygen atom is two "holes" short of a closed shell. In water, the two electrons of the two H-atoms plug into the holes; in CO_2, the four valence electrons of C plug into the holes of the two oxygen atoms. In water the hydrogen electrons are then not only attracted by the protons in the hydrogen, but also by the oxygen nucleus. In CO_2 the four valence electrons of C are attracted also by the oxygen nuclei. The resulting binding energies are of course multiples of a Rydberg; the decisive energy of atomic electrons. Surprisingly, it turns out that the binding energy in most simple molecules, such as water, carbon dioxide, glucose, and in hydrocarbons, is roughly 4.4 eV per bond or 100 kcal/mol. The total binding of CO_2 is 385 kcal/mol with its four bonds; it is 220 kcal/mol in water with its two bonds.

Now, the binding of oxygen in O_2 is less; it is 180 kcal/mol or about 60 per bond. This is understandable since it is not a plug and hole situation, nor are valence electrons available. Thus, when the oxygen atoms of air (where they are in the O_2 molecule) are bound into CO_2 and H_2O, as happens in the combustion process, we gain roughly 40 to 50 kcal/g per oxygen atom, taken from the air. That gives us an easy handle to find the energy gain in combustion.

Let us burn methane, the main constituent of natural gas: $CH_4 + 2O_2 = CO_2 + 2H_2O$. Two oxygen molecules of air are used for the combustion of one molecule of methane. Thus four medium bonds are broken and replaced by strong bonds in water and CO_2. Hence we should get 160 to 200 kcal for burning one mole of methane. The actual value is 190 kcal.

What about gasoline? It is a hydrocarbon chain. Consider one CH_2 link: $CH_2 + 3/2\ O_2 = CO_2 + H_2O$. We should gain three times the difference, which is 120 to 150 kcal/mol of CH_2, or 9 to 11 kcal/gram. Actually, it is 10 kcal/g.

The third example is glucose: $C_6H_{12}O_6 + 6\ O_2 = 6\ CO_2 + 6\ H_2O$. The gain should be 12 times the difference: 480 to 600 kcal/mol. The correct value is 590 (honey) which means 13.8 MJ/kg.

Let us use this result to estimate how long a bee can fly on one milligram of honey, the result of sucking at a few flowers. The bee does not fly like an airplane using the air flow passing by, but keeps itself suspended almost at rest, not unlike a helicopter. It produces a stream of air downwards to compensate for gravity. The bee is pulled down by the force mg, where m is its mass. To counter that force the bee produces a column of air of a cross section of σ m^2 moving downwards with a velocity v. Roughly speaking, σ is the area of the wings. The momentum of 1 m of that column is $\sigma \rho v$, where ρ is the density of the air. In one second the bee produces an air column of length v, so that the downward momentum produced per second is $\sigma \rho v^2$. A rate of change of momentum is a force; it balances the gravitational force so that we get $\sigma \rho v^2 = mg$, or $v^2 = mg/\sigma \rho$. The power P needed to produce that air stream is the force times the velocity. $P = mgv = (mg)^{3/2} (\sigma \rho)^{-1/2}$. We put $m = 10^{-2}$g, $\sigma = 3 \times 10^{-5}$ m^2 and $\rho \sim 1$kg/m^3. Then $P \sim 200 \mu$ J/s. A milligram of honey yields 14 J. It lasts for a few hours, including the efficiency of biological metabolism for mechanical work at 10% to 15%.

Another question is the amount of food needed for a person just to keep warm. Enrico Fermi began this estimate by making the remark, "I know from reading mystery novels that a corpse needs about half a day to cool from body temperature to room temperature." Assuming a body mass of 60 kg, we get from this information that the body needs about 2000 kcal per day to stay at body temperature; this is 100 W. Assuming all food is roughly as nourishing as honey, we find that just keeping warm takes about 600 g of food per day. Here we can count on full efficiency since all losses produce heat. This amount of energy would lift the person to 1.5 times the height of Mt. Everest, not counting metabolic and other losses incurred when climbing. It shows how much harder it is to produce heat than mechanical energy. Note that the human body produces 1.7 mW per gram, whereas the sun produces only 2×10^{-7} W per gram!

Mountains, Waves, and Ceilings ("Grand Canyon Suite")

Next question: The height of the mountains, water waves and leaky ceilings. Fig. 4 shows a very schematic mountain. The maximum height H of the mountain is reached when it becomes so heavy that it sinks into the base because of the plastic flow of the base material. Evidently when the mountain sinks the same amount of material must be displaced at the base as is lost on top, in order to make room for it. This displacement is carried out by plastic flow of the base material.

Let us transport one molecule down. We gain MgH in energy where M is the mass of the molecule and we must displace one molecule. The energy per molecule necessary to displace the material in the ground will be called ϵ_p, the plastic flow energy. It is hard to estimate, but there is an upper limit: the heat of fusion. The material of mountains is rock, that is, mostly SiO_2. You may deform it by heating it to the melting point, which is pretty high, several thousand degrees, and then melt it and then you can plastically deform it with practically no energy, and then you go down again and you have it in the deformed state.

Going up and down in temperature balances more or less as far as energy is concerned. Therefore, as a first orientation, we may put ϵ_p equal to the heat of fusion per molecule, which is 0.1 eV.

Now we equate ϵ_p with the gravitational energy, MgH, where M is the mass of the molecule (M = Am, with A = 60 being the molecular number and m the mass of the proton). We then get

$$H = \epsilon_p/Amg \sim 14 \text{ km.} \qquad (1)$$

This is too high an estimate because ϵ_p is certainly much smaller than the heat of fusion. The lattice defects and the domain structure of rock greatly reduce the energy of plastic deformation, say by a factor ten or more. On the other hand, mountains with slopes less steep than Fig. 4 can maintain a larger height because less material rests on a wider base. A pyramidal shape would give a factor 3. We arrive at a maximum mountain height of several kilometers relative to their surroundings (not sea level) which certainly corresponds to the facts.

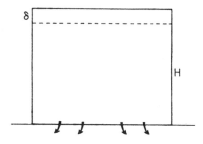

Fig. 4. The sinking of the mountain by the amount δ corresponds to the displacement of a layer of thickness δ from the top into the ground, and to a plastic flow of a comaparable volume in the ground.

Obviously the plastic deformation energy $\epsilon_p = \xi \epsilon_B$ is a small fraction ξ of the binding energy $\epsilon_B = 6.5$ eV of a molecule in SiO_2. ξ is of the order of 0.01 or smaller. The binding energy $\epsilon_B = \zeta$ Ry is a fraction $\zeta = 0.48$ of a Rydberg. Our expression for the height of the mountains can then be expressed in terms of Rydbergs:

$$H = \xi \zeta \text{Ry}/Amg \qquad (2)$$

What about the drops on the ceiling? When the ceiling is not tight, the leaking water forms a thin film covering the surface of the ceiling. This film is unstable. A slight accumulation at one point starts growing downwards as water flows into it from all sides, since this reduces the gravitational energy (see Fig. 5). When will the drop come off?—When the gravity force becomes larger than the surface tension that keeps the drop on the ceiling. The surface tension S is an energy per unit area or a force per unit length. Let us approximate the drop as a hemisphere of a radius R. Then the force F holding it up is the surface tension along the periphery where the drop merges with the film on the surface: $F = 2\pi RS$. When this force becomes equal to the gravity force $(2\pi/3)R^3\rho g$, the drop will fall. Here, ρ is the density of water. We then get for the radius of the drop (S = 73 mN/m in water):

$$R \approx (3S/g\rho)^{1/2} = 4.7 \text{ mm}. \tag{3}$$

The result is not exact since the form of the drop when attached to the ceiling deviates from a hemisphere, especially shortly before separation. However, it does give a size of the drops not far from the one we do observe all too frequently.

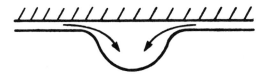

Fig. 5. The forming of a water drop from a thin water film below a surface.

Let us now turn to the water waves. When a light breeze starts blowing over a quiet surface of a lake, the wavelength λ of the initial waves is of the order of a few centimeters. "Willows whiten, aspens quiver, little breezes dusk and shiver," as the poet Tennyson says. We will not enter into the physics of wave production; suffice it to say that the wind transfers its energy first to those waves whose propagation velocity v is lowest. The expression for v is:

$$v = (g\lambdabar + S/\lambdabar\rho)^{1/2}$$

where S is the surface tension, ρ is the density of water, and $\lambdabar = \lambda/2\pi$. The first term comes from the gravity and the second term from the surface tension. It is evident that the longer $\lambdabar$ is, the more strongly gravity acts as a restoring force, and the smaller $\lambdabar$ is, the more the curvature of the surface causes a restoring force. The minimum of v occurs at

$$\lambdabar_m = (S/g\rho)^{1/2} = 2.8 \text{ mm}. \tag{4}$$

The corresponding minimum value v_m is 0.23 m/s. A wind with less than this speed would be unable to produce waves.

This is why lakes are so much like mirrors, even in the presence of weak winds. The first waves appear when the breeze surpasses v_m, and should have a wavelength of $2\pi\lambdabar_m$. Actually the minimum velocity v_m is somewhat smaller and the corresponding wavelength λ_m is somewhat larger, because we should have done our estimate by using the group velocity instead of the phase velocity of the waves, but the order of magnitude is the same. We should have found $v_m = 0.18$ m/s, and $\lambda_m = 44$ mm. Note that $\lambdabar_m$ is the same length as the radius of the falling drop as given by (3) apart from a factor $\sqrt{3}$ for the phase velocity calculation or a factor 0.68 for the group velocity calculation.

In order to compare these results with the mountain height, we express S and ρ in terms of molecular properties. The binding of a molecule at the surface is $(1-\xi')$ times the binding energy ϵ_B in the interior. ξ' must be near to 1/6, because a molecule in the interior can be considered to be bound to its six nearest neighbors, whereas at the surface there are only five. In water it is somewhat smaller: $\xi'=0.093$. Setting again $\epsilon_B=\zeta'$Ry, the surface tension S is the reduction of binding energy for the d^{-2} molecules in a m^2 of the surface. Here d is the distance between neighbor molecules. Then we get $S=\xi'\zeta'Ry\ d^{-2}$ and $\rho=A'm/d^3$, where $A'=18$ is the molecular number of water. Putting this into (4), we obtain

$$\lambda_m^2=R^2/3=(\xi'\zeta'Ry/gA'm)d=CdH, \text{ where } C=\xi\zeta A/\xi'\zeta'A'.$$

Since we do not know ξ too well, it is hard to give an exact value of C, but it is not too far from unity. In other words λ_m and the size of falling drops are roughly the geometrical mean between the maximum height of mountains and the intermolecular distance d. The greatness of mountains, the finger-sized drop, the shiver of a lake, and the smallness of an atom are all related by simple laws of nature.

Thermal Expansion ("The Anvil Chorus")

There is not much time left for thermal expansion. I will make it very short.

The length L of a rod of solid material increases with rising temperature. Call $(\Delta L)_1$ the increase when the temperature is raised by 1 K. Then the thermal expansion coefficient is defined by

$$\tau=(\Delta L)_1/L.$$

In order to estimate roughly the value of τ, we assume for a moment that ΔL is proportional to ΔT even for large changes of ΔL. Assume we deliver energy to the rod amounting to the binding energy ϵ_B per atom. ϵ_B is the energy necessary to liberate an atom from the material. If we did this, the substance would fall apart. Let us now very qualitatively equate "falling apart" with a doubling of the interatomic distances, hence a doubling of the length of the rod, $\Delta L=L$. Let us further assume that the linear dependence of ΔL on ΔT more or less holds even up to temperatures corresponding to ϵ_B per atom. ΔT is a measure of the energy given to an atom. For example, a rise of temperature by 1° corresponds roughly to an energy increase of the order of the Boltzmann constant k per atom, which is $\sim 10^{-4}$ eV. (The equipartition theorem ascribes the energy 3kT to a three-dimensional oscillator, so that the energy increase would be 3k. But we are interested only in the vibrations in one direction when we study the linear expansion of the rod.) If the linear dependence of ΔL with ΔT holds, the increase $(\Delta L)_1$ for $\Delta T=1°$ should be smaller than the increase $\Delta L=L$ when each atom gets ϵ_B, roughly by the ratio k/ϵ_B. Therefore, we get

$$\tau=(\Delta L)_1/L\approx k/\epsilon_B\approx 10^{-4}/(\epsilon_B)_{eV}$$

where $(\epsilon_B)_{eV}$ is the binding energy in eV which is of the order of a few eV for substances that are solid at room temperature. Indeed, the thermal expansion coefficients are between 10^{-4} and 10^{-5} per degree.

Now, my time is over. Let me just perhaps end with a warning. It is the same warning as Mr. Whitehead has said: Distrust simplicity. Seek simplicity and distrust it. And I would just like to tell an anecdote about Pauli. When I was an assistant to Pauli in Zurich, the experimental physicist there was Paul Scherrer. He always loved simple explanations. He came to Pauli and said, "Pauli, I have now a wonderful explanation for this effect" — I have forgotten what it was — "you know, this spin goes up and the other spin's down, and the action then comes out." And Pauli said, "Simple it is, but it is also wrong."

Well, with this I will end my talk.

Dr. Aubrecht: I am sure Professor Weisskopf would be willing to answer questions if there are any.

Question: These are nice explanations, but if you think about them they are drawing on an enormous wealth of knowledge, and they seem to be simplifying after the fact, rather than before, and would be confusing before.

Professor Weisskopf: Absolutely. There will be a warning in the preface of our book. This is not our advice on how to do physics, although you will hear probably during these days some of the high energy physics speculations that are probably even on a more tender footing. This is not the way one can do physics except the great geniuses who have the necessary intuition.

It is a way to understand physics that has already been done. It doesn't replace the actual calculations. Indeed, I wouldn't trust it if I hadn't seen the actual calculations. In other words, the purpose of it is purely pedagogical, namely to get the essential elements of the complicated calculations. This is the only purpose.

I consider the approach important today due to the advent of the computer and a general tendency towards complicated calculations. "Here is the Schrödinger equation, here is the computed result, and now I understand the physics."

Let me tell you another anecdote. When I meet a meteorologist, I always ask him why the wind blows from the west. Once I went to Jules Charney, a famous one—he died unfortunately—at MIT, and said why? He said "Well, very simple, come to my office," and there he had a printout, a computer printout where he'd put in the solar radiation, Coriolis force, and God knows what and he said, "Look at how all the arrows in our latitudes come out."

And then I said to him, "Jules, that is very nice, the computer understood it, but what about you and me?"

This is the sense of it.

Certain material in this article is excerpted from a series of articles appearing in *The American Journal of Physics*. It is reprinted here by permission.

Chris Quigg is Head of the Theoretical Physics Department at Fermilab, and Professor of Physics at The University of Chicago. He received the B.S. degree from Yale in 1966, and the Ph.D. from Berkeley in 1970. Professor Quigg has contributed to a broad range of topics in the theory of elementary particles and high energy collisions. Much of his work has been characterized by a commitment to the fruitful interplay between theory and experiment. He is the author of numerous review articles and summer school courses on a wide variety of subjects, and of a standard textbook on gauge theories of the fundamental interactions. His recent research has been concentrated on the scientific possibilities of the Superconducting Super Collider.

ELEMENTARY PARTICLE PHYSICS: DISCOVERIES, INSIGHTS, AND TOOLS[1]

CHRIS QUIGG

Fermi National Accelerator Laboratory
P. O. Box 500, Batavia, Illinois 60510

ABSTRACT

This is a lightly edited transcript of two lectures presented at the Conference on the Teaching of Modern Physics held at Fermilab in April, 1986. The informality of the spoken word has been preserved, but some of the immediacy of the interchange with the audience is inevitably lost.

LECTURE 1: THE FUNDAMENTAL CONSTITUENTS

What I would like to talk to you about this morning is ELEMENTARY PARTICLE PHYSICS, the science of the ultimate constituents of matter and the interactions among them. Like all of physics (but in an especially immediate manner), it tries to ask and answer the questions

- What is the world made of?

- How does the world work?

In common with other physicists, we hope that by beginning to understand the laws of Nature, by codifying them, by extending the domain over which they apply, we may be able to put our new knowledge to productive use.

The questions that we pose for ourselves (see Fig. 1) are

- What are the basic constituents of matter and energy?

- What are the forces by which these constituents interact with each other?

---

[1] Copyright © 1986 Chris Quigg

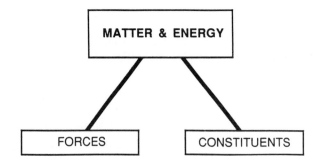

Figure 1: Goals of elementary particle physics.

What I will try to do in these two talks is to introduce you to the description of matter and energy to which we have come, and to emphasize both the simplicity and the tentativeness of that description. In the course of this, we will fill in some of the white space in Fig. 1. In a sense, it is easy to do this. It is easy because dramatic progress has been made over the last twenty years. The picture we have of fundamental physics is much simpler, much more comprehensive, and much more unified than it was a couple of decades ago. This has prompted some to say that a grand synthesis of natural law is at hand. It is unquestionably true that great progress has been made, and that the place at which we have arrived is at least a good starting point for the next great leap.

The reason we can explain our world view to students in relatively simple terms has to do with the emergence of something called THE STANDARD MODEL OF ELEMENTARY PARTICLE PHYSICS. The point of my lectures this morning will be to illustrate for you some of the prominent features of the Standard Model.

The Standard Model has a couple of aspects that I want to emphasize. One is the identification of a set of *elementary particles,* at least for our generation of scientists, called the *quarks* and the *leptons.* I'll spend much of this first lecture reminding you of some of the features of those constituents. On the other side of our chart, in trying to understand the interactions of those constituents, there has been the recognition of a grand principle and the development of a class of theories called gauge theories of the strong, weak, and electromagnetic interactions. I'll try to indicate to you in the beginning of the second lecture what is the strategy of gauge theories. We won't go through all the mathematical details of gauge theories, but as with most wonderful ideas, once someone has slogged through the details for you, you can explain it more or less simply, and I'll try to do that for you. Finally, the reason for the gleam in one's eye is that because of the simplicity of this picture, having identified the relatively small number of fundamental constituents and seen a nice mathematical framework in which to express their interactions, we see the promise of going further and gaining a more coherent understanding of all the forces of Nature. I'll try at the end of the second lecture to allude to that a little bit, and the thrust of where we go from here is what Howard Georgi is supposed to talk about in the next couple of days.[2]

Now, particularly because you are here at Fermilab, but also because I think it's important, I'd like to ask that in listening to me and in your working groups, you try to take into account the interplay between *discoveries* and *insights* and *tools,* or if you feel the need for labels, between experiment and theory and advances in technology (See Fig. 2.) One of

---

[2]A list of suggested readings appears at the end of Lecture 2.

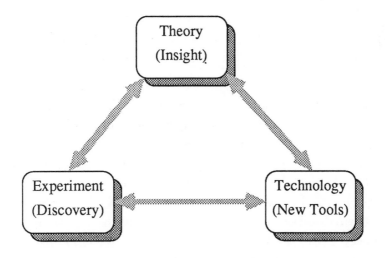

Figure 2: Synergism in basic research.

the things most disappointing to me when I look at my children's science textbooks is that there is usually simply an Aristotelean statement of "Mr. X or Ms. Y invented this or that, and this was the idea they got," and this is the way it is. The last person credited with using a technological innovation to learn something about the world is usually van Leeuwenhoek with his microscopes, and that took place three hundred years ago. The way things actually happen around here and at other great centers of science is that we do experiments, we make observations, we try to learn about things. The theory (which is not always done by theorists) leads us to catalog these observations, abstract from them, and so on. And in the long run these new insights into Nature give us the means for developing new technologies. Once a new technology is available, it is used immediately, often for the first time, in the pursuit of fundamental science to try to make new sorts of experience. You have the opportunity here to wander around and see — in addition to our buffalo herd — some of this state-of-the-art instrumentation in action. I would urge you to do that and to take that excitement of search and discovery home to your students.

One of the wonderful things Professor Weisskopf did this morning in showing those highly polished baubles of insight was to illustrate how in trying to understand things that are present in common experience we are led to retreat from common experience and make use of understanding that we gain on different levels. In particle physics we try to push always to the smallest (and we hope simplest) levels, hoping to find the most fundamental pieces of matter and the interactions among them. The whole history of science tells us that it ought to be possible to build up from those minimal parts to the larger complex systems we see around us.

In order to look at matter on fine scales and to see the interactions — to make them happen — we use particle accelerators and detectors, which together you may think of as the microscopes of high energy physics. We push to higher energies for two reasons. One is that these little things that are inside the deepest levels of matter are stuck together pretty firmly, and so to get inside and move them around and see what they do, you've got to hit them harder and harder. It is, in other words, a question of binding energy being larger as you go to deeper levels, and that requires that you hit things harder with projectiles of higher energy. The other reason we go to higher energy is related to the fact that you can listen to FM radio stations in underground parking garages, but can't listen to AM radio stations. That is, to see little things you've got to inspect them with probes of short wavelength. Short wavelength corresponds to higher energies.

Now to our main subject. The prerequisites for this lecture are the sum of human knowledge from Antiquity to twenty years ago, as represented in Fig. 3. The key idea illustrated here (and one of the enormous simplifications that physics has brought to us) is that we can explain and understand all natural phenomena in terms of a small number of fundamental forces. Since the 1930s these have been identified as the *strong force*, the *electromagnetic force* (itself the union of electricity and magnetism from a century ago), the *weak force* responsible for radioactivity, and *gravitation*. What we're going to try to do is to learn something about the properties of these forces, and to learn what are the most basic constituents upon which they act.

Thanks to a great number of experiments, principally over the last couple of decades, we have identified two classes of fundamental particles called the *leptons* and the *quarks*. I want to take a few minutes to tell you a little bit about them.

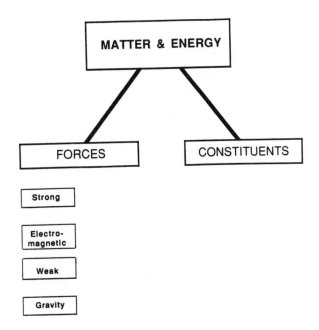

Figure 3: A starting point.

The first class is made up of particles like the electron. These are called *leptons* because the electron is a very light particle. Other members of the class turn out not to be very light, but the name persists. The leptons are particles which experience weak and electromagnetic interactions but not the strong force, not the force that binds protons and neutrons together in nuclei. We know of six such particles, shown in Fig. 4. The electron and its heavier cousins the muon and the tau lepton

LEPTONS (COLOR NEUTRAL)			
Particle name	Symbol	Mass (MeV/c^2)	Electric Charge
electron neutrino	ν_e	~ 0	0
electron	e or e^-	0.511	-1
muon neutrino	ν_μ	~ 0	0
muon	μ or μ^-	106.6	-1
tau neutrino	ν_τ	< 70	0
tau	τ or τ^-	1784	-1

Figure 4: Some characteristics of the leptons.

all carry the same electric charge; they all have spin-$\frac{1}{2}$; and, as far as we can tell, they are all pointlike. They have no extent, no gears and wheels running around inside. As far as we can tell by means of the resolution of our present "microscopes," which is down to a distance of 10^{-16} cm, these objects are just geometrical points. It is interesting to wonder whether, as we look more closely, they will develop structure inside. Are there little tiny things in there, or will the leptons remain forever truly elementary particles, structureless and indivisible? Together with the charged leptons there are three neutral particles called neutrinos, which experience weak interactions and form family patterns with the charged leptons, as we'll see in a moment.

All the known leptons can be made readily in accelerator laboratories, and they can be studied directly in the laboratory. When the charged leptons are produced at high energies, they fly out of the reaction for macroscopic distances. [The electron is absolutely stable, the muon lives for a couple of microseconds, and the tau lives for about a third of a picosecond.] We can measure their tracks by ionization and see where they have been, and so measure them readily. The neutrinos are more difficult to measure because they are neutral and don't cause ionization, but we can see the effects of their interactions when they hit other objects. A lot is known about them. The neutrinos, so far as we know, could be exactly massless, although what we have so far is upper limits on their masses. Because we can study the charged leptons in great detail, making beams of them and even storing them for long periods, we know quite a lot about their properties. The simplest of these is their mass, indicated in the chart in Fig. 4.

In observing the interactions of the leptons, we find that there are well defined families. The electron always goes in partnership with the electron's neutrino. That is to say that there are interactions which transform one into the other, but they always go back and forth. There is no interaction that we know that changes an electron into a muon or an electron into a muon's neutrino. We thus observe these rather rigid family patterns, which are suggestive that there is some deep relationship between the members.

The other class of particles we can study in the laboratory includes the proton and neutron. These are particles which experience the strong interaction (the nuclear force), in addition to the other forces. The proton and neutron are the most familiar. The pion, or π-meson, which is grossly speaking responsible for the nuclear force, is another. And then there are tables and tables ... Just yesterday I received in the mail this year's edition of the Particle Data Tables which runs to 350 pages and has everything that you want to know about all the hundreds of species of these *hadrons*. That's quite a thick book just listing numbers and references and properties.

Now, unlike the leptons, which all were of one general kind, all spin-$\frac{1}{2}$ particles, these are particles that have integer spins, half-integer spins, small spins, large spins. All of them are composite particles. You can see that by scattering electrons from them, for example. You find that they are big and squishy inside, and typically have a size of about 10^{-13} cm. At a certain resolution, the proton resembles a Nerf basketball.

Hadrons range in stability from the proton, which has a lifetime of 10^{31} years or more, down to the Δ (Delta) and other resonances, which have lifetimes on the order of 10^{-24} to 10^{-25} seconds. The lifetime of the proton, you will notice, is many orders of magnitude longer than the age of the Universe, which is of order 10^{10} years. So obviously we have not derived the limit by watching one proton for a very long time — there *isn't* that much time — but by watching many protons for a much shorter time, on the order of a year.

The hadrons make up a great zoo of particles, in which we can recognize a certain taxonomy. A large step to bringing order and understanding to this diverse collection of beasts came in the mid-1960s with the proposal that these hadrons, these composite objects, were made up of a small number of more fundamental objects called *quarks*. Like the leptons, the quarks would be spin-$\frac{1}{2}$, pointlike particles. And we now know, as I'll try to convince you in the next few moments, that these quarks really exist, and that they are smaller than about 10^{-16} cm.

The essential distinction between the quarks and the leptons, and indeed between the quarks and most of the other constructs that we use in science, is that we don't get to see the quarks in the laboratory. We have not been able to isolate them. As a matter of fact, we now have a strong conviction that you can't isolate them. Because of that it's helpful, I think, to spend some time reminding you why we believe in quarks. Since they are not seen directly, one is entitled to ask whether this whole story about the quark model is not just so much making of myths. So what I'd like to do in the next few minutes is to try to evoke for you some of the experimental bases for our belief in quarks. The evidence will have to be circumstantial because we can't remove a quark from a hadron and hold it in our hands, but there's so much of it, it's so consistent, and it's so overwhelming that you will be led ineluctably to the belief that quarks are real!

Why do we believe in quarks? The first motivation for quarks came from observing the family patterns of the hadrons, the neutrons, protons, pions, and other things, which had been discovered up through the early sixties. As you know, in atomic spectra we observe degenerate multiplets in which energy levels with different magnetic quantum numbers, say, have exactly the same energy in the absence of magnetic fields. Only by applying perturbations (in the form of magnetic fields) do you break that degeneracy and learn about all the individual levels that are there.

That line of analysis of atomic spectra (which led to the introduction of group theory into physics), that way of thinking of degenerate multiplets, carries over to other situations and is used again and again in our attempts to understand the fundamental constituents. The first new setting is the observation that the proton and neutron seem very much alike. Both particles live in the atomic nucleus. They have almost exactly the same mass. One happens to be charged; the other isn't. The similarity led to the idea of a family partnership between them, to the idea of *isospin*.

In the same way, one could look at the particles which had been discovered in the early sixties and notice family partnerships among them. One of the great heroic enterprises of that period was to try to figure out what were the multiplets, which particles went together, and so on. Well, that's a long and fine story. The end of that long and fine story is that there's a symmetry group called $SU(3)$ (which you'll hear about in Chris Hill's lecture this afternoon), and that all of the particles known at that time could be classified as members of $SU(3)$ families.

A puzzle to be explained was that whereas for angular momentum (or the rotation group) you can build up arbitrarily large multiplets, the $SU(3)$ clans seemed to be limited to families of a few small sizes. In the case of the particles like the pions, the so-called *mesons*, the families contained either one member or eight members. And in the case of particles like the proton, called *baryons*, all the families had one or eight or ten members.

So a challenge after the establishment of $SU(3)$ symmetry was to understand why only a few of these family sizes were special. The way you can do that is by saying that the hadrons, which we already know to be composite because of their finite size, are composite in a very special sense. There is a fundamental triplet of quarks (which we now call *up*, *down*, and *strange*) three flavors of quarks if you like, and there are simple rules for combining these three fundamental entities into the mesons and the baryons, the pion-like particles and the proton-like particles.

If you make the rule that a meson is one quark and one antiquark joined together by a force to be understood later, then the arithmetic of $SU(3)$ tells you that a family of three members times a family of anti-three members gives resulting families of either one or eight members. That's good; that's the result that you wanted to get. And if you say that particles like the proton are made of three quarks joined together, it turns out that, by the arithmetic of $SU(3)$, you can only make families of one or eight or ten members, again the desired result. Having found that this arithmetic works, you must then ask whether quarks are real, and what are the forces that allow these combinations to form and prevent more complicated combinations like six quarks or 27 quarks from joining together.

What we find is that we are led by the success of this picture to try to give it a deeper meaning, and to understand on a dynamical level why these things happen.

If you say that there are quarks inside the proton, then there ought to be some way of learning that they are there. One piece of evidence which makes that plausible is found by studying the scattering of an electron beam from a target. Here, in Fig. 5, is a standard experiment. You take an electron beam of known energy, and allow it to hit a target. The target might be a piece of carbon, a bottle of hydrogen, whatever you like. And then you observe the direction and energy of the scattered electron and, if you wish, you can observe something about the recoil particle or

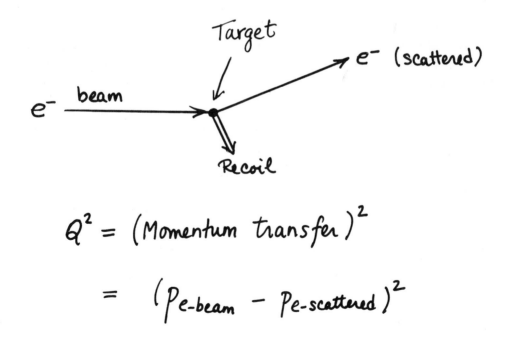

$$Q^2 = (\text{Momentum transfer})^2$$

$$= (P_{e\text{-beam}} - P_{e\text{-scattered}})^2$$

Figure 5: Electron scattering kinematics

particles.

The point of this exercise is to see what happens as you vary the angle and energy of the scattered electron, and to understand what that reveals about the inner structure of the target material. Let's proceed by analogy, by looking at the historical precedent. Take as a target a carbon nucleus, really a carbon fiber, scatter electrons from it, and require that the carbon nucleus remains intact after the scattering, so we are studying the reaction

$$\text{electron} + \text{Carbon nucleus} \rightarrow \text{electron} + \text{Carbon nucleus.} \qquad (1.1)$$

If you hit the carbon nucleus very hard, because it's a loosely bound collection of protons and neutrons or maybe of alpha particles, it is likely to fly apart. By requiring that it stay together, you are selecting a very rare occurrence. This is called the form factor effect. If you require that the carbon nucleus remain intact, you find that the rate at which this process occurs decreases rapidly as the amount of energy you deliver to the carbon nucleus increases. This is illustrated in Fig. 6(a).

On the other hand, if you relax the constraint that the carbon nucleus must come off intact and just say that you are going to observe the outgoing electron without regard to what came out with it, then you find that the cross section is almost independent of how hard a blow is delivered (dot-dashed line in Fig. 6(a)). The reason for this difference is that you're seeing the scattering of the electron from the individual protons inside the carbon nucleus, and at a certain resolution those protons behave as structureless particles.

So in the old days, in doing nuclear physics scattering experiments you could deduce the idea that there must be relatively structureless, electrically charged objects inside the nucleus by seeing the slow variation of this inelastic scattering rate. Of course, you could also knock the protons directly out of the carbon nucleus and verify your conclusion. If you pursue this, you can change to a situation in which your target is an individual proton, as shown in Fig. 6(b), where I've changed the scale of my abscissa by a couple orders of magnitude. Whereas I was hitting my carbon nucleus with 0.06 units of punch, I'm now hitting the proton 100 times harder.

On this scale, the proton itself doesn't like to remain intact. We see the structure of the proton reflected in the fact that the cross section or

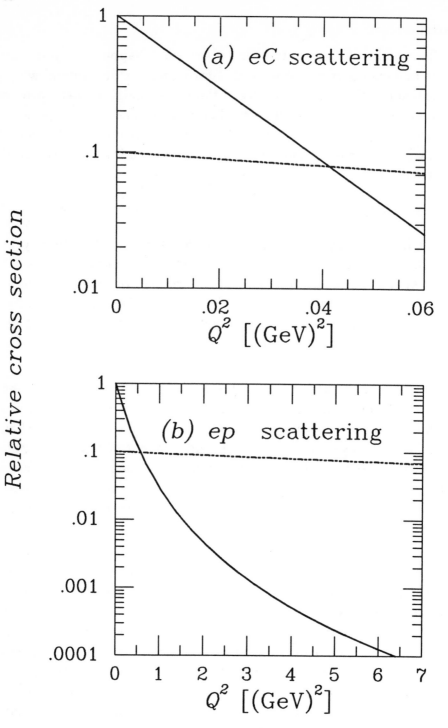

Figure 6: Elastic (solid lines) and inelastic (dot-dashed lines) cross sections for (a) eC scattering; (b) ep scattering.

rate for the reaction

$$\text{electron} + \text{proton} \rightarrow \text{electron} + \text{proton} \qquad (1.2)$$

falls off rapidly. That's because the proton tends to become excited or to produce new particles when it is hit hard. On the other hand, if we relax the constraint that the proton come off intact, we find that there is once again a contribution to the cross section which is essentially independent of how hard you hit the proton.

Just as we interpreted the proton as being something hard and point-like and electrically charged inside the carbon nucleus, it's tempting to conclude that there is something hard and pointlike and electrically charged inside the proton, and that is a role which could well be played by the quarks. Experiments which first showed this were done at Stanford Linear Accelerator Center in 1967 and 1968, and immediately led people not to accept, but to take seriously the idea that quarks really were inside the proton.

If quarks can't be knocked out of the proton, how do we know anything about the properties of quarks? Let me evoke just a few of the ways that we learn about quarks. The quark electric charges are unusual, compared to common experience: the up quark has charge 2/3; the down and strange quarks have charge $-1/3$. These are measured in units in which the proton's charge is $+1$ and the electron's charge is -1. These assignments come in the first instance from the group theory of $SU(3)$, but you can seek more direct ways of determining them.

One of these more direct ways is to look at the decay rates for spin-one particles made out of a quark and an antiquark, the so-called *vector mesons*, particles that resemble heavy photons, and which decay into pairs of electron and positron (anti-electron). The way this happens in the quark model is that the quark and the antiquark which make up the vector meson can annihilate each other, if they find themselves at the same place, in a burst of electromagnetic energy we call a *virtual photon*, which later on will disintegrate according to the laws of quantum electrodynamics into the electron-positron pair.

Now, you can calculate this decay rate. In fact, Professor Weisskopf did it first. But we don't have to do that; we can normalize one rate to the other, as follows. The rate at which the decay occurs is determined by two basic things, as indicated in Fig. 7. One is the probability for

$$\text{Decay Rate} \propto Q_q{}^2 \, |\psi(0)|^2$$

Figure 7: Decay of a vector meson into an electron-positron pair, in the quark model.

the quark and antiquark to get together and annihilate in the first place. In nonrelativistic language, this is related to the probability for them to meet at a point — so that's given by the quantum-mechanical wave function squared at the origin, *i.e.* for zero separation between the quark and antiquark. [That's this factor $|\psi(0)|^2$ in Fig. 7.] I'm going to make the gross assumption that for the vector mesons I want to talk about, that probability is the same, that they have more or less the same structure. So that's one factor which must be present, but which I'm going to pretend has no effect.

The other thing that enters is the strength of the electromagnetic coupling between the quark and antiquark, the rate at which they combine to make photons. The electromagnetic strength is just governed by the *charge* of those objects, and so the overall rate is proportional to the charge squared of these things. You can look at things called the *rho, omega,* and *phi* mesons and measure their decay rates into electron and positron pairs. You will find that the ratio of those rates is exactly in the proportion suggested by these funny charge assignments. There are numerous other ways of making that test, as well.

One of the most striking pieces of evidence that quarks are real came later with the discovery of families of particles made of two kinds of still heavier quarks called the *charm* quark and the *bottom* or *beauty* quark. And here, in Fig. 8(a) I show you the spectrum of particles composed of the charm quark and anticharm quark. You see that there are various levels, with different values of angular momentum. They make atom-like transitions from one state to another, so that if this spectrum were unlabelled and you were asked to identify it, it would be natural to say

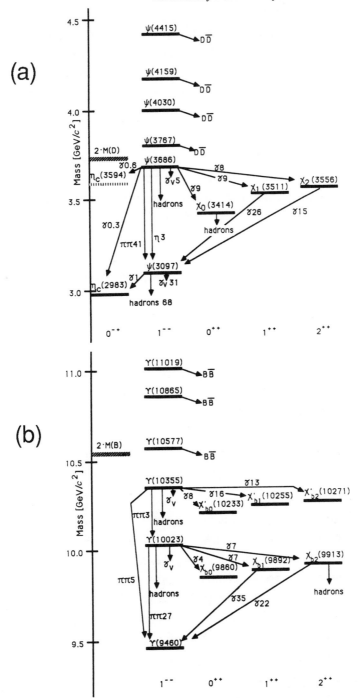

Figure 8: (a) The charmonium states (first members observed in 1974);
(b) the upsilon family (first members observed in 1977).

this is an atomic spectrum of some kind. We find it both here, for the *charmonium* states, and also for the heavier *upsilon* particles made up of *b*-quarks, shown in Fig. 8(b). The number of states and the order of levels are exactly in agreement with the idea that fundamental spin-$\frac{1}{2}$ objects are put together — one particle plus one antiparticle — to make them up.

Still another piece of evidence for the reality of quarks — and again it's because we cannot see them directly that we have to keep making indirect arguments and asking, "Does the world behave as if there really were elementary quarks inside the hadrons?" — comes from looking at the reaction

$$\text{electron} + \text{positron} \rightarrow \text{hadrons (mostly pions).} \qquad (1.3)$$

There are large facilities in which we make storage rings for electrons and positrons and bring them into head-on collision.

In the quark model, we believe that the way this reaction happens is sort of to run backwards the decay reaction we just looked at: the electron and positron come together and make a virtual photon which then disintegrates into a quark and antiquark. We don't observe the quark and antiquark; by some process which is still a little mysterious to us (although we believe we understand it in principle), the quark and antiquark materialize into well-collimated sprays of pions and other hadrons.

Let's study these reactions at high energies. Here (in Fig. 9) is a projection onto a large detector about two meters in diameter at an accelerator laboratory in Hamburg, Germany. The beams were perpendicular to the plane of the page, and you see going out from the collision point one spray of pions here, one spray of pions there. It is difficult not to be led to the conclusion that one spray represents the direction of the outgoing quark and the other the direction of the outgoing antiquark. The routine events that we see at high energy do seem to display and "remember" the directions of the quark and antiquark.

Indeed, you can go further. Knowing that the quark and antiquark are spin-$\frac{1}{2}$ particles like the muon, you can say that the angular distribution of these sprays, the rate at which you see them, with respect to the beam direction, ought to be the same as the angular distribution of the reaction

$$\text{electron} + \text{positron} \rightarrow \text{muon} + \text{antimuon.} \qquad (1.4)$$

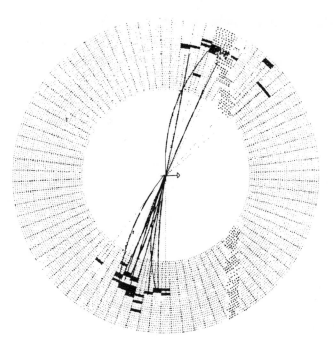

Figure 9: A two-jet event produced in 30 GeV electron-positron annihi-lations.

And that behavior is precisely what is observed.

Well, all this is part of the evidence for the reality of quarks, and as we say in France, it's a good story. It's a good story, but it's not completely consistent. It's not completely consistent because in building models of physical phenomena, it has paid off over the years to respect the grand principles that have great force and wide applicability. One such is the Pauli exclusion principle, which tells us how to build up the periodic table of the elements. The Pauli principle has served us well there. We can show in quantum theory that it must be true, and so it should serve us well for quarks, too.

The problem is that if you make the simplest quark model you can think of for the baryons, for particles like the Delta resonance, the Pauli principle seems not to be respected. Let me just remind you of how that goes. This first resonance, the Δ^{++}, which weighs 1232 MeV/c^2, has charge +2. In the quark model we make it out of three up quarks:

$$\Delta^{++} \sim uuu. \tag{1.5}$$

It's the lowest-mass particle of that kind, and so you expect on general grounds that each of the up-quarks is in an *s*-wave relative to any other: there's no orbital excitation between them. In order to get the total spin of the particle equal to 3/2, all three of the quarks must align their spins in the same direction. And similarly, the isospin, the *up-versus-down*-ness of the quarks, has to be aligned. That is similar to the statement that they are all up quarks. All this means that if a make an interchange of any two of the up quarks in this particle, the wave function is unchanged — symmetric. It's symmetric in space because the quarks are in relative *s*-waves, and in spin and isospin because we have completely symmetric configurations for both of those quantum numbers.

We are taught in quantum mechanics courses that bound-state fermion wave functions, wave functions of particles with half-integer spin, are supposed to be *anti*symmetric when we exchange everything in sight. So we are faced with two logical possibilities. One logical possibility is that the quark model is fundamentally flawed. We have come to a contradiction and either we have to give up the Pauli principle or abandon the quark model.

The other possibility, which seems like the easy way out, but turns out to be extremely profound, is that everything in sight isn't everything there is. There is some new degree of freedom that we haven't thought of yet, and in terms of that new degree of freedom, the three up quarks are *not* identical particles, but in fact can be distinguished. Then we can, if we like, make the wave function antisymmetric in terms of the dintinguishing characteristics. This new degree of freedom now is named *color*. We say that each quark flavor: up, down strange, and the others, comes in three distinct colors: red, green, and blue, if you like, and we require any hadron to be neutral in color. So a proton must be made of a red, a green, and a blue ("white") and a quark and antiquark must be of the same color and anticolor to form a meson.

This seems too easy, to invent something you've never seen before and couldn't see as an excuse for complying with the Pauli principle. Is there not some way to show that this additional attribute, color, is present? Let us return to the very simple reaction of electron-positron annihilation into hadrons to see if we can find evidence for the new degree of freedom. We used this reaction to argue that hadrons were emitted in jets, and that those reflected the production of quarks. Now I'm going to use that fact to make a model in which I can calculate the rate of hadron production, assuming that the things initially produced are quarks. Again, I know how to calculate these rates in all their glory, but I don't want to do that. As I told my students yesterday in the middle of a disastrous calculation on the blackboard, I only do arithmetic in public to make them feel more secure.

I've already commented [see page 42] on the similarity between muon pair production and quark pair production. I'm going to use the rate for muon pair production as the unit of cross section. At any energy, the rate at which muons are produced is the unit called *one*. That's a convenient name because this rate will be proportional to the charge squared of the muon — it's an electromagnetic interaction. The charge squared of the muon is 1, so the cross section is 1.

The quark model lets me make up quarks or down quarks or strange quarks, which then materialize as they choose into hadrons. But I can calculate the rate just by saying that the probability for the quarks to materialize into hadrons is unity. Once I've made the quarks, they will turn themselves into hadrons, and for the moment I don't have to know how that happens. The probability of making up quarks in our convenient units is the charge squared of the up quark, which is $(2/3)^2 = 4/9$. To make down quarks, it's $(-1/3)^2 = 1/9$. And to make strange quarks it's $(-1/3)^2 = 1/9$. So if I add up the three different ways I can make hadrons, I find that the cross section for making hadrons should be

$$\sigma(\text{hadrons}) = 2/3, \qquad (1.6)$$

in units in which the muon pair cross section is one.

That's assuming that there is only one kind of up quark, one kind of down quark, and one kind of strange quark. If I now accept the color hypothesis and say that there are red, green, and blue up quarks or down quarks or strange quarks, then I have not three diagrams of the kind shown in Fig. 10, but in fact nine diagrams, all leading to distinct final states. And so the prediction that I make for the cross section will be, not 2/3 but three times that, or

$$\sigma(\text{hadrons})\big|_{\text{color}} = 2. \qquad (1.7)$$

Now, we may go off and do an experiment (or in fact a whole series of experiments) to see which of these predictions, if any, is true. Here in

45

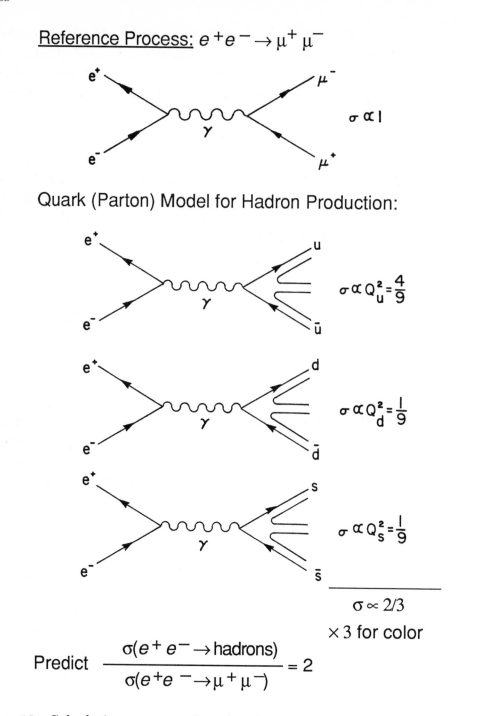

Figure 10: Calculating cross sections for electron-positron annihilations into hadrons.

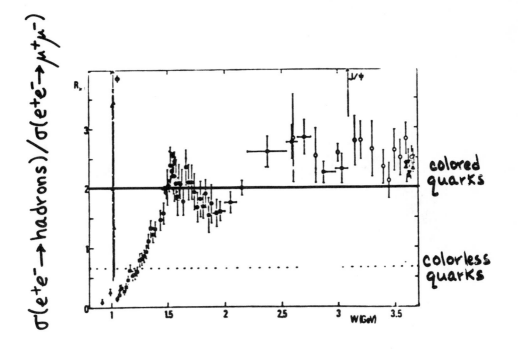

Figure 11: The ratio of hadron production to muon pair production.

Fig. 11 is the ratio of the rate of hadron production compared to the rate of muon pair production, as measured in electron-positron annihilations. At low energies there are individual resonances, the rho and omega (which are not shown on this plot), phi (which is shown here at about 1 GeV), some wiggles, and then after a while the ratio settles down to some approximately constant number. The quark model said the ratio should be a constant, so that's good. And the measured constant is within shouting distance of two, our prediction with colored quarks. It is humiliatingly far from the prediction of 2/3 in the case of colorless quarks. So this is a piece of evidence that the color degree of freedom is present.

As we move up to higher energies, we can make other flavors of quarks like charm quarks and beauty quarks. What happens there is shown in Fig. 12, where the energy scale ranges all the way up to 40 GeV. You can see that from about eleven billion electron volts up to 40 billion volts the cross section is constant and equal to a number close to 11/3. A prediction of 11/3 is precisely what you get by taking three times the

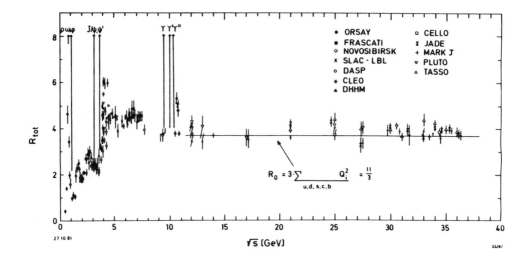

Figure 12: The ratio of hadron production to muon pair production at higher energies.

charge squared of up and down, strange and charm, and beauty. [Charm has charge $+2/3$, beauty has $-1/3$.] And so you see that there is very good agreement between the colored quark prediction and experiment, and there would be terrible disagreement, in the absence of color.

There are a number of other ways of getting at the color quantum number and convincing yourself that it is there, but they are all in this same spirit of counting up degrees of freedom in a more or less direct way. Our knowledge of the quarks is summarized in Fig. 13.

This brings us to a rough knowledge of the fundamental constituents. We have discovered particles which, at the current limit of resolution, are structureless and indivisible. For the quarks there are two and a-half families known, pending the observation of the top quark. [The indirect evidence for its existence is overwhelming.] And for the leptons, there are the three families we have discussed earlier. As we near the end of this lecture, then, our world view has advanced to the state of knowledge represented in Fig. 14. The quarks experience the strong, electromagnetic, weak, and gravitational interactions, and the leptons

QUARKS (COLOR TRIPLETS)			
Particle name	Symbol	Mass (MeV/c^2)	Electric Charge
up	u	310	2/3
down	d	310	$-1/3$
charm	c	1500	2/3
strange	s	505	$-1/3$
top/truth	t	$\gtrsim 22{,}500$	2/3
bottom/beauty	b	5000	$-1/3$

Figure 13: Some characteristics of the quarks.

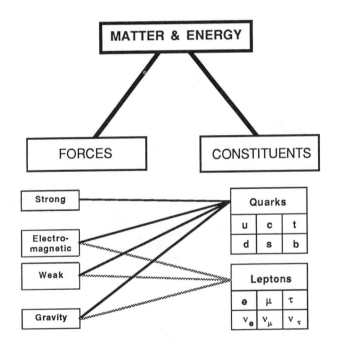

Figure 14: Progress toward the Standard Model.

the last three, but not the strong interaction.

In the second lecture I want to concentrate on the left-hand side of our diagram, but in the few minutes that remain before you rush out to drink coffee, I want to say a few words about experiment. This is offered as a stimulus to thought, and obviously not as the definitive treatment of the subject.

To a good approximation there is a single experiment done in high energy physics. It is shown in Fig. 15. A beam enters from the left and interacts with a target, from which a product emerges. If you are Lord Rutherford, the beam is alpha particles and the target is a gold foil. The product may be the same as the incident beam, or something different. The detector is often depicted in textbooks as a tin cup into which little things fall and collect.

Now, the point of doing these experiments is to try to study what is going on inside the target, and to see what are the manifestations of the interactions between the beam and the target. We have the possibility

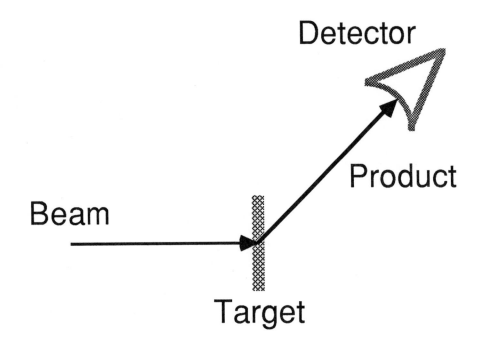

Figure 15: The Experiment.

of changing targets, of varying the properties of the beams by changing species or energies, and of observing different products. In the reactions that we study at Fermilab, it's often the case that the number of products is on the order of a hundred or so, and we want to learn as much as we can about all of those.

What is the goal of a detector system? The goal of a detector system is to measure all you can about everything that happens in the event; that is to say, to measure all the characteristics of all the particles produced in an event. This places the following requirements on a detector: you have to cover as much space as you can, as much of the angular range as possible, in all three dimensions, so you don't miss anything. But for reasons that we'll discuss immediately, you want to have high spatial resolution. If you had just one large tin cup that registered everything coming out of a collision and didn't distinguish where it was coming out, that would be less interesting — because it gives less information — than having a lot of little tin cups and counting who went here, who went there, and so on.

You would also like to be able to identify particle characteristics. And finally, one of the great challenges, particularly now, is to try to select events of interest, the special things that you want to study, from the routine background. There is a saying in particle physics that yesterday's sensation is today's calibration and tomorrow's background.

You want to do all of this keeping the cost of construction, operation, and data reconstruction within reasonable bounds. What are reasonable bounds? There is a detector you can see in a big orange building down the road, which does all these things. The price of that detector (the Collider Detector, CDF) is about $50 million. With respect to data reduction, the coin of the realm here is a VAX-11/780 computer, and the data analysis for that experiment is estimated to require 50 such computers running full time.

Let me now say a few words about the principles that underlie detection, the ways that we can think of learning things about these produced particles.

Charged particles lose energy by ionization as they pass through matter. Of course, there is a whole science and technology built up of how they ionize, which has to do with electrodynamics, the properties of materials, *etc.*, which is itself very interesting and good physics. What you want to do is to measure the position and magnitude of the ionization trails to learn something about where the particles went and what they were.

In some cases, if you measure how long it took to go from here to there, as you do in elementary physics labs, you can measure the velocity of particles and therefore infer something about their identity.

Magnetic fields deflect charged particles into curved orbits. By measuring the curvature of the orbit you can, knowing the properties of the magnetic field, determine the particle's momentum.

Beyond that, different kinds of radiation can be emitted by particles under different conditions. One of the most useful so far in particle physics is the Cherenkov radiation emitted by particles which pass through a medium faster than the speed of light in that medium. A shock front builds up radiation with characteristic opening angle and intensity patterns, and by measuring the intensity of the radiation and the angle with respect to the particle direction, you can make inferences about the energy of the particle, its mass, and other characteristics. Coherent radiation is also emitted by particles crossing the interface between materials (transition radiation), and by particles passing through magnetic fields (synchrotron radiation).

Neutrinos are wonderful particles to detect. They interact so feebly that they are almost not there,[3] and so you infer their presence by the fact that you didn't see something. Pauli's original reason for inventing the neutrino was that there seemed to be missing energy in radioactive beta-decay. In the same way, we can try to sum up all the momentum carried by particles produced in a high-energy collision, and if there is a big lump missing off in that direction, then you say, "Ah, a neutrino or something like a neutrino went off in that direction," because you believe in momentum conservation. So nonobservation can be a good way of observing, provided you can be sure that you would have observed something else, had it been there.

Among the particles that do something interesting when they pass through matter, electrons and photons are special because they produce characteristic electromagnetic showers, converting all the original energy

---

[3]See "Cosmic Gall," in John Updike, *Telephone Poles* (Alfred A. Knopf, New York, 1969), p. 5.

of the particle to ionization and then the relaxation of excited atoms. By recording the deposit. d energy, you can do electromagnetic calorimetry. You can observe the development of the shower and, by adding up all the energy, learn what the energy of the electron of photon was.

Hadrons passing through matter lose some energy to ionization, but also have strong interactions with the nuclei and so they will start nuclear cascades, nuclear showers. If you make a large enough block of instrumented steel, you can again collect all the energy from the incident hadron.

Finally, muons are an exception to these sorts of patterns because they can go through huge thicknesses of material without losing much energy. They radiate much less than electrons do because they are so much heavier, and they do not induce nuclear showers. As a result you can identify muons by making a big block of material and watching what charged particles come out the other side. In the case of the Fermilab neutrino beam, we have about a kilometer or so of steel and earth in the way just to absorb all the muons which otherwise would contaminate the beam.

Using all these principles we can arrive at the idea of *layered detectors*. What you try to do is to exploit different characteristics of the various physical principles of detection to do different things. Close in you need a detector which has very good spatial resolution and can sustain high rates because lots of particles are emerging from a small volume. There is a special class of detectors called *vertex detectors* used close in. Next there are charged particle tracking chambers which trace the progress (often through a magnetic field) of particles coming out from the collision point.

Combined with this, or sometimes in addition to this, there is often an attempt made to identify particle types by using some of the coherent radiation schemes. After all that nondestructive tracking has been done with only a little material in the way of the outgoing particles, you then put lots of material of various sorts in the way to do the calorimetry, contrived so that anything that penetrates the entire detector must be a muon, which you may wish to measure again.

Here (Fig. 16) is a picture of the Collider Detector, which I hope you will take the time to see while you are here at Fermilab. Note from the sketch that a typical person is one-fourth to one-fifth the size of the detector. An exploded view of half of the detector is shown in Fig. 17. There is

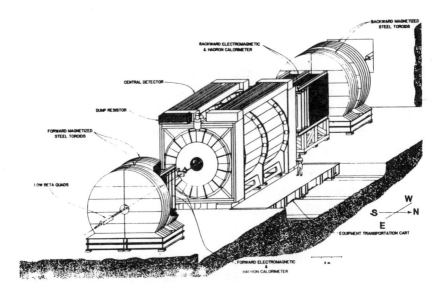

Figure 16: CDF, the Collider Detector at Fermilab.

a highly sophisticated vertex tracking device around the interaction point. Then you find, all immersed in a superconducting solenoid, the central tracking, an electromagnetic shower calorimeter, hadron calorimetry, the magnet yoke (which is iron), so that the particles which penetrate to the outside should be muons. That's all in the central region. The same sorts of pieces are found as you go toward the forward direction. At each location and for each task, you try to choose the best detector in terms of performance, reliability, cost, and so on.

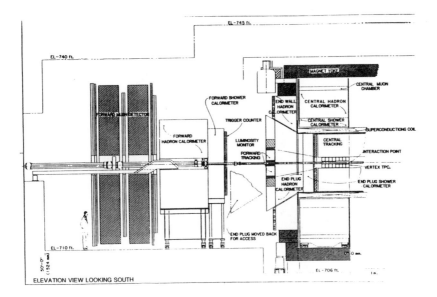

Figure 17: Exploded view of the Fermilab Collider Detector.

I do hope that while you are here you will spend some time looking at these detection devices and trying to understand a little bit about them. After the break, we will move on to the strategy of gauge theories.

LECTURE 2: THE IDEA OF GAUGE THEORIES

In this second talk, I want to focus on the interactions and to explain a bit of the motivation for gauge theories, and the basic elements of the gauge theory strategy. What we shall see is that symmetries in Nature, when we recognize and use them properly, can be used not only as restrictions that guide the formulation of theories, but also as tools that help us construct the theories directly.

Let us now recall the theories currently in use to describe the strong, weak, and electromagnetic interactions. The first of these, and in many ways the prototype for quantum theories, is *Quantum Electrodynamics,* or *QED*. This is the most successful of physical theories: it works, essentially without modification, from distances on the subatomic scale down to nearly 10^{-16} cm out to enormous distances on the interplanetary scale. When you consider that the theory is built upon experiments first done by Cavendish and others on the scale of half a meter or so, the success of the extrapolation is really quite striking.

Quantum electrodynamics is in part the model for, and is incorporated in, the theory of weak and electromagnetic interactions brought to its final form by Weinberg and Salam 20 years ago. The resulting theory describes at the same time the weak and electromagnetic interactions. Although for the moment it is not nearly as well tested as QED itself, the electroweak theory has many very precise experimental successes. It anticipated a new kind of radioactivity called neutral weak currents, required the existence of the charmed quark, predicted the recently discovered carriers of the weak interactions, W^+, W^-, and Z^0, and (to the level at which we have been able to do experiments) gives a precise and quantitative description of everything we see in the electroweak realm.

A theory that we'll discuss at somewhat greater length is *Quantum Chromodynamics,* a theory of the strong interactions. It is called "chromo" because it is based on the idea that the *color* property of quarks which distinguishes them from leptons and enabled the quark model to survive the Pauli principle functions in some sense as a strong charge. And so the theory is called *QCD* in imitation of QED.

QCD is based on the color symmetry of the quarks in a way we'll review a bit later. For a variety of reasons, not least of which is that the strong interactions are strong and theoretical physicists are only good at calculating the consequences of feeble interactions, QCD has not yet been tested as precisely as the other interactions. It does give us lots of insight into the systematics of high energy collisions and the spectrum of hadrons. It predicts force-carrying particles called *gluons,* and in some restricted realms there are some quantitative successes which are rather impressive.

I'm now going to explain where gauge theories come from, and the strategy involved in deriving them. So far as we can tell, gauge theories provide the basis for correct, useful descriptions of all the fundamental interactions. They have a number of properties which we'll talk about later on. The reason for talking in general terms about how we construct gauge theories is that it's very easy to make up theories, and it's particularly easy to make up wrong theories. If you can find some guiding principles, they may restrict your search for different classes of theories. Now, you have to be careful not to restrict yourself too much, but if you pick a guiding principle like energy conservation or Lorentz invariance or some such, which is supported in great detail by lots of experimental data, and say provisionally that you will only look at theories which satisfy that principle, then you've saved yourself the trouble of looking at a lot of theories which have no chance of being correct. In the same spirit, if you can find and attach yourself to a principle which will lead you only to make theories from the class of those that might possibly be right, that's a good thing, at least in terms of economy of effort.

The strategy of gauge theories goes roughly like this. We recognize a symmetry in Nature. This afternoon you will be reminded that for many sorts of symmetries (continuous symmetries like rotation invariance, translation invariance, and so on), there is a deep connection with conservation laws. Rotation invariance is intimately related with the conservation of angular momentum, for example. By recognizing conservation laws, by seeing symmetries in Nature, we are led to build equations of physics that respect the symmetries in question. Having done that, we then try to impose the symmetry in a stricter form. I'll show you immediately by means of an example what I'm trying to say here, but for purposes of giving an outline let me proceed without explaining. When the new requirement is imposed, it will happen that the equations of physics from which we began must be modified in order to accommodate the stricter form of the symmetry. This can be done in a mathematically consistent way only by introducing new sorts of interactions, and new particles to carry those interactions.

There is an opportunity for blunder here. If I pick a symmetry that I think I see in Nature and I go through this program, I may well arrive at a theory which is mathematically self-consistent but which, because I was inept in my choice of the symmetry, doesn't describe the world we live in. The literature is littered with the corpses of such theories, and I will spare you examples of them.

Now, you may ask, "What is he trying to say? What does all that mean?"

To give an example, I have to beg your indulgence. The indulgence is to suppose that we know quantum mechanics but not electromagnetism. Now, from the times I've taught graduate courses in electricity and magnetism, I know that half of that statement (at least) is likely to be true. And from the times I've taught quantum mechanics, I have my doubts about the other half of the supposition.

I'm going to begin with quantum mechanics and lead us to electromagnetism.[4] What I need to know about quantum mechanics is that the

---

[4]This may seem to be a fake, because it's not the way electromagnetism was invented. But it should have been!

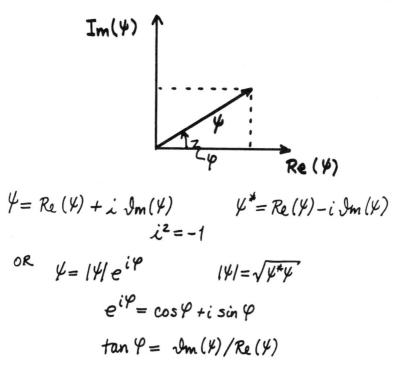

$$\psi = Re\,(\psi) + i\,Im(\psi) \qquad \psi^* = Re\,(\psi) - i\,Im(\psi)$$

$$i^2 = -1$$

OR

$$\psi = |\psi|\,e^{i\varphi} \qquad\qquad |\psi| = \sqrt{\psi^*\psi}$$

$$e^{i\varphi} = \cos\varphi + i\,\sin\varphi$$

$$\tan\varphi = Im\,(\psi)/Re\,(\psi)$$

Figure 18: Argand diagram representation of the quantum mechanical wave function $\psi(x)$.

quantum mechanical state of a system is described by some complex wave function called $\psi(x)$. This is a complex function with a real part and an imaginary part and, if I like, I can describe it as a vector in an Argand plot as in Fig. 18. As you know, you can characterize that in various ways, as shown in the sketch. Corresponding to the wave function ψ there is the complex conjugate ψ^* of the wave function, its reflection about the real axis. Of the various representations for the wave function given in Fig. 18, the one most convenient for our purposes will be to write the wave function as

$$\psi(x) = |\psi(x)|\exp i\phi. \qquad (2.1)$$

Now, everyone knows that in quantum mechanics observable quantities, things you can measure in the laboratory, are expressed as *expectation values* or *scalar products* which are integrals over some appropriate region of space, of a volume element times the complex conjugate of the wave function times a Hermitian operator O times the wave function,

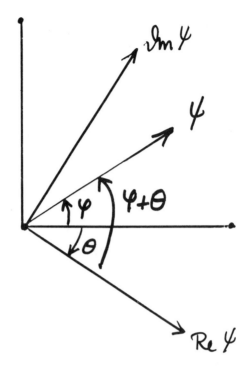

Figure 19: New definition of the real and imaginary axes.

symbolically

$$\langle O \rangle = \int_{(V)} dV \, \psi^*(x) \, O \, \psi(x). \tag{2.2}$$

We can verify by heavy-handed means that the quantity we're going to measure is unchanged if we redefine the phase. I come along and say that the real and imaginary axes in Fig. 18 are very fine for you, but I don't like them. I'm going to change to the new coordinate system shown in Fig. 19. In terms of those coordinates, I can of course measure the real part and the imaginary part of ψ, or express ψ in terms of a new set of polar coordinates. Rotating the definition of the real axis down by an angle theta is equivalent to multiplying ψ by $e^{i\theta}$:

$$
\begin{aligned}
\psi &\rightarrow e^{i\theta}\psi \\
\psi^* &\rightarrow e^{-i\theta}\psi
\end{aligned}
\; ; \tag{2.3}
$$

ψ^* gets rotated in the opposite sense.

In terms of the new ψ and ψ^*, our observable becomes

$$\langle O \rangle = \int_{(V)} dV \, \psi^*(x) e^{-i\theta} O e^{i\theta} \psi(x)$$

$$= \int_{(V)} dV \, \psi^*(x) O \psi(x). \tag{2.4}$$

The factors $e^{-i\theta}$ and $e^{i\theta}$ eat each other up, giving back *one*, so the quantity we are calculating is unchanged by the operation of changing coordinates. It is the same before and after I've made the change of phase indicated in Fig. 19. That is to say that the absolute phase of the quantum mechanical wave function is *arbitrary*. It is not something to which measurements can be sensitive.

Now, in fact this sort of phase symmetry has a deep connection, if you formulate it properly in detail, with the conservation of electric charge. From phase symmetry of precisely this kind you can derive the fact that the electric charge must be conserved.

For the moment, I'm not going to focus on that, but only to admire the fact that I could make this change of convention. Just to put it in symmetry language, I can say that ordinary quantum mechanics is invariant under *global phase rotations,* phase rotations in which the convention is changed by the same amount at every seat in the Fermilab auditorium. Here, in Fig. 20(a), is where we started out. Each of you agreed with me that this would be our direction for the positive real axis, the original convention for zero phase ($\phi = 0$). Later on, we all agreed together that the direction shown in Fig. 20(b) would define the direction of zero phase. We found that physics didn't change when we made that rotation.

Now, some of you might object and say, "Why should *you* be able to tell *me* what *my* phase is? Couldn't we be more democratic and choose a different phase convention independently at every point in space?" Not in a haphazard fashion: you might want to have some common harmony with your neighbors, but might it be possible to have a position-dependent definition of the zero of phase? Would that be all right? Do the laws of quantum mechanics admit that sort of symmetry, a more general symmetry than the phase invariance we have just investigated?

This freedom to choose a position-dependent phase convention would mean that instead of multiplying my wave function by a fixed rotation, $e^{i\theta}$, I would multiply it by a position dependent phase, $e^{i\alpha(x)}$. Does that

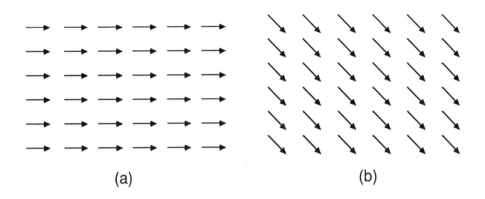

Figure 20: (a) Original convention for zero phase of the quantum mechanical wave function; (b) new convention for zero phase. Each arrow represents a seat in the Fermilab auditorium.

work? Well, let's suppose we are talking about the quantum mechanics of a free particle. We can check whether the new symmetry is respected either by talking about observable quantities, as we have just done, or by plugging the transformation law into the Schrödinger equation and asking whether the same equation holds before and after the phase change. Let me do the exercise in terms of observables; what we'll find is that the quantum mechanics of a free particle is not invariant under local phase rotations.

How can we see that? There are observables in the world like momentum, and there are pieces of the Schrödinger equation itself, which involve derivatives or gradients. What we can do is to calculate how a gradient changes when you make a position-dependent phase change on the wave function. Here is what happens: as

$$\psi(x) \rightarrow \exp i\alpha(x) \cdot \psi(x), \tag{2.5}$$

the gradient becomes

$$\nabla\psi(x) \rightarrow \exp i\alpha(x) \cdot [\nabla\psi(x) + i\psi(x)\nabla\alpha(x)]. \tag{2.6}$$

Unlike the wave function, the gradient is not simply multiplied by a phase. The fact that $\alpha(x)$ has a position dependence means that its gradient is nonvanishing, and that gives rise to the second term on the right-hand side of Eqn. (2.6).

That extra term means that if I try to calculate an expectation value like

$$\psi^* \nabla \psi \to \psi^* e^{-i\alpha(x)} e^{i\alpha(x)} \left[\nabla \psi(x) + i\psi(x) \nabla \alpha(x) \right], \qquad (2.7)$$

I do not recover the original value. So the answer to the question, "Does ordinary quantum mechanics admit a local variation of phase of the wave function?" is that it does not.

Now, especially after Chris Hill's lectures this afternoon in which symmetry will be made a part of your being, you might ask yourself, "Couldn't I change the equations of physics a little bit because that symmetry seems so nice and appealing?" In other words, is it possible, if only as a little mathematical homework problem, to make some changes in the equations so that everything will work out and the modified equations admit this more general phase symmetry?

The answer to that question is yes, but only if you introduce an *interaction*, a specific kind of interaction called a *gauge field*. What we need to introduce will be the electromagnetic field, or something like it. Let me show you the arithmetic rather schematically. I'll then remind you that the answer we get to is something you already know, and you will be prepared to take the gauge theory leap of faith.

The solution is that I'm going to introduce (in three vector notation) an electromagnetic vector potential $\mathbf{A}(x)$, and I'm going to make the following rule: when I rotate the phase of the wave function by an amount[5]

$$\psi(x) \to \exp iq\phi(x) \cdot \psi(x), \qquad (2.8)$$

I'm going to change my newly introduced electromagnetic vector potential by shifting it my an amount

$$\mathbf{A} \to \mathbf{A} - \nabla\phi. \qquad (2.9)$$

How do I know to do that? I know to do that because I went through the equations and asked, "What do I have to do so they come out right after phase rotation?" And this is the answer.

Now, in addition to all that, I make an agreement with myself that everywhere in the laws of physics — in the definition of observables, in

---

[5]In terms of our earlier notation, I have renamed α to be $q\phi$, where q is supposed to be the electric charge. That's to suggest that the theory we will derive is electromagnetism. If I like, I can choose another charge and derive another theory.

the Schrödinger equation, and so forth — everywhere I see a gradient I will replace it with something named *the gauge covariant derivative,*

$$\mathcal{D} \equiv \nabla + iq\mathbf{A}. \tag{2.10}$$

Those of you who are good at juggling factors of i and $\hbar$ will notice a resemblance between this expression, to which I've been led by fiddling the equations of physics, and the familiar replacement of classical electrodynamics in which the momentum $\mathbf{p}$ becomes $\mathbf{p} - q\mathbf{A}$. This is a source of comfort and reassurance.

What remains is to verify that this new object, the generalized gradient, when acting on the wave function goes into simply a phase factor times itself under the combined transformations (2.8) and (2.9). As a little homework exercise, you can check that $\psi^* \mathcal{D}\psi$ is invariant under local phase transformations.

And so, I've invented a theory. I've had to change the gradient, redefine the momentum operator, *etc.* But I've invented a theory in which the appearance of the equations is identical before and after local phase rotations, and all the observables we can imagine will be the same before and after the local phase rotations.

The fact that you've seen the final results before invites you to believe — and it's even true — that the theory we've derived in this way is exactly the theory of electromagnetism. If we do the same steps in a covariant, relativistic way, the theory we derive is precisely quantum electrodynamics.

So that's the arithmetic of it, and that's the general strategy of it. We can carry out the same kind of analysis for other theories or for more complicated theories. The arithmetic becomes more involved, but the strategy is always the same. The encouragement for trying the strategy in other settings comes from noticing that we can recover the idea of QED by starting with a symmetry and proceeding along these simple lines.

The phase symmetry is just the gauge invariance of quantum electrodynamics; the shift we made in the vector potential is the freedom textbooks normally call gauge invariance. Something to be emphasized to students is that gauge invariance means more than just the ability to choose arbitrarily the zero of a potential. It has, as we have just seen, a deep connection with symmetry through quantum mechanics.

Now that we've looked at one example, we can ask what are the general consequences of this strategy. *Global symmetry,* in which we make a continuous transformation (like a phase rotation) everywhere by the same amount, leads to a conserved current, a conserved charge. In the example we have considered, this is the electric charge. The *local symmetry* implies in addition that there must be an interaction. It had to be mediated by a spin-one vector field. It turns out that it had to be a massless field. And furthermore, at least if you follow your nose, the interaction between that new force and matter turns out to be a form traditionally known as "minimal coupling."

In this light we can think of electrodynamics as the gauge theory [the theory built upon this phase invariance or gauge invariance[6]] built upon the group of phase transformations, the group of rotations in a plane called the unitary group $U(1)$. Can we do the same for other continuous groups? Do they have to be commuting ("Abelian") groups, or not? The answer is that you can always construct a theory, for any continuous gauge group. Some of them will have more complicated properties, but you can always make the construction.

That completes the first topic for this lecture, how to construct a gauge theory. I told you a moment ago that electrodynamics, grossly speaking, in the form of Maxwell's equations, is valid not only down

---

[6] Why do we use the term gauge invariance? The original argument in the style we have just explored was given in papers written around 1921 by Hermann Weyl. At that time, the known forces were electromagnetism and gravity, and so it was a natural impulse to try to give a unified basis for the two. Gravity had something to do with geometry, and so it was natural, I suppose, to try to think of a geometrical basis for electromagnetism. Weyl's contribution, for which the general strategy is exactly the one I used today, was to say, suppose that there is a scale invariance of the world, so that the laws of physics have to be the same as the scale, or measure of length, changes from point to point. Requiring the equations of physics to have this invariance, he found the necessity of inventing an interaction and hoped to identify this interaction with electromagnetism. It turns out, as we've just seen, that you need a phase change rather than a scale change to recover electromagnetism, but the idea — the strategy — has persisted. Since quantum mechanics wasn't invented until a few years after he'd made this proposal, we can hardly blame Weyl for not understanding the importance of phases at the time.

In Weyl's original papers he used a German term, *Eich,* meaning calibration or gauge. Following correspondence with Fock, London, and others, after the invention of quantum mechanics, Weyl changed his program to one of phase invariance, but retained the old term. Gauge invariance had caught on, and it is the term we still use today.

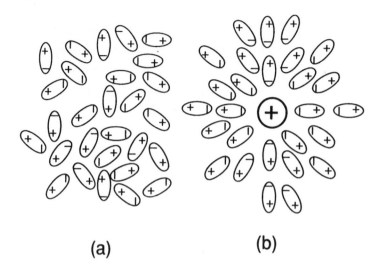

(a)   (b)

Figure 21: Bipolar molecules in a dielectric medium: (a) disordered state; (b) ordered (polarized) state in the presence of a test charge.

to very short distances (around 10^{-16} cm), but also applies out to very large distances. The best measurement that I know about comes from measuring the rate of falloff of the magnetic fields of the large planets, Jupiter and Saturn. It indicates that Maxwell's equations hold out to a distance of about $4 \cdot 10^{10}$ cm. There is indirect evidence that pushes the range of validity out another twelve orders of magnitude.

What is interesting for electrodynamics, and later for the theory of the strong interactions, is that in spite of the theory's enormous range of validity, there are well-understood modifications to the simple behavior in a polarizable medium. This is the phenomenon of charge screening. Let me make a model for such a medium. Here, in Fig. 21(a), is a polarizable medium in which molecules behave as little boats with a positively charged end and a negatively charged end. In the absence of some external source of charge, the boats are distributed without macroscopic order. Of course, the oppositely charged ends attract each other in such a way that, grossly speaking, in any lump of this medium you will find a net charge zero: the charges cancel out, on balance.

What happens if I stick some positive charge right in the middle of the medium? As long as the little boats are free to move about, they will orient themselves so that the negatively charged ends of the boats are attracted to the positive test charge. Schematically, the pattern will be as shown in Fig. 21(b).

The effect of this is that if I imagine probing the test charge, measuring the charge in the medium by inserting a hypothetically nonperturbing probe, the charge my probe feels will be less than the charge carried by the test charge. This is because the total charge enclosed in a circle centered on the test charge, with radius given by the distance from the test charge to the probe, is less than the test charge itself.

In order to see the full strength of the test charge in a molecular substance, I must approach the test charge closely — so closely that my probe is within the molecular scale. Once my probe is there it sees the whole charge, unscreened by the molecules in the medium. That's a gross way of indicating that the effective charge, the charge I measure, increases at short distances.

Because of quantum mechanical effects and the possibility that the vacuum can fluctuate into pairs of electrons and positrons for very short times, the same thing occurs in the vacuum. The vacuum that we live in is not an empty thing, but something in which pairs of electrons and positrons are coming and going all the time. While they are here, they can be polarized by a local test charge. The effect of that *vacuum polarization* is precisely the same as in a dielectric medium, to screen a charge and to make the effective charge larger at short distances than at large distances.

I now want to move on, building on the idea of gauge theories, to the force between quarks as another example of how we take a symmetry and build a theory from it. We noticed that every flavor of quark (*e.g.* up, down, strange, and charm) came in three distinct varieties called colors. And in order to make the mesons, we had to have antiquarks which came in anticolors. We named these colors red, green, and blue, but we could have chosen A, B, and C.

Physics isn't supposed to change if we change the names. The color symmetry means that when I interchange the names red, green, and blue, or reassign them in some continuous fashion, nothing should change. So I should build my laws of physics to have that property. That is to say that the interactions of a red quark and a green quark and a blue quark should all be the same. That suggests that I might be able to build a theory in which that freedom to name red, green, and blue is respected ocally: a gauge theory of the color force, QCD.

Now, in this case, because we have three kinds of charge instead of only ne, as in the case of electrodynamics, the arithmetic is more complicated,

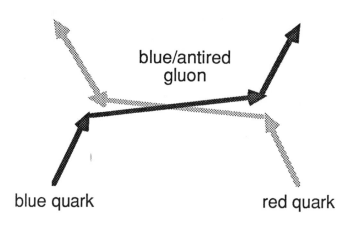

blue/antired
gluon

blue quark

red quark

Figure 22: Quark-quark scattering in QCD.

and so I won't work it out in public. But you can make such a theory, a theory in which there are interactions between the quarks mediated by massless spin-one particles. We call these force particles gluons because they glue the quarks together. The strategy we have followed is to think of color, the attribute that differentiates the quarks from the leptons, as the charge of the strong interactions, and to build a theory based on local color symmetry.

What do these interactions look like? Fig. 22 shows the scattering of a blue quark and a red quark. They interact by exchanging a blue–antired gluon, and emerge as a red quark and a blue quark. Notice that in this example the gluons themselves carry color charge, in fact one color charge and one anticolor charge. Since the gluons are colored, they will have strong interactions mediated by gluons. You can construct these interactions just by drawing colored pictures. Here, in Fig. 23, is a green–antiblue gluon scattering from a green–antired by exchanging a blue–antired gluon.

Now, you've never seen a quark but you all believe that they exist. How can I convince you that the gluons exist? By doing a variation on one of the ways that I convinced you that quarks exist. Back on page 42 we talked about electron-positron annihilation into hadrons proceeding through the the formation of a quark–antiquark pair. The quarks materialized into hadrons, mostly pions, which remembered the direction of their parent quarks. That gave us the two-jet events of the kind illustrated in Fig. 9.

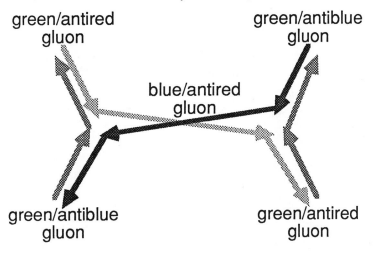

Figure 23: Gluon-gluon scattering in QCD.

In much the same way, now that we have invented gluons interacting with the quarks we may imagine that sometimes one of the outgoing quarks radiates a gluon,

$$
\begin{aligned}
\text{electron} + \text{positron} \;\rightarrow\; &\text{quark} + \text{antiquark} \\
&\;\;\llcorner\!\!\rightarrow \text{quark} + \text{gluon,}
\end{aligned}
\tag{2.11}
$$

just as an outgoing muon may radiate a photon in

$$
\begin{aligned}
\text{electron} + \text{positron} \;\rightarrow\; &\text{muon} + \text{antimuon} \\
&\;\;\llcorner\!\!\rightarrow \text{muon} + \text{photon.}
\end{aligned}
\tag{2.12}
$$

When that happens, I expect my two-jet event to change into a three-jet event. One quark jet splits into a quark jet and a gluon jet.

At high energies in electron-positron annihilations, three-jet events are quite common. Fig. 24 shows a picture of one in the same detector in which we saw the two-jet event in Lecture 1. You can see one fully developed jet, and two smaller jets. The fully developed jet may represent the debris from the quark. Then the smaller jets are the offspring of the antiquark and the gluon. The frequency at which these events are seen and the detailed properties of the events are all consistent with the idea that the mechanism for generating them really is a quark, an antiquark, and a gluon in the semifinal state before the hadrons materialize.

Now I want to talk about polarization effects and the effective charge of the quarks. There will be similar screening effects to the one we discussed for electrodynamics. In this case, since we have three kinds of

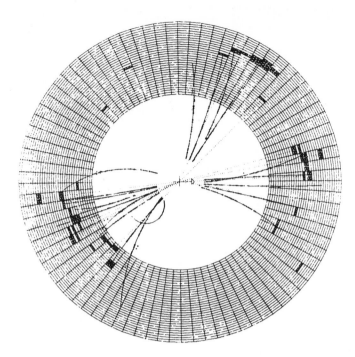
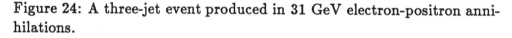

Figure 24: A three-jet event produced in 31 GeV electron-positron annihilations.

charge, I can imagine that the molecules in my analogy are little triangu-
lar objects which have a red corner, a green corner, and a blue corner. In
the absence of a test charge, they will be oriented in some neutral, disordered way. If I insert a test charge, say a red quark, that will attract the
blue and green corners of the surrounding "molecules," and repel the red
corners. The resulting arrangement is shown schematically in Fig. 25.

Just as before, if I ask at a certain radius how much redness lies within
a circle, the result will be less than the redness of the test charge because
some of it is screened out or cancelled by the antiredness from the blue
and green corners of the triangles. There is a color charge screening in
this case, which tells you that the effective charge tends to become larger
as you probe on shorter distance scales. This is entirely analogous to
what we saw in QED.

The difference in this case is that there is something else that can
happen, because the gluons carry color. Because the gluons carry color,
quarks can *camouflage* themselves and hide their color. Fig. 26(a) shows
our test charge, the red quark. We now send some emissary in to say,

Figure 25: Colored molecules polarized by a test charge.

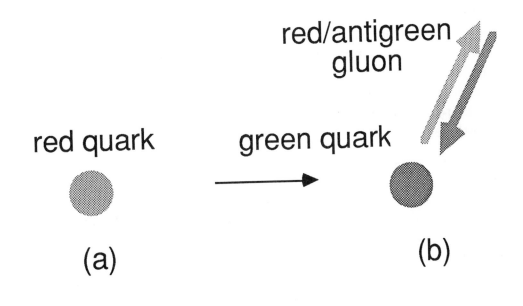

Figure 26: Camouflage.

"Hello, are you red?" While our emissary is on the way in to ask whether this is a red quark, the quark can fluctuate quantum mechanically into a quark and a gluon. And if it chooses to, it can fluctuate into a green quark and a red–antigreen gluon. The red–antigreen gluon goes out and takes a walk in quantum mechanics space, as indicated in Fig. 26(b).

Our probe arrives and says, "Hello, are you red?" And the quark says, "No, I'm green. Go away." So because of the fluctuations made possible by the fact that gluons can carry color you find, if you look too closely, *less* red charge than you thought was there. In order to see the full red charge, you've got to look on a bigger scale, the scale of the promenade of the gluons.

We have two effects going on: one, the normal screening effect as in electrodynamics; the second, the camouflage effect made possible because unlike the photons, which don't have an electric charge, the gluons do have a color charge.

There's a competition between these two effects, and in the theory we believe to be true, QCD, camouflage wins. The consequence of that is that the strong force, as measured by the effective color charge, becomes weaker and weaker at short distances. If you look closer and closer you find that the strong charge is getting tinier and tinier. What this means is that for practical purposes if you find quarks close together in a small space inside a bubble or a little balloon or a bag, they behave almost like independent particles. Because of the camouflage effect, as long as the quarks remain close together, each one hardly feels the color charge of the others.

On the other hand, if you try to separate two quarks by a large distance, then each is able to see more clearly the full charge of the neighboring (but no longer very close) quark. And so the strong force becomes more formidable as you go to large distances. We believe that this effect, properly implemented, is responsible for the fact that we can talk about the quarks as being quasi-free particles within protons, but we can't extract the quarks from the protons. The net *antiscreening* of color charge gives us the possibility of understanding that apparent paradox.

Let me now say just a few words about the theory of weak and electromagnetic interactions. The symmetry that we recognize here is the family symmetry between, say, the electron and its neutrino or the muon and its neutrino. This is a family pattern which seems to be perfectly respected for the leptons and very well respected for the quarks. What we do is to take that family symmetry and combine it with the phase invariance that we saw was a good thing in electromagnetism.

When you do that cleverly, you find that the resulting theory is a rather agreeable one in which the force carriers are the photon, two carriers of the charge-changing weak interactions, W^+ and W^-, plus a fourth force carrier called Z^0. The first three were expected on the basis of previous observations, but there was no evidence for charge-preserving weak interactions at the time the theory was formulated.

The rest, as they say, is history. The new kind of weak interaction which would have been mediated by the Z^0 was in fact discovered in experiments first at CERN, then here and at Brookhaven in 1973. The properties of the new interaction were refined by experiments over the next five years, and had precisely the character outlined by the electroweak gauge theory. Now, those of you alert to the newspapers may recognize that I asserted to you a few moments ago that the carriers of these forces had to be *massless* particles, and yet you have read that the particles carrying the weak interaction, the W and the Z, weigh 100 times as much as a proton.

And so there is something which had to be understood. The great contribution of Mr. Weinberg and Mr. Salam was to understand how to use a phenomenon called *spontaneous symmetry breaking* to change the force carriers to massive particles. Unfortunately, if I'm going to get to the University of Chicago in time for my afternoon class, I won't have time to tell you about that. Ask someone in the discussion section to explain how it works.

This then is the Standard Model (Fig. 27). Let us put aside gravity, since gravitation is generally a weak perturbation on particle physics. We have a few elementary forces, all of the same mathematical character. They are all mediated by spin-one particles whose properties we understand rather well, and which are given to us in large measure by the symmetries that generated the theories. We have a few (although perhaps not few enough) elementary particles. Putting together these elements we should be able to understand everything!

Now, the mathematical similarity of these theories and the observational similarity of quarks and leptons — the fact that apart from the color quantum number they seem to be so similar — invites us to ask,

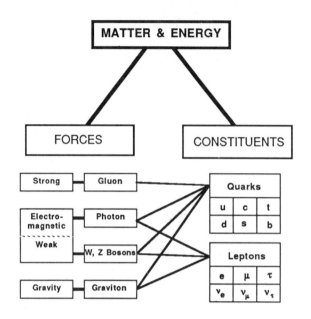

Figure 27: The Standard Model of Particle Physics.

"Is is possible to put the quarks together with the leptons? Is it possible to give a common basis to all these theories?"

The answer to that, at least in principle, is that it seems like a good idea to try, and that we know how to do it by constructing examples of *unified theories,* whether or not these theories turn out to be true. Howard Georgi will tell you more about this opportunity tomorrow.

Where do we stand, then, at the end of all this? We stand in a pretty good place. We have arrived at a fairly simple scheme. It has broad applicability, and if we ask how well are gauge theories tested, it is fair to say that there are no experimental embarrassments. There is no single piece of data that is both true and says the whole idea of this theory or that must be wrong. That is very important.

There are lots of predictions which we have to make sharper by doing the theoretical calculations better, and many others for which the experiments have been very difficult or outside the range of our instruments, and those must be tested better. QED, of course, is the standard by which we judge other theories; it is really very good. For the electroweak theory, the tests are becoming very quantitative at the level of tenths of a percent, but we still need to test it further. And for quantum chromodynamics, a few tests are becoming quantitative and we are learning how to do better.

For unified theories in which we try to put everything together, as you'll see in Howard's lectures, the questions that we're asking still are at the level of yes and no: Is this a good idea? Do its essential consequences actually follow in Nature? We're not really at the level of comparing prediction with observation for numbers.

Let me now take just a couple of minutes to talk about problems. I've sketched for you today this edifice (Fig. 27) of elementary particles and forces, and tried to convey to you a certain enthusiasm for the style of arguments that led us here, and a certain respect for the success of the theories in making predictions about the world around us. When confronted with that success we may briefly celebrate with our colleagues who invented the ideas and made the observations from which they sprang. But then you have to ask yourself, if this theory works so well, why is it working so well? Is it really internally consistent? What are the reasons that it goes together so well? That kind of questioning represents a whole range of activities now going on.

In addition to that, once you've solved every problem in sight, you can look at the few problems you didn't solve which are now made approachable by virtue of the last problem you solved. Let me mention a couple of those.

The theory of the weak and electromagnetic interactions helps us understand why quarks and leptons have masses. If you work out the formalism in more detail than I did this morning, you find that going through the arguments about recognizing a symmetry and then hiding, or spontaneously breaking, the symmetry, that you're given little spaces in the equations, little empty boxes where the mass of the electron belongs, and the mass of the muon, and the mass of the up quark, and so on. That's very good because until the invention of that theory, you didn't have those little boxes to write the numbers in — you didn't know how the masses could come about.

So that's progress. It's incomplete progress because nobody tells you from first principles what numbers to write in the boxes. At the moment, it is still information you have to take from experiment. It would be nice if we had a more complete theory in which we were told not merely that here's a box to put a number in, but here's the number to write in it or here's how to compute the number that goes in there.

Another annoyance: we have several sets of quarks and leptons, but for ordinary experience all we need is the electron and the up and down quarks that make up protons and neutrons. Crudely speaking, that's enough to account for us, so why should we need these other things? There are indications from the internal consistency of the electroweak theory that quarks and leptons do go together in some way. But why they come together in the way we observe, why there are three sets of them, whether there are more, all is outside the scope of present theories.

A lot of the complaints we have about the standard model have to do with *arbitrariness*, the general problem of having boxes that need numbers written in them. So, to engage in a little self-flagellation, let's count the parameters of the standard model. It doesn't matter that I haven't told you yet what some of them are; you'll get the picture. There are three coupling strengths for the strong, weak, and electromagnetic interactions, six quark masses, three numbers that describe how the weak interactions of the quarks cross family lines, something called the CP-violating phase, two parameters of something called the Higgs potential, three masses for the electron, muon, and tau, and one number called a vacuum phase (you don't care what that is). But if you add them all up, it's a big number, namely 19.

If I go further and make a unified theory of the strong, weak, and electromagnetic interactions, I get some interrelations between parameters, but in order to build such a theory I need to introduce some new parameters. So the number is still around twenty. That seems not completely satisfying.

The other thing you can do is to count up the number of fundamental fields. Leaving aside the discovery of the top quark (which will come sometime soon), there are 15 quarks if you count all the colors, six leptons, one photon, three intermediate bosons W^+, W^-, and Z^0, eight colored gluons, a Higgs boson, and a graviton. [That last one is to show I'm not hopelessly reactionary.] This too is a total which exceeds the number of fingers and toes of a single theoretical physicist.

Well, there are lots of speculations about how to make our present theories more complete, and how to go beyond them, and all of us are hard at work on that. In addition to theoretical work, the other thing we need is clearly to get more experimental information, and to do this at the highest possible energies — the shortest possible distances. Part of the beauty of the current framework is that it is good enough, it needs to be taken seriously enough, that we can trust it to tell us when it doesn't work any more. In the case of the electroweak theory, the frontier is particularly well defined. From general arguments about the structure of the theory as it now stands, from every invention we've made to go beyond the standard electroweak theory, there is an indication that new and important clues have to be found in collisions of the fundamental particles at energies around 1 TeV, 10^{12} electron volts.

Because of this, when I'm not standing here in the Fermilab auditorium, one of the ways I occupy my time is in trying to convince the taxpayers of the United States that they should build for us an instrument to explore the 1 TeV scale. The device we have in mind is a large superconducting proton-proton collider. We want to have energies of 20 TeV per beam so that quarks and gluons and other things inside the proton will themselves carry several TeV into the elementary collisions. We use superconducting magnets to make a strong magnetic field to confine the protons in a relatively small circle as we're accelerating them, and also to lower the power consumption. The present design calls for magnets of about 6.5 tesla.

How big is this device? Well, you all know the formula for the radius of curvature of a charged particle moving in a magnetic field. You may not know it in the appropriate engineering units, which are[7]

$$\text{Radius} = \frac{10}{3}\text{km} \cdot \frac{\text{Beam Momentum}}{\text{TeV}/c} \div \frac{\text{Magnetic Field}}{\text{Tesla}}. \qquad (2.13)$$

And so for a 20 TeV beam in 5 Tesla magnets, the radius of curvature would be about 13 km. If you make allowances for straight sections in which to do the experiments and the acceleration, this is a device which is about twenty miles in diameters. It is a large undertaking, and we are taking care to propose it in a sensible and responsible way.

I show you in Fig. 28 that it is not completely out of scale with human experience and human structures. At the left of the picture you can see the size of the Fermilab ring, a four-mile circle you can jog around during your visit. The largest circle shows the size of the supercollider we would like to build. The irregular loop is the Washington Beltway. You can see

---

[7]In Congressional Units, (10/3) km = 2 miles.

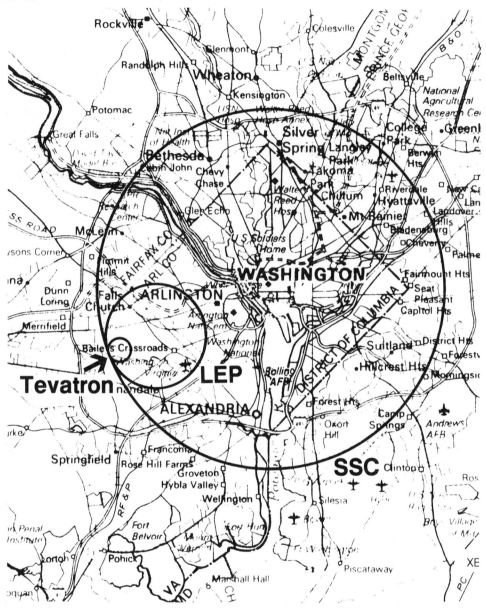

Figure 28: The Superconducting Super Collider and two smaller colliders, LEP at CERN and the Tevatron at Fermilab, superimposed to scale on the environs of Washington, D.C.

that they are about the same size. If only they had built that highway in the right shape, we would already have a site for our next accelerator!

Suggested Reading

C. Quigg, "Elementary Particles and Forces," *Scientific American* **252**, (4) 84 (April, 1985).

S. Weinberg, *The Discovery of Subatomic Particles*, W. H. Freeman, San Francisco, 1983.

J. Mulvey (editor), *The Nature of Matter*, Oxford University Press, Oxford, 1981.

F. E. Close, *The Cosmic Onion: quarks and the nature of the universe*, Heinemann Educational Books, London and Exeter, N. H., 1983.

C. Sutton, *The Particle Connection*, Simon and Schuster, New York, 1984.

C. T. Hill, "Quarks and Leptons: it's Elementary," *The Science Teacher,* September and October, 1982.

Physics through the 90s: Elementary Particle Physics, National Academy Press, Washington, 1986.

E. W. Kolb and C. Quigg, "Exploring the Universe from Quarks to Cosmology," *The Physics Teacher*, December, 1986.

C. Quigg and R. F. Schwitters, "Elementary Particle Physics and the Superconducting Super Collider," *Science* **231**, 1522 (March 28, 1986).

K. Gottfried and V. F. Weisskopf, *Concepts of Particle Physics, vol. 1*, Oxford University Press, Oxford, 1984.

D. H. Perkins, *Introduction to High-Energy Physics, second edition*, Addison-Wesley, Reading, Massachusetts, 1982.

I. J. R. Aitchison and A. J. G. Hey, *Gauge Theories in Particle Physics*, Adam Hilger, Bristol, 1982.

C. Quigg, *Gauge Theories of the Strong, Weak, and Electromagnetic Interactions*, Benjamin/Cummings, Reading, Massachusetts, 1983.

Christopher T. Hill received his Bachelor's and Master's degrees from MIT in 1972 and completed a Ph.D. in Elementary Particle Physics in 1977. He has been a post-doctoral researcher at the University of Chicago's Enrico Fermi Institute, and Fermilab. In 1978 he was elected Arthur H. Compton Lecturer at the University of Chicago and delivered a series of ten lectures on elementary particles to the general public in the spring of 1979. These lectures were very successful and led to ongoing general lectures at the Adler Planetarium, Fermilab, and other forums.

In June 1985 he became a permanent staff member at Fermilab. His research has covered many topics, from consequences of grand unified theories for the masses of quarks and leptons, to the nature of high energy cosmic rays. Most recently he has investigated the observational consequences of super-conducting cosmic strings.

He has been very active in the development of educational programs at Fermilab including the much imitated "Saturday Morning Physics' for high school students. As co-author of a proposal to create a "Summer Institute" for high school science teachers he helped to raise $400,000 to fund the program for a five-year period. Dr. Hill is presently a member of the Board of Directors of the "Friends of Fermilab," an organization specializing in the development of science education programs in conjunction with Fermilab and local educational institutions.

Symmetry in Physics [1]

Christopher T. Hill
Fermi National Accelerator Laboratory
P. O. Box 500, Batavia, Illinois 60510

Abstract

We present methods for introducing the concept of symmetry into the introductory physics curriculum.

1 Introduction

The concept of symmetry is fundamental to our understanding of the physical world. It is where we can discern true or approximate symmetries, such as those involved in the basic forces of nature, that we profess any real understanding. Where nature displays little or no apparent symmetries, such as in the spectrum of elementary quarks and leptons, we find ourselves most befuddled. Moreover, all thinking in modern theoretical physics is aimed at understanding the possible role of deep mathematical symmetries in nature. The realization of the importance of symmetry to the understanding of the laws of physics is a modern concept, belonging almost entirely to the twentieth century and beginning largely with Einstein and the special theory of relativity.

Why is the concept of symmetry essentially totally absent in the introductory physics course? Symmetry is probably the greatest component of what we mean when we speak of the "beauty of physics", yet the student of physics does not begin to see this underlying motif until rather late in the usual curriculum. Perhaps it is here that we do ourselves the greatest disservice in denying a peak into this structure to the casual physics student. Having delivered the "Symmetry" lecture in the Fermilab Saturday Morning Physics program for the past six years I've found that as a conceptual framework it can be introduced to the introductory (high school) physics students in a substantive and meaningful way. The student must be led to discover the physical manifestations of symmetries after exploring the mathematical concept *without burdensome abstraction*. This may lay the basis for further study of group theory, having provided a concrete realization of the ideas in geometric examples and the ways in which nature is constrained by symmetries. During a unit on conservation of momentum, energy and angular momentum, the underlying origin of these principles as a consequence of the fundamental symmetries of space and time can be demonstrated and the content of Emmy Noether's famous theorem connecting these can be motivated without attempting a proof (see Section 4).

In this brief article I will outline a set of basic mini-units which can be injected into the standard curriculum at various points without largely disrupting the latter. Pausing to contemplate an elegant symmetry argument in the course of analyzing a tedious physics problem can contribute much to enliven the subject, even for beginners. And, *please* view this as a small beginning, but by no means a conclusion to this subject; you are heartily encouraged to develop it further yourself!

2   HOW DO WE THINK ABOUT SYMMETRY?

Mathematicians solve many problems in geometry and topology by turning them into *equivalent algebraic problems*. This approach to understanding symmetry as a subject unto itself begins approximately with the 19th century French mathematician, Galois [1], who in his short, tragic life laid the foundation and fundamental applications of what we call "group theory", a mature branch of modern mathematics (the biography in Scientific American of ref.(1) is highly recommended reading).

We will not develop group theory here in its general form, but rather think concretely about the symmetries of a very simple geometric object... ...the equilateral triangle. This is the simplest nontrivial example and the results for any student the first time through this introductory exercise are often very surprising.

Prepare two transparencies as in Fig.(1) and Fig.(2) each featuring an equilateral triangle, both of the same size. The transparency of Fig.(1) has the three axes of symmetry labeled as I, II and III, while the transparency of Fig.(2) has the vertices labeled as A, B, and C.

Transparency (1) is laid down on the projector table and the students are informed that this is a *reference triangle* which must be considered

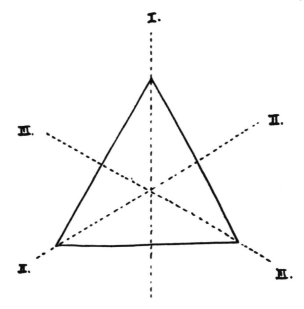

Figure 1: The reference triangle

to be glued in place and has the purpose of serving as a reference grid, or a kind of "coordinate system"; once laid in place we will not move it again. Transparency (2) on the other hand is an *experimental triangle*; we will be overlaying the reference triangle with the experimental triangle. Our problem is to *find all possible* **distinguishable** *ways in which the experimental triangle can be lifted up and brought down on top of the reference triangle.* The vertices of the experimental triangle are labeled to allow us to identify the distinguishable ways in which this can be done.

We begin by overlaying the experimental triangle on the reference triangle with the vertices reading ABC clockwise around the experimental triangle. This will be called the **initial orientation**. Our problem now is to discover a way in which we can pick up the experimental triangle and bring it back down on top of the reference triangle so that the vertices read something other than ABC clockwise. Each such operation is called a **symmetry operation** and our problem is to find all possible distinguishable symmetry operations of the equilateral triangle. How do we proceed?

Some student will no doubt suggest rotating the experimental triangle until the vertices now read CAB clockwise from the top. This certainly corresponds to a symmetry operation, which is a rotation through 120°.

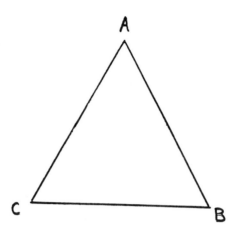

Figure 2: The experimental triangle

We shall designate this first discovery as $\mathbf{Rot}_{120°}$ and it should be written down on the blackboard as such (there will be six such operations and this should be written *second* from the top of a column).

Now return to the initial orientation. What else? It is obvious at this point that a rotation through 240° is another symmetry operation which yields the result BCA. However, it is important to emphasize that we should always return to the initial orientation before performing the next operation (this is a bit like pressing the **CLEAR** button on a pocket calculator before doing the next calculation). Thus we discover a second distinguishable symmetry operation which we designate $\mathbf{Rot}_{240°}$

Q: *Why is this a distinguishable rotation?*
Because the vertices now occur in the sequence BCA, we see that the triangle has been moved to a new orientation, distinguishable from the initial orientation ABC or from the $\mathbf{Rot}_{120°}$ orientation of CAB.

Q: *Why do we distinguish between a symmetry operation and an orientation of the triangle?*
Here there is an important distinction. The symmetry operation takes the triangle from *any* given initial orientation and

maps it into a new orientation; the position of vertices, A,B, and C, defines a particular, absolute position and orientation of the triangle in space. We are really interested here only in the symmetry operations, but not the absolute positions and orientations in space. That is, we are really only interested in the *relative* positions and orientations that the triangle can be brought to by a symmetry operation starting with any arbitrary initial position and orientation. We only introduce the labeling of vertices to keep track of them, starting always in the initial orientation ABC, with the triangle positioned on the table (we could equally well have labeled sides of the triangle). The symmetry operation can be performed on any equilateral triangle, of any size, drawn with any color ink, in any initial position and orientation. It captures the essence of symmetry, but contains no additional information about any particular triangle or its absolute position and orientation in space.

Are there other symmetry operations to consider? A student may suggest a rotation by $-120°$. But we now see that this takes the triangle to BCA (from the initial orientation, of course) and therefore this is not a new operation, i.e. it is not *distinguishable*. We may thus write the equation: $\mathbf{Rot}_{-120°} = \mathbf{Rot}_{240°}$. Again, we do not care about the *path* that takes us from one orientation to another; we only care about the new orientation *relative to* the initial orientation. That's what defines a symmetry operation. Thus $\mathbf{Rot}_{-120°}$ and $\mathbf{Rot}_{240°}$ are the same.

At this point a student may suggest a rotation through 360°. Is this a symmetry operation? We see that it maps the triangle from the initial orientation ABC back to the initial orientation ABC. Consequently, it *is* a symmetry operation, but a very special one. For one, it is equivalent to doing nothing at all. As such we shall refer to it as the "do nothing operation", or *the identity operation*. We shall denote it at the top of our blackboard list by the boldface $\mathbf{1}$. Secondly, note that the identity element is a symmetry operation of any object; even an amoeba has the symmetry of the identity symmetry. Thirdly, we note that a rotation through 360° is equivalent to a rotation through any *integer multiple* of 360°, e.g. 720°, $-360°$, etc. All are equivalent to the "do nothing" operation, $\mathbf{1}$.

Q: *Then isn't our* $\mathbf{Rot}_{120°}$ *equivalent to* $\mathbf{Rot}_{120°+360°\times N}$ *where N is an integer?*

Yes.

What is the analogous statement for $\mathbf{Rot}_{240°}$?

We now have three symmetry operations; are there more? In fact, the student will generally suggest performing a *reflection* about one of the three axes of the reference triangle. We begin with the initial orientation and consider "skewering" the experimental triangle (as if we had a barbecue skewer) along one of the axes of symmetry indicated on our reference triangle. For example, skewering along axis I, we then pick up the triangle and flip it and we arrive at the new orientation, ACB. We denote this symmetry operation as a *reflection about axis I* or as $\mathbf{Ref}_I$. Similarly, we return to the initial orientation and consider the other two symmetry operations, (a) the reflection about axis II, or $\mathbf{Ref}_{II}$ which yields the vertex position BAC and (b) the reflection about axis III, or $\mathbf{Ref}_{III}$ which yields the vertex position CBA. Thus we now have a list on the blackboard of six of the symmetry operations which has the form:

Table I. The six symmetry operations of the equilateral triangle.

Notation	Operation	Vertices
1	"do nothing" or identity	ABC
$\mathbf{Rot}_{120°}$	rotate by 120°	CAB
$\mathbf{Rot}_{240°}$	rotate by 240°	BCA
$\mathbf{Ref}_I$	reflect about axis I	ACB
$\mathbf{Ref}_{II}$	reflect about axis II	BAC
$\mathbf{Ref}_{III}$	reflect about axis III	CBA

Are there any other symmetry operations? At this point many students recognize that we have discovered essentially the six permutations of three objects, i.e., the six permutations of the three vertices of the triangle. That raises an interesting question:

Q: *Are the symmetries of all such objects, such as squares, pentagons, hexagons, cubes, etc. given by the permutations of their vertices?*

In fact the answer is *no*. It doesn't work that way for the square as we can easily see. Suppose we have a square with vertices labeled ABCD. A true symmetry operation of the square is a rotation through 90° and gives DABC, which is also a permutation of the vertices. However, is there a symmetry operation which can give the vertex ordering BACD, which is certainly a valid permutation of ABCD? (Think in terms of an experimental square on a transparency; what would we have to do to the transparency to get BACD starting from ABCD?) Clearly, this is not a symmetry operation of the entire square because we would have to *twist* the experimental square to get the vertices into this position, but then the sides would not overlay properly! Thus, while all symmetry operations of geometric objects are indeed permutations, not all permutations are symmetry operations of geometric objects. We have actually discovered our first example of a subgroup; the square is a subgroup of the group of permutations of four objects. The equilateral triangle is simpler and it does have only six symmetry operations, the ones we've listed above, which are equivalent (isomorphic) to the permutations of three objects.

Thus far our exercise has been almost trivial, but now we make the great observation of Galois and his colleagues. We now ask, can we obtain additional symmetry operations by combining together two of the operations previously obtained? That is, let us take two of our six operations, say $\mathbf{Rot}_{120°}$ and $\mathbf{Ref}_{II}$, and first perform one of them on the experimental triangle (try $\mathbf{Rot}_{120°}$) and *without returning to the initial orientation* perform the other operation ($\mathbf{Ref}_{II}$). We see that if we begin in the initial orientation that $\mathbf{Rot}_{120°}$ leads to CAB and then following with $\mathbf{Ref}_{II}$ we obtain the orientation ACB. But ACB is not a new orientation of the triangle, and it corresponds to $\mathbf{Ref}_I$ as seen by our table. We have therefore discovered an interesting result: *first performing* $\mathbf{Rot}_{120°}$ *and following it by* $\mathbf{Ref}_{II}$ *yields the* result $\mathbf{Ref}_I$.

Let us write an equation for this result:

$$\mathbf{Rot}_{120°} \otimes \mathbf{Ref}_{II} = \mathbf{Ref}_I. \tag{2.1}$$

Here we have introduced a symbol, $\otimes$, which represents the action of combining the symmetry operations in the order indicated (without returning to the initial orientation in between). It is easily seen that the

$\otimes$ combination of any pair of our symmetry operations (which we also refer to as "elements") produces another of the elements. We say that our set of elements is **closed** under the operation $\otimes$. Thus, in a sense the combining of two symmetry operations is something like *multiplication of numbers*. In this sense the "do nothing operation" is the *identity*:

$$1 \otimes X = X \otimes 1 = X. \qquad (2.2)$$

Q: *Why do we call this "multiplication" rather than "addition"?*

The answer is really one of convention. Multiplication and addition have very similar mathematical properties; the identity element in addition is 0, while in multiplication it is 1. The inverse of 4 under addition is -4, while under multiplication it is $\frac{1}{4}$. Hence, the positive and negative integers **close** under addition while the rationals close under multiplication. Note however that there is an important difference between addition and multiplication: 0 has no multiplicative inverse, i.e. infinity makes no sense mathematically. Denoting the combination of symmetry operation by "multiplication" is also a consequence of the fact that matrices can *represent* the group elements and matrix multiplication can *represent* the $\otimes$ operation as we shall see subsequently.

Thus we have made a very important observation: *the symmetry operations form an algebraic system with an operation consisting of performing successive operations.* This algebraic system is called a **group**. The symmetry operations are the analogues of the rational numbers under this group multiplication. We refer to the symmetry operations as **group elements** or simply as **elements**. We present the complete multiplication table of the symmetry group of the equilateral triangle, designated as S_3, in Table(II). The entries in this table should be verified by performing several of the cases with the experimental triangle and reference triangle transparencies. Table(II) is to be read like a highway mileage map; if we choose to perform the product $X \otimes Y$ we first find the row labeled on the left by X, then the column labeled on the top by Y and we look up the corresponding entry in that row and column for the result.

Table II. S_3 Group multiplication table

	1	**Rot**$_{120°}$	**Rot**$_{240°}$	**Ref**$_I$	**Ref**$_{II}$	**Ref**$_{III}$
1	**1**	**Rot**$_{120°}$	**Rot**$_{240°}$	**Ref**$_I$	**Ref**$_{II}$	**Ref**$_{III}$
Rot$_{120°}$	**Rot**$_{120°}$	**Rot**$_{240°}$	**1**	**Ref**$_{III}$	**Ref**$_I$	**Ref**$_{II}$
Rot$_{240°}$	**Rot**$_{240°}$	**1**	**Rot**$_{120°}$	**Ref**$_{II}$	**Ref**$_{III}$	**Ref**$_I$
Ref$_I$	**Ref**$_I$	**Ref**$_{II}$	**Ref**$_{III}$	**1**	**Rot**$_{120°}$	**Rot**$_{240°}$
Ref$_{II}$	**Ref**$_{II}$	**Ref**$_{III}$	**Ref**$_I$	**Rot**$_{240°}$	**1**	**Rot**$_{120°}$
Ref$_{III}$	**Ref**$_{III}$	**Ref**$_I$	**Ref**$_{II}$	**Rot**$_{120°}$	**Rot**$_{240°}$	**1**

There are several important properties of this group multiplication able that are shared by all groups:

- A group is a set of elements and a composition law, $\otimes$, such that the product of any two elements yields another element in the set.

- Every group has an identity element satisfying eq.(2.2).

- Each element of the group has a unique inverse element. That is, given an element X there exists one and only one element, Y (which may even be X itself), such that $X \otimes Y = Y \otimes X = 1$.

- Group multiplication is *associative*. That is, given X, Y and Z we have $X \otimes (Y \otimes Z) = (X \otimes Y) \otimes Z$. In words, first perform Y and follow by Z and remember the result (call it W). Now return to the initial orientation and first do X and follow by W. This result will be the same as having first done X followed by Y then followed by Z. This is the meaning of associativity and you should carefully think it through to make sure you understand it.

- Each element of the group occurs once and only once in each row and each column of the multiplication table. This can actually be proved as a theorem from the preceding statements. This is a powerful constraint on the mathematical structure of the group; essentially the group multiplication table forms a kind of "magic square".

- **Group multiplication is not necessarily commutative**[2]. That is, $X \otimes Y$ need not equal $Y \otimes X$!!!

This last result, namely that group multiplication is not commutative, is really quite remarkable. Here we have discovered a simple system of six elements with a multiplication law and the system is not even commutative. For example, (*this should definitely be shown explicitly with the triangles*) consider first performing $\mathbf{Rot}_{240°}$ and following by $\mathbf{Ref}_{II}$, that is, calculate $\mathbf{Rot}_{240°} \otimes \mathbf{Ref}_{II}$. You should obtain the result $\mathbf{Ref}_{III}$. On the other hand, consider first performing $\mathbf{Ref}_{II}$ followed by $\mathbf{Rot}_{240°}$, i.e., compute $\mathbf{Ref}_{II} \otimes \mathbf{Rot}_{240°}$. The result is $\mathbf{Ref}_{I}$. Summarizing:

$$\mathbf{Rot}_{240°} \otimes \mathbf{Ref}_{II} = \mathbf{Ref}_{III}$$

$$\mathbf{Ref}_{II} \otimes \mathbf{Rot}_{240°} = \mathbf{Ref}_{I}$$

Thus, although ordinary multiplication is commutative, e.g. $3 \cdot 4 = 4 \cdot 3$, group multiplication need not be. When a group has commutative multiplication it is said to be an **abelian group**, after the mathematician Abel. The general group, such as the equilateral triangle group, is noncommutative, or **nonabelian**.

We finish this discussion with an important example as to how group mathematics underlies the structure of our physical world. One may wonder how noncommutative mathematics can have anything at all to do with nature, or physics. A simple demonstration will show this.

Take a textbook and hold it in front of you with the binding down as though you were going to open it up on a table. Now extend your right arm parallel to your chest and parallel to the floor (like a right turn signal) and let this be the positive x–axis. Now extend your arm straight out in front of you; let this define the positive y–axis. We wish to rotate the textbook by 90° about the positive x-axis and follow this by a

---

[2]This discussion should be considered for introduction into a mathematics class as a unit for this reason. The concept of the commutative property of ordinary addition or multiplication is almost vacuous without showing a counterexample, namely a system in which it doesn't hold! Our equilateral triangle symmetry group affords such a simple example. Unfortunately, there are no simple examples of nonassociative systems, even though they do exist.

rotation through 90° about the y–axis. The rotations should always be performed in the sense of a right hand screwdriver. Perform the two successive rotations and note the book's position. Now return the book to the initial orientation and perform first the rotation about the y–axis followed by a rotation about the x–axis. You will find that the book ends up in two different positions. The symmetry group consisting of rotations through 90° is a noncommutative group. The continuous group consisting of all rotations of objects in three dimensions (the full symmetry of a sphere) is thus noncommutative. It is known as **O(3)** and it governs the physics of angular momentum and spin.

The subject of group theory is an entire branch of mathematics in which many people have specialized and undertake ongoing research. The continuous groups, possessing an infinite number of operations that vary continuously with "angle" parameters, like the rotations of a sphere about a given axis through any angle, were first completely classified early in the 20th century by Cartan. Remarkably, only very recently have all possible discrete symmetry groups been classified. This job was made difficult by the existence of certain "sporadic" groups, such as the "monster group" with $\approx 8 \times 10^{53}$ elements. The classification of the discrete groups constitutes one of the longest and least comprehensible theorems in mathematics (2).

The application of group theory to physics is a rich and fundamentally important subject, significantly different than pure mathematical research into groups and their properties. While mathematicians may struggle to classify the discrete groups, such as the monster, nature embodies only a small subset of all possible mathematical symmetry groups. It is remarkable, however, that as we probe deeper into the shortest distances and most elemental properties of matter we seem to discover evidence of ever more sophisticated symmetry groups at work. Nature seems to read books on group theory!

Exercises and food for thought:

1. Construct the symmetry group of the square and its associated multiplication table. Verify the properties discussed above for the

general group.

2. Construct the permutation group of four objects and its associated multiplication table. Is it isomorphic to the square's group?

3. A subgroup of a group is a subset of elements which themselves form a group. Clearly, each subgroup must contain the identity. Can you identify some subgroups of the equilateral triangle group? What is the largest subgroup of the equilateral triangle symmetry group (not counting the entire group itself [each set is a subset of itself]; technically we want the largest *proper* subgroup).

4. If the square in the preceding problem is squashed into a rectangle identify the surviving subgroup (see the preceding problem) which describes the remaining symmetry. (As a preliminary exercise consider an isosceles triangle in which vertex A is lifted along axis I in Fig.(1); what is the resulting symmetry group? Is it a subgroup?)

5. If an infinite floor is tiled with equilateral triangles we have a *lattice*. Is the symmetry group of the equilateral triangle also a symmetry group of the lattice? (*Yes, but it is only a subgroup; It is called the "point group" of the lattice*). Is the full symmetry group of the lattice equivalent to the group of a single triangle? (*No. It includes in addition the set of translations along the axes which bring the entire lattice down on top of itself*). Does a rotation commute with a translation? This problem illustrates how groups enter solid state physics in which they are of paramount importance.

3   PHYSICAL SITUATIONS INVOLVING SYMMETRY

In the previous lecture we introduced the formal notion of the symmetry group. Specifically what has it to do with physics? We present here a number of problems which illustrate the power of symmetry arguments in the solution of physics problems. Our last examples, (5) and (6), that of the modes of oscillation of a system of three coupled masses, goes the farthest in illustrating the power of group theory (and group representation theory).

Figure 3: Symmetric arrangement of three masses, all at rest

Example 1.

In Fig.(3) we have a symmetric arrangement of three masses following the equilateral triangle with a single mass situated at the center. What is the force of gravity exerted upon the center mass due the other three? Of course, we must be careful in stipulating that the system really is fully symmetric. Thus, each particle is at rest and has no internal degree of freedom (such as angular momentum, or quadrupole moment of mass) which violates the symmetry; or at least we seek an approximate solution in the limit in which such complications can be ignored.

Therefore, it is obvious that the force, by symmetry, vanishes for the particle in the center. This involves the symmetry considerations of the preceding section as well. Suppose that by adding up the individual force vectors due to the three masses at the vertices we had obtained the nonvanishing result shown in Fig.(4). Now this answer is clearly wrong, but why is it wrong from the point of view of the symmetry group? Consider performing a typical group operation on this result, e.g., perform the operation $\mathbf{Ref}_I$. This maps the system into itself, but it maps the answer into a new one shown in Fig.(5). The same system cannot produce two different results, so we have shown that the result is wrong. Actually, consideration of $\mathbf{Ref}_I$ alone does not eliminate a result lying on the axis I, however we may then consider, e.g., $\mathbf{Rot}_{120°}$ to dispose

Figure 4: Hypothetical result for the force exerted on center mass

of that case. Clearly, the only result which is invariant under any of the group operations is *zero*.

This trivial example illustrates a very important aspect of nature: If a physical configuration possesses a given symmetry then the dynamics of the system will possess the symmetry[3]. Thus, our result in this case does not depend upon the kind of force law involved so long as the symmetry constraint is in effect. The central particle could be a pion surrounded by (spinless) nuclei interacting through the strong nuclear force and the conclusion would be the same: the force must vanish by symmetry!

Example 2.

In Figure(6) we have yet another configuration which this time does not possess the full symmetry of the equilateral triangle. Now what is

---

[3] There is a caveat here of great importance: the system must be *stable* in the symmetric configuration. Many systems, such as ferromagnets, though they are described by rotationally invariant equations of motion, are unstable in rotationally invariant states. These systems undergo "spontaneous symmetry breaking" at low temperatures. The ferromagnet develops nonzero magnetization below the Curie temperature. The problem shown above actually exhibits this phenomenon. If the small mass at the origin in the above problem was slightly displaced away from the center it would then experience a nonzero force pulling it farther *away* from the center and the small symmetry breaking fluctuation is amplified. Replacing all the masses by positive charges eliminates the instability.

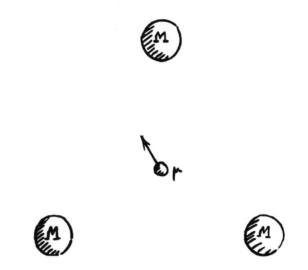

Figure 5: Action of **Ref**$_I$ on answer obtained in Fig.4

the force exerted upon the central mass?

Of course, there is a residual symmetry here. This system is described by the subgroup of the full equilateral triangle symmetry group consisting of the elements **1** and **Ref**$_I$. Thus, the resulting force vector must lie along the axis I. Symmetry does not tell us what the sign of the force is. For gravity the force is away from the center while for electromagnetism it depends in an obvious way upon the choice of charges. This typifies the situation scientists often face in understanding a new phenomenon. A symmetry may be present which goes along way toward controlling the physics, while some unknown underlying dynamics may be present which determines the quantitative outcome.

In the present case it is instructive to consider the small displacement from the center, a, of the mass μ. The general inverse square law gives:

$$F = \frac{\alpha}{(r-a)^2} - \frac{\beta}{(r+a\cos\theta)^2} \tag{3.1}$$

where α and β are given by the precise form of the force law (i.e. gravity versus electromagnetism, etc.). Here θ is 60° as is seen by the the geometry of the situation. Symmetry tells us that $F = 0$ when $a = 0$, hence that $\alpha = \beta$. For $\frac{a}{r} \ll 1$ we may consider the first terms in the series

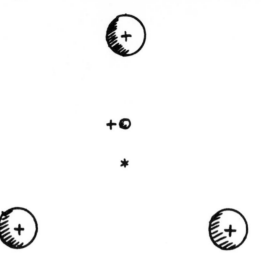

Figure 6: Asymmetric arrangement of masses, all at rest

expansion of eq.(3.1) and thus:

$$F \approx \frac{3a\alpha}{r^3} \qquad (3.2)$$

using $\cos\theta = \frac{1}{2}$. For gravity $\alpha = GM\mu$ and the force is up, away from the center and the configuration is unstable. If the vertices correspond to charges, Q, and center to charge $-q$, then $\alpha = kqQ$ and the force is toward the center and the central mass is in a stable potential.

Example 3.

One of the most conspicuous atomic transitions is the $2P \rightarrow 1S$ "dipole transition" in, e.g., atomic Hydrogen. The $1S$ orbital is the groundstate quantum mechanical motion of an electron in the Hydrogen atom and is the shape of a perfectly rotationally symmetric "cloud" if the atom is in free space. The $2P$ levels are not rotationally symmetric and "point in a direction" like a vector, and form, therefore, a "triply degenerate" state, i.e. the electron can at any instant be in one of three independent P–orbitals, $2P_x$, $2P_y$, or $2P_z$, corresponding to the three independent directions in space. The three 2P states are degenerate, i.e. have the same energy, because of rotational symmetry. If the electron is in the $2P_x$ state we can just rotate the atom at no cost in energy (or just rotate our reference frame) and thus put the electron into a $2P_y$ or $2P_z$

state. There are also higher orbitals such as the D, F etc. with more orientational information than the S or P (these are like tensors).

If the atom is placed in a strong magnetic field we have broken the rotational symmetry. Suppose the magnetic field points in the z–direction. Then we still have rotational symmetry in the perpendicular x–y plane. Therefore, the $2P_x$ and $2P_y$ orbitals will remain degenerate, but the $2P_z$ orbital will develop a slightly different energy. This energy splitting is proportional to the magnetic field strength and the observed transition photons from $2P_z \to 1S$ and $2P_x$ or $2P_y \to 1S$ will have slightly different energies. This is known as the **Zeeman effect** and is one of the principal methods for determining the presence and strength of magnetic fields in the sun and other astronomical objects.

Incidently, putting atoms into crystal lattices also breaks up the degeneracy of atomic levels due to the interatomic forces and their symmetries which follow the symmetries of the crystal lattice. Can you think of a way to get the $2P_x$, $2P_y$ and $2P_z$ to each have different energies? If the hydrogen atom could be placed in a perfect cubic crystal would the $2P$ levels be degenerate? (*Ans: yes*) What about a non–cubic lattice?

Example 4.

Figure(7) shows yet another configuration of large masses, M, arranged on the vertices of a symmetric hexagon, however the topmost vertex has a mass $m \neq M$. Given this each vertex is a distance a from the center, find the force experienced by the center mass μ (this should be done in less than one minute).

It is clear that the resulting force must be directed along the vertical axis of symmetry and thus can depend only upon the uppermost and lowermost masses. Since the force vanishes when $m = M$, it must depend only upon the difference $m - M$ (gravity depends only linearly upon the "pullers"). Also, when $m \gg M$ the force must be in the $\vec{up}$ direction. Hence, *without any computation at all*, we arrive at the answer:

$$F = \frac{G\mu(m - M)}{a^2}\vec{up} \qquad (3.3)$$

These are only a small handful of simple, illustrative problems. I urge you to develop more of them (particularly *clever* ones). The following

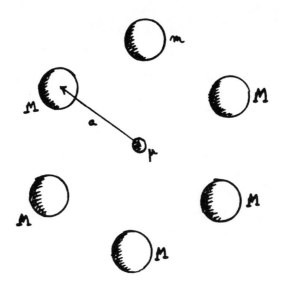

Figure 7: Hexagonal arrangement of masses; the topmost mass is different. What is the force on μ?

discussion amplifies the theory of symmetry groups to get at a very powerful relationship between physics and symmetry through the properties of *group representations*.

Example 5.

We consider the question, "Can we find any standard mathematical objects which close under multiplication and satisfy the same rules of combination as any given symmetry group?" Such a set of objects forms what is known as a **representation** of the group. The problem of classifying all of the representations of symmetry groups is an extremely important and very rich subject and forms what is known as **representation theory**. It also has a great many physical applications.

For any group there always exists a **trivial representation** which consists of letting each element be represented by the number 1. Thus we write:

$$
\begin{aligned}
\mathbf{1} &\rightarrow 1 \\
\mathbf{Rot}_{120°} &\rightarrow 1 \\
\mathbf{Rot}_{240°} &\rightarrow 1 \\
\mathbf{Ref}_I &\rightarrow 1 \\
\mathbf{Ref}_{II} &\rightarrow 1 \\
\mathbf{Ref}_{III} &\rightarrow 1
\end{aligned}
\qquad (3.4)
$$

This forms a representation in the sense that the product of any two elements as given by Table(II) produces a result consistent with the product of any of the representative elements. Thus $\mathbf{Rot}_{240°} \otimes \mathbf{Ref}_{II} = \mathbf{Ref}_{III}$ is consistent with $1 \times 1 = 1$; but of course this is trivially the case. The trivial representation thus contains no information about the group.

Are there any other representations using only numbers? It should be fairly clear to you upon thought that the absolute magnitude of each number representing a symmetry element must be unity (for example, since $\mathbf{Ref}_I \otimes \mathbf{Ref}_I = \mathbf{1}$ the number representing $\mathbf{Ref}_I$, call it α, must satisfy $\alpha^2 = 1$; also the rotations map unit vectors into unit vectors. Indeed, if we consider *only the three element subgroup* consisting of $\mathbf{1}$, $\mathbf{Rot}_{120°}$ and $\mathbf{Rot}_{240°}$, we could always write down a *complex representation*:

$$
\begin{aligned}
\mathbf{1} &\rightarrow 1 \\
\mathbf{Rot}_{120°} &\rightarrow \exp\left(2\pi i/3\right) \\
\mathbf{Rot}_{240°} &\rightarrow \exp\left(4\pi i/3\right)
\end{aligned}
\qquad (3.5)
$$

however, when we include the reflections this ceases to be a possible representation of the full equilateral triangle group).

There is a representation of the full equilateral triangle group which is nontrivial, yet involves only 1 and -1. It is:

$$
\begin{aligned}
\mathbf{1} &\rightarrow 1 \\
\mathbf{Rot}_{120°} &\rightarrow 1 \\
\mathbf{Rot}_{240°} &\rightarrow 1 \\
\mathbf{Ref}_I &\rightarrow -1 \\
\mathbf{Ref}_{II} &\rightarrow -1 \\
\mathbf{Ref}_{III} &\rightarrow -1
\end{aligned}
\qquad (3.6)
$$

This representaion recognizes the difference between those operations which can be done without lifting the experimental triangle (the rotations) and those that require lifting and flipping (reflections). Another way of saying it is that rotations are direct operations which we may perform upon the system while reflections require a mirror or a **parity transformation**. We shall call this the **parity representation**. We see that the correspondence with the multiplication table(I) may be checked by considering several examples:

$$
\begin{aligned}
\mathbf{Rot}_{120°} \otimes \mathbf{Rot}_{240°} &= \mathbf{1} & \rightarrow \quad 1 \cdot 1 &= 1 \\
\mathbf{Ref}_I \otimes \mathbf{Rot}_{240°} &= \mathbf{Ref}_{III} & \rightarrow \quad (-1) \cdot 1 &= (-1) \\
\mathbf{Ref}_{II} \otimes \mathbf{Ref}_{III} &= \mathbf{Rot}_{120°} & \rightarrow \quad (-1) \cdot (-1) &= 1.
\end{aligned}
\tag{3.7}
$$

Both the trivial representation and the parity representation are examples of **unfaithful representations** because the same representative (1 or -1) corresponds to two or more elements (e.g. -1 corresponds to $\mathbf{Ref}_I$, $\mathbf{Ref}_{II}$ and $\mathbf{Ref}_{III}$). Are there any faithful representations of the equilateral triangle symmetry group?

In fact, there is one more representation that is faithful, but it cannot be given in terms of numbers. It requires 2×2 matrices and the multiplication operation is now matrix multiplication. The matrix representation may be written as:

$$
\begin{aligned}
\mathbf{1} \quad &\rightarrow \quad \begin{pmatrix} 1 & 0 \\ 0 & 1 \end{pmatrix} \\[2mm]
\mathbf{Rot}_{120°} \quad &\rightarrow \quad \begin{pmatrix} -\frac{1}{2} & -\frac{\sqrt{3}}{2} \\ \frac{\sqrt{3}}{2} & -\frac{1}{2} \end{pmatrix} \\[2mm]
\mathbf{Rot}_{240°} \quad &\rightarrow \quad \begin{pmatrix} -\frac{1}{2} & \frac{\sqrt{3}}{2} \\ -\frac{\sqrt{3}}{2} & -\frac{1}{2} \end{pmatrix} \\[2mm]
\mathbf{Ref}_I \quad &\rightarrow \quad \begin{pmatrix} -1 & 0 \\ 0 & 1 \end{pmatrix} \\[2mm]
\mathbf{Ref}_{II} \quad &\rightarrow \quad \begin{pmatrix} \frac{1}{2} & \frac{\sqrt{3}}{2} \\ \frac{\sqrt{3}}{2} & -\frac{1}{2} \end{pmatrix} \\[2mm]
\mathbf{Ref}_{III} \quad &\rightarrow \quad \begin{pmatrix} \frac{1}{2} & -\frac{\sqrt{3}}{2} \\ -\frac{\sqrt{3}}{2} & -\frac{1}{2} \end{pmatrix}
\end{aligned}
\tag{3.8}
$$

Again, we see that the representation gives results consistent with Table(II) for a few sample cases:

$$\mathbf{Rot}_{120°} \otimes \mathbf{Rot}_{240°} = \mathbf{1}$$

$$\rightarrow \begin{pmatrix} -\frac{1}{2} & -\frac{\sqrt{3}}{2} \\ \frac{\sqrt{3}}{2} & -\frac{1}{2} \end{pmatrix} \cdot \begin{pmatrix} -\frac{1}{2} & \frac{\sqrt{3}}{2} \\ -\frac{\sqrt{3}}{2} & -\frac{1}{2} \end{pmatrix} = \begin{pmatrix} 1 & 0 \\ 0 & 1 \end{pmatrix}$$

$$\mathbf{Ref}_I \otimes \mathbf{Rot}_{240°} = \mathbf{Ref}_{III}$$

$$\rightarrow \begin{pmatrix} -1 & 0 \\ 0 & 1 \end{pmatrix} \cdot \begin{pmatrix} -\frac{1}{2} & \frac{\sqrt{3}}{2} \\ -\frac{\sqrt{3}}{2} & -\frac{1}{2} \end{pmatrix} = \begin{pmatrix} \frac{1}{2} & -\frac{\sqrt{3}}{2} \\ -\frac{\sqrt{3}}{2} & -\frac{1}{2} \end{pmatrix} \qquad (3.9)$$

$$\mathbf{Ref}_{II} \otimes \mathbf{Ref}_{III} = \mathbf{Rot}_{120°}$$

$$\rightarrow \begin{pmatrix} \frac{1}{2} & \frac{\sqrt{3}}{2} \\ \frac{\sqrt{3}}{2} & -\frac{1}{2} \end{pmatrix} \cdot \begin{pmatrix} \frac{1}{2} & -\frac{\sqrt{3}}{2} \\ -\frac{\sqrt{3}}{2} & -\frac{1}{2} \end{pmatrix} = \begin{pmatrix} -\frac{1}{2} & -\frac{\sqrt{3}}{2} \\ \frac{\sqrt{3}}{2} & -\frac{1}{2} \end{pmatrix}$$

The reader is invited to check other cases. Of course, this kind of representation is possible because matrix multiplication is itself *noncommutative*. In fact, the set of all matrices of a given order, e.g. 2×2, with *nonvanishing determinant* form a group with respect to matrix multiplication (why must we stipulate nonvanishing determinant?).

Are there any higher matrix representations? Of course, we can always combine the two numerical representations (these are 1×1 matrices) and the 2×2 case to form higher dimensional matrix representations. For example, consider the set of 3×3 matices consisting of a 1 in the upper left-hand corner and a 2×2 matrix from our set of matrix representations in the lower right-hand corner, with zeros everywhere else. This *is* a representation, but it contains no new information not already contained in the cases examined above. It is known as a **reducible** representation because it is a set of matrices that are **block diagonal**, and each block contains one of the three basic representation described above. *The three representations we've discussed above are the only* **irreducible** *representations of the equilateral triangle symmetry group.* In general it is not trivial to decide if a given representation is reducible. This is because we may take a group of block diagonal matrices and multiply on the left by some matrix, S, and on the right by S^{-1} and we still preserve the group multiplication table, but the resulting matrices no longer *appear* to be block diagonal. Nonetheless, such a representation is equivalent to the block diagonal reducible one and is itself reducible. So the general problem is to determine whether a given representation is equivalent to a block diagonal one by right multiplication by some S and left multiplication by S^{-1}. If no such S exists, then the representation is irreducible and therefore interesting. This problem is then solved and constitutes the central subject of group representation theory. We refer the reader to any good book on group theory for a discussion of representation theory.

Example 6.

We consider now a physical system as shown in Figure(8) which consists of three equal masses attached together by springs. This may be viewed as a kind of molecule and we will assume that it can move only in the 2-dimensional plane for simplicity (such a system could be fabricated out of air-pucks and springs for use on an air table as a demonstration). The system possesses in its equilibrium rest state the symmetry of the equilateral triangle; it will be governed by the symmetry group of the triangle in a very interesting way. The system can undergo several kinds of motion, consisting of uniform center–of–mass translations, rotations and internal oscillations. Indeed, this is a property shared by all molecular systems and the heat capacity of a gas essentially counts the various states of motion of the systems comprising the gas. First, we may count the number of independent motions of the system, e.g. how many numbers must be given to specify the exact state of the system at any time? These are called **degrees of freedom** of the system. Since we have 3 masses and each mass requires 2 coordinates (an x- and a y- coordinate) we thus have $2 \times 3 = 6$ degrees of freedom. The tying together of the masses by springs does not change the counting because the springs are free to stretch and given the 6 coordinates we can calculate the length of

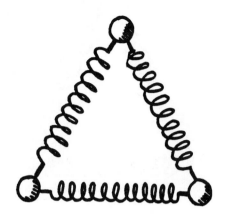

Figure 8: "Molecule" consisting of masses and springs.

each spring (if, on the other hand, the masses were attached together by 3 rigid rods which are not free to stretch we would have $6 - 3 = 3$ degrees of freedom).

Describing the motion by giving the 6 coordinates does not tell us much about the system's motion as a whole. Therefore, we wish to describe the system in terms of basic motions that we can separate qualitatively. In fact, this separation also reduces the mathematical complexity of analyzing the motion of the system. These are called the **normal modes** of the system's motion.

First, suppose the system is initially at rest in its equilibrium position. Clearly, two modes of motion are just given by uniform translation of the center of mass in the x– or y– directions. Thus, we may dispose of **2** degrees of freedom (d.o.f.) and analyze the remaining 4 d.o.f. by holding the center of mass of the system fixed in space.

With the center of mass fixed we may consider a uniform oscillation of the system as a whole as shown in Fig.(9). This is known as a **breather mode** since the system simply expands and contracts but remains always in the fully symmetrical shape. Thus, if we act upon the breather mode with one of our symmetry operations, e.g. simply perform $\mathbf{Rot}_{120°}$ or $\mathbf{Ref}_{II}$ on the system at some arbitrary instant of time, we see that we

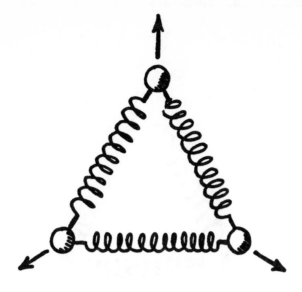

Figure 9: The breather mode

remain in exactly the same state of motion. *Therefore, the breather mode corresponds to the trivial representation of the group.* It has a characteristic frequency, ω_0 given by the mass, M, and the spring constant, K. Next, we may consider a uniform rotation about the fixed center of mass of the system, Fig.(10). Let us consider clockwise rotation with angular frequency ω_r. If we act at some instant upon the system with the symmetry group elements, $\mathbf{1}$, $\mathbf{Rot}_{120°}$ and $\mathbf{Rot}_{240°}$, we find that the motion will remain the same. On the other hand, acting with the elements $\mathbf{Ref}_I$, $\mathbf{Ref}_{II}$ and $\mathbf{Ref}_{III}$ we see that the motion becomes counterclockwise, i.e. these operations map the frequency into $-\omega_r$. *Therefore, uniform rotation corresponds to the parity representation of the group.* Therefore, we are left with 2 remaining d.o.f. Consider the motion shown in Fig.(11). Here one of the vertices moves out along an axis of symmetry while the other vertices are attracted toward the axis. These three motions, corresponding to the three vertices, are not *independent*, i.e. if we add together all three motions with equal strength we obtain no motion at all $(\alpha + \beta + \gamma = 0)$. However, if we act upon any of the motions we obtain one of the others. For example, $\mathbf{Rot}_{120°}(\alpha) = \beta$, or $\mathbf{Ref}_{II}(\alpha) = \gamma$. How do we rewrite these three modes in terms of two so that they close under the action of the symmetry elements? An answer is the following:

$$mode\ 1 = \alpha - \beta \qquad mode\ 2 = \alpha + \beta - 2\gamma \qquad (3.10)$$

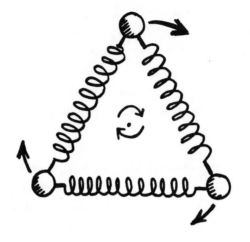

Figure 10: Uniform rotation

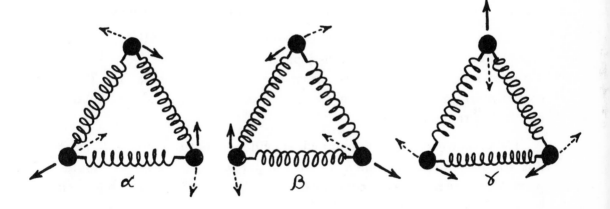

Figure 11: The doublet modes

Now under the action of any symmetry element these two modes go into linear combinations of themselves. *Thus, these two modes correspond to the doublet representation of the symmetry group.* Because they are related to each other by symmetry elements, they *must have the same frequency of oscillation.* Of course, this frequency will not be the same as that of the breather mode. The normal modes forming a representation of the group are said to be **degenerate** in their oscillation frequency.

Example 7.

In the mid 1960s the strongly interacting particles could be placed into multiplets of a continuous symmetry group, $SU(3)$ [3]. One of the representations of $SU(3)$ has eight components, and is known as an octet (the $SU(3)$ symmetry elements can be represented as 8×8 matrices in an irreducible way; the eight members of the octet mix amongst themselves under an $SU(3)$ transformation). The eight spin–0 mesons fit into one multiplet, the eight spin–$\frac{1}{2}$ baryons into another, and so on (see Fig.(12)). There is also a *10* component representation into which the spin–$\frac{3}{2}$ baryonic resonances fit (in fact, one, the Ω^-, was missing at the time $SU(3)$ was discovered and it was correctly predicted by the theory). The particles in the multiplets were not degenerate indicating that the $SU(3)$ symmetry was not exact, but the pattern was clearly established.

The puzzle was that the smallest representation of SU(3), namely a triplet consisting of three spin–$\frac{1}{2}$ particles with charges $(\frac{2}{3}, -\frac{1}{3}, -\frac{1}{3})$ were not seen directly (in these units the electron charge is -1). These particles are known respectively as the "up" quark, the "down" quark, and the "strange quark".

Today, however, we have compelling evidence that the quarks do exist but are permanently confined within the particles we see in the laboratory. All of the mesons are composed of quark and anti–quark, while each baryon contains three quarks. $SU(3)$ symmetry is not exact because the strange quark mass is much larger than the up or down quark masses (this is analogous to breaking the equilateral triangle symmetry by making one of the vertex masses much different than the others; the doublet modes would cease to be degenerate). Incidently, there is another totally distinct $SU(3)$ symmetry associated with the force holding the quarks together inside of the mesons and baryons. This latter $SU(3)$ is called the "color symmetry" and it is an exact symmetry (it is also a *gauge group*, which is a subject beyond the level of our present discussion).

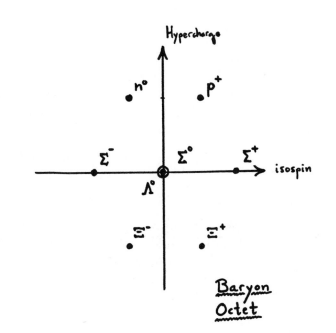

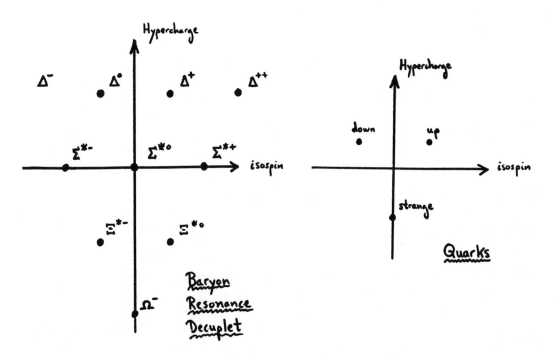

Figure 12: $SU(3)$ baryon octet, resonance decuplet and quark triplet

4 SYMMETRIES OF THE LAWS OF PHYSICS AND NOETHER'S THEOREM

What are the symmetries of the basic laws of physics? What do mean by a symmetry in a law of physics? Just as in the case of the equilateral triangle a symmetry operation is an action we perform upon a geometric configuration such that it remains unchanged, we mean now an operation we can perform upon an experiment such that the outcome of the experiment is unchanged. However, the operation must hold for *any conceiveable experiment* if it is to qualify as a symmetry of the *laws of physics*.

Let us imagine an effort to map out the symmetries of the laws of physics. We suppose that we have a vast region of perfectly empty space and an enormous amount of time. For example, we go into a void in the Universe that measures $\approx 10^5$ parsecs on each side with a laboratory (fig.(13)). The laboratory moves through the void carrying out various experiments. For example, the laboratory measures the quantities:

Physical Quantities:

1. The masses of the electron, proton, mesons, W–bosons, etc.

2. The electric charges of these particles.

3. The speed of light.

4. Planck's constant.

5. The lifetimes of various particles and nuclear levels.

6. Newton's universal constant of gravitation.

These quantities are measured to enormous precision and the values are plotted against various configurations of the laboratory:

Configurations:

1. The position in space of the laboratory.

Figure 13: Laboratory for measurement of symmetries in physics.

2. The orientation in space of the laboratory.

3. The time of the measurement.

4. The velocity of the laboratory.

After many billions of years the results of the study are completed. They may be summarized simply: *The physical quantities measured by the laboratory have no dependence upon the configuration of the laboratory.* In other words, changing the position, orientation, time, or velocity of the laboratory does not influence the outcome of experiments conducted in the laboratory! *These are symmetries of the laws of physics.*

In fact, such an experiment can be performed. For example, here at Fermilab we might try to determine if some basic physical quantity, such as the lifetime of the charged pi–meson, depends upon the orientation of the lab. One simply performs the experiment at different times of the day and looks for correlations between time of day and measured lifetime; the rotation of the earth takes care of the reorientation of the laboratory (of course, we must be careful about systematic errors in such an experiment; maybe the power company switches to a different generator in the evening which somehow contaminates our pion beam and gives us a fake signal. What are other potential systematic errors and how do you avoid them in the design of an experiment?). Such an experiment is reminiscent of the famous Michelson–Morley experiment which failed to show any dependence of the speed of light upon the absolute state of motion of the earthbound laboratory through an "ether–filled" space. As the consequence of symmetry is simplicity, this experiment, once properly interpreted by the special theory of relativity, washed away the concept of the ether from physics.

There are other compelling indications from astrophysics that our Universe is the same everywhere and for all time. The physical processes occuring in distant stars and more distant galaxies produce the same spectral lines as in laboratory measurements on earth. Such measurements reach back 10 billion years to the early Universe where even Quasi-stellar objects reveal spectral lines of Hydrogen equivalent to those we see today[4]. Also, the measurements are independent of direction is

---

[4] Of course, these lines are redshifted due to the general recession of these objects, but the redshift is a universal multiplicative effect and the relative frequency ratios of lines are not affected

space. In fact, it is hard to imagine such a homogeneous and isotropic Universe if the laws of physics themselves are not independent of space, time, direction and state of motion.

But can we be absolutely sure that these symmetries are exact and that lurking well below the sensitivity of our experiments there are not small violations of translational invariance or perhaps the fundamental constants change very slowly as the Universe expands? The answer is no. We may prefer the esthetic simplicity of the belief in absolute symmetry, but we can be no more sure than our best experiments can determine. Yet these space–time symmetries are so nearly exact that we can proceed to *understand* nature by insisting that they are truly exact. The result is a completely self–consistent picture of physical laws.

Changing the position, orientation, time, or velocity of the laboratory does not influence the outcome of experiments conducted in the laboratory! In fact, these symmetries form the *Poincaré Group* of space–time symmetries of the laws of nature — the basic space–time symmetry group of the laws of physics. They hold over cosmological as well as microscopic and subnuclear scales. What are the physical consequences of these symmetries? This connection is given by one of the most important *theorems* in theoretical physics, known as **Noether's Theorem** [4]:

For every continuous symmetry of the laws of physics there exists a corresponding conservation law.

Since the translational symmetry operations can act in any one of three directions in space we find that there is a conserved quantity known as **momentum** which forms a three component vector. Since temporal symmetry operations act in one direction of time we find that there exists a conserved quantity known as **energy** which forms a scalar in Newtonian physics. Relativity unites space and time and in so doing melds energy and momentum together into one quantity called a 4–vector.

Rotational symmetry operations can be performed in any of three independent directions and thus there exists a vector quantity known as **angular momentum**. In relativity the rotations combine together with the three independent Lorentz transformations and angular momentum becomes associated with a six component *tensor*.

Thus Noether's theorem gives us the remarkable connection between symmetry and dynamical, physical conserved quantities:

$$
\begin{array}{ccc}
\text{momentum} & \longleftrightarrow & \text{space translations} \\
\text{energy} & \longleftrightarrow & \text{time translations} \\
\text{angular momentum} & \longleftrightarrow & \text{rotations}
\end{array}
\tag{4.1}
$$

Furthermore, even electric charge, baryon number, quark color and other conserved quantities are associated with symmetries in a deeper and more abstract manner.

As the arts and music have moved in this century farther away from symmetry, indeed adopting *antisymmetry* as a structural element, it is remarkable that symmetry has been increasingly understood by physicists as fundamental to the formulation of the laws of physics. How many times have we glimpsed an equilateral triangle's simplicity yet missed it's inner complexity and logical beauty? So too it is with nature. Perhaps her deepest secrets lie hidden before our very eyes!

FOOTNOTES AND REFERENCES

1   *The Short Life of Evariste Galois*, T. Rothman, Scientific American **April** (1982) 136.

2   *The Enormous Theorem*, D. Gorenstein, Scientific American, **Dec** (1985) 104.

3   *The Eight-Fold Way*, M. Gell–Mann and Y. Neeman, *Benjamin, New York* (1964).

4   See e.g. *Women in Mathematics*, L.M. Osen, *MIT Press* (1974) 141.

Howard Georgi was born in 1947 in San Bernardino, California. He received a B.A. from Harvard in 1967 and a Ph.D. from Yale in 1971. Returning to Harvard in 1971 as a postdoc, he was a National Science Foundation Postdoctoral Fellow and later a Junior Fellow in the Harvard University Society of Fellows. In 1976, he joined the teaching faculty at Harvard as an Associate Professor and was awarded an Alfred P. Sloan Foundation Fellowship. After promotion to Professor in 1980, he was elected to the American Academy of Arts and Sciences. In 1981, he rejoined the Society of Fellows as a Senior Fellow. He has written numerous articles in particle physics and two books, *Lie Algebras in Particle Physics,* and *Weak Interactions in Modern Particle Theory.*

Unification and Particle Physics

Howard Georgi
Department of Physics
Harvard University
Cambridge, MA 02138

INTRODUCTION

Let me begin by saying how honored and delighted I am to speak to you all today.

I am supposed to talk about GUTs and the present state of elementary particle physics. I should say at the outset that I am not sure that it is necessary, desirable or possible to teach elementary particle physics in a systematic way at the high school or beginning-college level. At any rate, it is much more important to do a good job of teaching mechanics. But I think that a course at any level can be enlivened by occasional reference to the bizarre and fascinating world of elementary particles. Therefore, as I outline our present understanding of the theory of elementary particles, I will pause at various points to present vignettes from particle physics that make sense taken out of context and can be used to illustrate more basic ideas.

I was also going to advertise Chris Hill's talk, which I thought would be after mine. The concept of symmetry is crucial in modern particle physics. Very often, when a practicing particle physicist says that he or she "understands" something, what is meant is that the answer can be quickly and simply obtained using a symmetry argument. But symmetry arguments can, I think, be used more than they are at all levels. After Leon invited me to this conference, I decided that I should find out something about how physics is actually taught in high school. I went to see the physics teacher in my local high school (Masconomet Regional, in Topsfield, Massachusetts) and got a copy of the textbook used in the honors physics class, College Physics by Franklin Miller, Jr. I found that in most respects it was a complete and well-written book. Serious efforts had clearly been made to keep it up to date. But symmetry was simply not mentioned. I then phoned my own high school physics teacher (Tony duBourg at The Pingry School, in New Jersey). He confirmed that symmetry was not discussed in any systematic way in their curriculum either. That's a small sample, but I see a pattern. I am going to talk a lot more about symmetry, as I am sure that Chris Hill has already done.

SYMMETRIES

Symmetry is the idea that if the laws that describe some system look the same when the system is transformed in some way, then we can use the properties of the transformation to determine the properties of the system. It is helpful to distinguish between *continuous* symmetries, such as rotations, that depend on a set of continuous parameters (i.e., angles), and *discrete* symmetries, such as reflection in a mirror (called a "parity" symmetry by particle physicists), that cannot be varied continuously. You use both of these, for example, to analyze the magnetic field due to a toroidal current distribution. Using a reflection symmetry argument, you can show that the magnetic field is azimuthal. Using a rotation symmetry, you can argue that the magnitude of the magnetic field is independent of the azimuthal angle. Then you can write down the magnetic field, even in a funny-shaped toroid, without doing much work at all.

Continuous symmetries are particularly important, partly because of a general argument called *Noether's Theorem*. The statement is that for every continuous transformation that leaves the physics of a system unchanged, there is a quantity that is conserved, that is that does not change with time. You are all familiar with examples of this theorem. Angular momentum is conserved in systems in which the laws are unchanged by rotations. Momentum is conserved in systems in which the laws are unchanged by translations. Both of these examples involve ordinary three dimensional space. Particle physicists have learned that the laws of particle physics look approximately unchanged under "rotations" in other "spaces" as well. In general, the conserved (or almost conserved) quantity associated with such a rotation is called the *charge* associated with the continuous symmetry. For example, protons and neutrons, along with a family of other such particles, are called baryons. Baryon number, which essentially counts the number of baryons, is an example of a charge that might be conserved in our world—alas, there is no experimental evidence that it is not. It has the value 1 for all baryons. If it is conserved, then by Noether's theorem it is associated with a symmetry in which the quantum mechanical wave function of every baryon is rotated by an angle in the complex plane. More familiar electric charge is similar, although, as Chris Quigg told you yesterday (and I will discuss later), it is even more interesting.

Isospin is an example of a symmetry that is certainly only approximate. It rotates protons into neutrons, leaving the strong, nuclear force between the two nearly unchanged. The associated charges, also called isospin, are only approximately conserved, meaning that in certain circumstances, isospin changes can be ignored.

Discrete symmetries also play an important role in particle physics, although I will not discuss them much here. For example, the discrete symmetry called TCP which seems to be a symmetry of any reasonable theory implies that all particles have antiparticles with exactly the same mass and opposite values of all the charges associated with continuous symmetries. For example, antiprotons and antineutrons and all other antibaryons carry baryon number -1. TCP is a combination of a parity symmetry with interchange of particle and antiparticle and a change in the direction of time.

Symmetries played a crucial role in the development of our understanding of the strong interactions that play a role inside the atomic nucleus. Baryon number and electric charge are associated with what are called "U(1)" symmetries, the "1" because they act on only one type of particle at a time and do not mix up different particle states. Isospin is a so-called "SU(2)" symmetry, because in its simplest representation, it mixes up only two states, the proton and neutron, for example. Isospin, however, also shows up in larger representations. For example, the *triplet* of pions, pi plus, pi zero and pi minus (π^+, π^0 and π^-), are mixed up by isospin rotations.

Gell-Mann found that he could classify all the strongly interacting particles (generically called *hadrons*) known in the 60s in terms of a larger symmetry, "SU(3)", that, in its simplest form, mixes up three states at a time. The isospin SU(2) and also the electric charge of the hadrons were contained within SU(3). He further discovered that he could understand the isospin and SU(3) properties of all the known hadrons if he assumed that they were built out of three types of quarks, u, d and s with electric charges 2/3, -1/3 and -1/3 (times the charge of the

118

proton) and their antiquarks, ($\bar{u}$, $\bar{d}$, and $\bar{s}$) with charges -2/3, 1/3 and 1/3. These quarks are the three fundamental objects that are mixed up by an SU(3) transformation. For example, the proton is a bound state of two u quarks and and a d quark, while the neutron is a bound state of two d quarks and an u quark. Isospin SU(2) consists of a subset of the SU(3) transformations that just mix up two of the three quarks involved in SU(3), the u and the d. In this picture, the isospin rotation that interchanges the proton and the neutron can be understood in terms of an isospin rotation that interchanges the u quark and d quark.

This may sound like nothing more than a change of language, but in fact there are two advances hidden in this simple idea. One is that when we know how the quarks change under an isospin or SU(3) rotation, we can automatically determine how the isospin rotation works on all the particles that are built out of these quarks. Even more important, the quark picture suggests a dynamical reason for the existence of the isospin symmetry and a specific dynamical picture of its breaking. If the strong forces that bind quarks together into strongly interacting particles are independent of the type of quark, then the only things that distinguish one quark from another are their masses and their properties under weaker interactions. It turns out that the quarks really do exist, as Chris Quigg told you, and are bound together by a force something like electromagnetism. It is called QCD or the color interaction and I will discuss further in a moment. It has the crucial property of independence of quark type. Thus, isospin is a good approximate global symmetry because the u and d quarks have nearly the same mass. SU(3) is more badly broken because the s quark is somewhat heavier, but the s quark is still close enough in mass to the u and d quarks to make Gell-Mann's SU(3) a convenient and useful tool.

In fact, it is important to note that there is nothing very fundamental about an approximate symmetry such as isospin. It is a great convenience in many calculations, but we now "understand" it in a more fundamental way by looking directly at the underlying quark dynamics. In fact, it is easy to make mistakes by taking the symmetry to be a thing more fundamental than it is. Nature has made it easy to make such mistakes by supplying a series of quarks heavier than the u, d and s—the c (for charm) with charge 2/3, the still heavier b (for bottom or beauty) with charge -1/3, and the still undiscovered but undoubted t (for top or truth) with charge 2/3. For these heavy quarks, it is pointless to talk about symmetries that rotate them into one another. The mass differences between them are simply too large. But we understand their properties directly from the underlying theory, so we do not need the symmetry.

THE PION

Another of the key developments was the understanding of the strange case of the pion. The π mesons are the lightest of the particles that participate directly in the strong interactions, the lightest objects built out of quarks and antiquarks. They are produced by the millions at FNAL when high energy particles collide. The π^{+}, for example, is a bound state of a u quark and a $\bar{d}$ antiquark. Its mass is less than 1/6 the proton's mass. An obvious question is, "Why are the pions so light?" In a simple quark model, the obvious assumption to make about the u and d quark masses is that they are about 1/3 the mass of the proton. But then why doesn't the pion have a mass about 2/3 the proton mass?

In fact, the masses of all the light, strongly interacting particles can be understood by putting these quark masses together, but there is another important contribution to the physics—the interaction between the quark spins. Quarks, like electrons, carry spin angular momentum, $\hbar/_2$, and they act like little magnets, not only ordinary electromagnetic magnets, but also magnets under the much stronger color force. The color magnetic force makes the state in which the quark and antiquarks spins are oppositely directed (paired) have lower energy than the state in which they are aligned. In fact, this is exactly what we see. The pions have quarks spins paired and are much lighter than the analogous bound states of the same quarks with spins aligned (called the ρ mesons).

Note that the same phenomenon, the dependence of the energy on the relative orientation of the magnetic dipoles, occurs also in heavy quark systems such as "charmonium", a bound state of c and its antiquark $\bar{c}$. It also occurs, with ordinary electromagnetic dipoles in the bound state called "positronium" of an electron and its antiparticle the positron. The "ortho" state with the spins aligned has the two magnetic dipoles oriented in opposite directions (because the charges of the positron and electron are opposite) while the "para" state has spins oppositely directed and the magnetic dipole moments are parallel. Two magnetic dipoles in the form of current rings have lower energy if they are sitting on top of one another with moments aligned (although electric dipoles work the other way). Because the divergence of $\vec{B}$ is zero even inside the almost point-like electron and positron, they behave like current loops, and the para state has lower energy than the ortho state. I like to use this in discussions of electromagnetism as an example of the difference between electric and magnetic dipoles. Then it is easy to make contact with the analogous situation in particle physics.

But there is something more to the lightness of the pions. There is good reason to believe that most of mass of the quark that we "see" in the mass of the proton or the ρ is a dynamical effect of quark confinement, that the u and d quarks in the underlying QCD theory actually have masses much smaller than 1/3 the mass of the proton. Let's consider the limiting situation in which the u and d quark have no mass at all. This theory actually has a larger symmetry than isospin symmetry. The QCD force looks the same even if two separate isospin rotations are made, one on each of the spin states of the quarks. A quantum mechanical spin 1/2 particle has two spin states. If the particle is moving, we can always take the two spin states to be the right handed and left handed states with spin parallel and antiparallel to the direction of motion. The handedness of a spinning particle is also called "chirality." If the particle has a mass, these two spin states are closely related because we can flip the handedness simply by slowing the particle down and reversing its direction with the spin unchanged. But if the particle is massless, it always moves at the speed of light. Then the left handed and right handed spin states act like completely independent massless particles. Thus there are two independent isospin rotations that leave the QCD theory unchanged, one for the left handed quarks and one for the right handed quarks. The combination in which both states are rotated together is ordinary isospin. The other independent combination, in which the left handed and right handed quarks are rotated in opposite ways is called "chiral" isospin symmetry.

SPONTANEOUS SYMMETRY BREAKING

A quark confined within a massive particle like the proton, even a quark that is massless in the underlying theory, cannot have left handed and right handed spin states that are independent. Because the massive particle inside which it is confined must travel at less than the speed of light, the quark cannot be moving at the speed of light in a straight line. It must be bouncing around in there. When it bounces, its direction and therefore its handedness changes. The independent isospin symmetries for the left handed and right handed quarks seem to have disappeared. But there is still a remnant of this symmetry. Since the underlying theory has the symmetry under independent isospin symmetry, the left handed u quark, for example, can be paired, when confined inside the massive strongly interacting particles, with a right handed u quark, or with a right handed d quark, or any rotated combination of the two. There are an infinite number of possible ways of pairing up the left handed quarks with the right handed quarks to produce the dynamically induced quark masses. But once the decision is made, it is made everywhere and for all time, because the original isospin symmetries were only global symmetries. We usually associate this decision with the "vacuum" state, the "empty" space in which all the elementary particles live. If the u and d quarks were exactly massless in the underlying QCD theory, there would be an infinite number of possible vacuum states, all physically equivalent, but only one picked out in our world. The chiral symmetry is broken by the vacuum state.

We call this situation *spontaneous symmetry breaking*. The symmetry is broken, not by any intrinsic interaction, but by an arbitrary choice of one among many possible lowest energy states of the system. An amusing example of spontaneous symmetry breaking that is familiar to all of you is obtained by considering the transverse waves in an infinite stretched rope. This system has lots of symmetry. Let us eliminate rotations from considerations by stretching the rope in the x direction and looking only at oscillations in one transverse direction, the y direction. There remain reflections (the nontrivial one, in the plane perpendicular to the rope, just tell you that waves traveling to the left have the same properties as those traveling to the right) and two kinds of translations—along the rope and in the transverse direction. The translations along the rope are associated with the very existence of traveling waves that carry momentum (remember Noether's theorem).

I want to concentrate on the translations in the transverse direction. These are rather different from those along the rope. After a longitudinal translation, the infinite rope looks just the same. But although the laws describing the motion of the rope are the same after a transverse translation, the rope looks different, because it has moved up or down. The actual position of the rope picks out a definite value of y. The symmetry of transverse translations is spontaneously broken. The analogy with chiral symmetry and the pi is the following. Imagine that each y value of the quiet stetched string represents some particular combination of right handed u quark and right handed d quark that is paired with the left handed u quark inside hadrons. Because all values of y are equivalent, we can simply define y=0 to be the sensible situation in which the left handed u quark is paired with the right handed u quark, as in an ordinary particle mass.

One might think that since our rope is infinitely long, as the physical vacuum is infinitely big, that the presence of other possible orientations would be irrelevant, because you can never translate the whole infinite system, only a finite part of it. But in fact, the presence of the infinite number of degenerate vacuum states, even though you can't reach them, implies that the system has an important property. The system supports waves whose freqency goes to zero as their wavelength goes to infinity. This is clear, because when the wavelength is very large, the disturbance just looks more and more like one of the equivalent positions for the quiet string so the restoring force and therefore the frequency of the mode goes to zero as the wavelength gets very long.

In a relativistic quantum theory, such waves correspond to massless particles whose energy goes to zero as their momentum goes to zero. The spontaneous breaking of a global symmetry always leads to such massless particles. These are called Goldstone bosons in honor of Geoffrey

121

Goldstone who gave one of the first clear derivations of this fact in particle physics language (although it was probably known to Archimedes in the example of the rope). If the u and d quark, in the QCD theory, were exactly massless, the proton mass would not be very different from what it is in our world, but the pions would be exactly massless Goldstone bosons. In our world, the quarks have a small mass, so the Goldstone theorem is not exact. The pions have a small mass as well.

The "Goldstone theorem," the fact that any spontaneous symmetry breaking is associated with a massless particle, is a very general phenomenon. That's nice, because it means that you can learn something without much input. But there is a flip side; it's such a general phenomenon that the presence of a Goldstone boson doesn't tell you much about the dynamics. It tells you that the underlying theory has a symmetry and that the vacuum breaks it spontaneously, but it tells you nothing about the mechanism of the symmetry breaking. If I changed the physics of my string, by putting beads on it at regular intervals for example, that would change the relation between wavelength and frequency, but the system would still have the Goldstone boson, the property of having zero frequency at zero wavelength, because that follows just from the existence of the spontaneously broken symmetry. Likewise, the presence of a Goldstone boson doesn't tell you what the analog of the y position of the string is or what it is that forces it to have a definite value that does the spontaneous breaking.

All of this is reasonably well understood in QCD, but there are still very important theoretical questions associated with the pions. No one has found any simple picture that combines the two facts about the pions that I just discussed: that they are quark - antiquark bound states and partners of the ρ mesons; and that they are Goldstone bosons. This is a very interesting project in quark chemistry.

GAUGE SYMMETRIES

So far, all the symmetries I have discussed are what physicists call *global* symmetries. What makes a global symmetry "global" is that the transformations under which the laws of physics are unchanged must be the same everywhere in space and for all time. A *gauge symmetry,* on the other hand, is a rotation that can depend on when and where you are. An example of a gauge symmetry is the familiar gauge invariance in electrodynamics. I'm sure that you are all familiar with this as the fact that if the magnetic field, $\overrightarrow{B}$, is written as the curl of a vector potential, $\overrightarrow{B}$ =curl $\overrightarrow{A}$, the vector potential can be changed by the gradient of a function, $\overrightarrow{A} \rightarrow \overrightarrow{A} + \overrightarrow{\nabla}\phi$ without changing $\overrightarrow{B}$. In quantum mechanics, this change is associated with a rotation of charged particle wave functions in the complex plane, depending on the function ϕ.

Another example of a gauge symmetry is the symmetry of the QCD interactions. Many different experiments show that each type of quark comes in three distinct quantum states called colors (red, blue and green), but the laws of QCD look the same if the colors are mixed up in a way that depends on when and where. This is the color SU(3) gauge symmetry. This situation is very different from a symmetry such as isospin, or Gell-Mann's SU(3). It makes sense to say that I have a proton sitting here, because the isospin symmetry is only approximate. I can tell the difference between a proton and a neutron. But even if the isospin symmetry were exact, so that it wouldn't matter whether my particle sitting here is a proton or a neutron, I could still tell whether another particle sitting somewhere else is the same or different. If I define my particle here to be a proton, that one definition settles the question of proton vs. neutron everywhere and for all time.

But gauge symmetry is different. If I define my quark sitting here to be blue, I can compare its color with the color of another quark sitting at the same place at the same time, but I still cannot tell whether a quark over there is red, blue, or green, or whether my quark here will still be blue a minute from now. The question doesn't even make any sense, because I can always make a rotation that leaves my blue quark blue but changes the colors of the quarks everywhere and everywhen else. Of course, this kind of concept of color is not consistent by itself.

When I make a rotation that changes the colors here and there differently, something between here and there must also change to make the physics look the same, otherwise nothing could depend on color at all. That something is the "gluon field" that carries the force between quarks in the same way that the electric field carries the force between electrical charges. The fact that QCD looks the same under color rotations that depend on space and time determines the properties of the gluon field. Thus "gauge symmetry" is a dynamical principle. It is a name for the rules by which one constructs the quantum field theories that describe forces like electromagetism and the color force that holds quarks together. Along with the principle of *renormalizability*, that ensures that the theory makes sense even at very short distances, the gauge symmetry completely determines the nature of the theory up to a single dimensionless *coupling constant*, like the famous $\alpha = 1/137$ of quantum electrodynamics. The concept of a "coupling constant" is important, so I will describe it in some detail.

The idea starts with Coulomb's law: the force between two charged particles is proportional to the product of their electric charges and inversely proportional to the square of the distance between them. In cgs units, the force looks like

$$F = q_x q_y / r^2$$

where q_x and q_y are the charges and r is the distance between them. This equation defines what we mean by the charges in this system of units. The product of the two charges is equal to the force times the square of the distance between the charges. In particular this means that in this system of units, the product of the charges has the units of a product of an angular momentum and a velocity. But the velocity of light, c, and Planck's constant, $\hbar$, are obviously built into the structure of the world. Thus, if we form the combination $q_x q_y / \hbar c$, we have a dimensionless measure of the strength of the interaction between q_x and q_y. It is independent of the units in which we measure force and distance. This would not be very interesting except for the experimental fact that charge is quantized. All particles that can be isolated in the laboratory seem to have an electric charge (as measured by Coulomb's law) that is an integral multiple of the proton charge, e, so it makes sense to use the proton charge as the standard against which all other charges are measured. Thus, taking out factors of the proton charge, we can write

$$q_x q_y / \hbar c = Q_x Q_y e^2 / \hbar c \ ,$$

where Q_x and Q_y are dimensionless (integer) "charges", the ratios of q_x and q_y to the proton charge and $e^2/\hbar c$ is a dimensionless measure of the strength of the electromagnetic interactions. This is the coupling constant:

$$\alpha = e^2 / \hbar c.$$

The experimental value of α, about $1/137$, is much smaller than 1. It is small enough to allow a straightforward treatment of the quantum theory that makes use of *perturbation theory*, a series expansion of the predictions of the theory in increasing powers of the small parameter α.

In the discussion below, I will absorb the proton charge e into the dimensionless coupling constant a and describe the charge of any particle by its ratio, Q_E, to the proton charge. The E stands for electric charge. As I discussed earlier, the fact that this charge is conserved is related to the existence of the gauge symmetry, by Noether's theorem.

I should emphasize that while the quantization of electric charge is an experimental fact, the Quantum electrodynamics theory does not require it. The theory would still make sense if there were also a particle with electric charge $Q_E = \pi$, for example.

In quantum field theory, the language of relativistic quantum mechanics, forces and their fields are associated with particles. The electromagnetic field is associated with the photon, the particle of light. The color field is associated with 8 "gluons". Collectively, these objects are called gauge bosons, "bosons" because they are particles with spin angular momentum that is an integral multiple of $\hbar$ and "gauge" because the fields with which they are associated are determined by gauge symmetry. The gauge symmetries guarantee that both the electromagnetic force and the color force have infinite range because the electric field or color field due to a particle at one point must extend out to arbitrary distances in order to compensate for gauge rotations that can have space dependence that goes arbitrarily far out. This, in turn, implies that the corresponding particles have no mass. Only a massless particle can mediate a long range force.

The coupling constant of a theory with gauge symmetry can be thought of as the probability that a charged particle, when accelerated, will spit off one of the gauge bosons. The coupling constant of QCD is larger than that of electromagnetism, at least at ordinary distances.

The gauge bosons are spin one particles, that is they carry an intrinsic angular momentum $\hbar$. A basic result of quantum mechanics is that a massive spin one particle has three spin states, corresponding to spin component in some arbitrary direction $\pm \hbar$ or 0. But a massless gauge boson has only two spin states. A particle moving at the speed of light cannot have any angular momentum transverse to its direction of motion (a handwaving way to see this is to note that Lorentz contraction causes an object spinning around an arbitrary axis at low speed to look more and more as if it is spinning along the axis of motion as the speed approaches that of light). The two spin states of a massless gauge boson are $\pm \hbar$ along the direction of motion. In fact, these two spin states just correspond to the two states of clockwise or anticlockwise circular polarization of light.

QUARK CONFINEMENT

There is an important mathematical difference between electromagnetic gauge invariance and color gauge invariance that has a dramatic effect on the physics. Electromagnetic gauge invariance is a U(1) gauge symmetry that acts on only one state at a time, while color is an SU(3) gauge symmetry that mixes up different (though physically equivalent) states. The mathematical term is that the color gauge symmetry is *nonabelian*, which just means that the order in which you make a series of color rotations makes a difference in the final result. That means that the associated charges cannot all be represented as numbers. They are matrices that need not commute when multiplied in different orders.

Electromagnetic gauge symmetry, on the other hand, is *abelian*, which means that the order of the charge rotations doesn't matter. The physical consequence of this difference lies in the properties of the gauge bosons. The photon is electrically neutral, it does not carry electric charge. The nonabelian nature of the color gauge symmetry ensures that the gluons themselves have color. This, in turn, makes the dynamics of color very different from the dynamics of electromagnetism. Quantum mechanical effects that Chris Quigg talked about yesterday cause the color force to get weaker at short distances (this is called *asymptotic freedom*) and stronger at large distances (while electromagnetism works the opposite way). One consequence of this is color confinement. Because the color force gets strong at long distances, the quarks within protons and neutrons cannot be pulled out and isolated. The force between a quark and an antiquark (say) goes to a constant value at large separations. The quark and antiquark are confined by the gauge interactions.

There is an interesting way to view quark confinement that can be connected with rather simple physics. Consider, for a moment, a large, parallel plate capacitor carrying a non-zero charge. The electric field is almost entirely inside, perpendicular to the plates and determined by the charge density on the plates. When the plates are pulled apart, so long as the parallel plate approximation remains valid, the electric field and therefore the force exerted on the plates remains constant. We believe that the color force between quarks and antiquarks at long distances works something like this. The color analog of electric field is organized by the quantum mechanical effects of the nonabelian SU(3) gauge interactions into a long tube, within which the electric field lines are parallel, like those in a parallel plate capacitor. When this tube is stretched, by pulling on the quarks, energy must be fed into the system at a constant rate, to build more of the electric field tube. This gives the constant force at long distances.

If the tube of color electric field were infinitely thin, it would be what is called a relativistic string. Such an object is, in a sense, a perfect spring, because when it is stretched, all of the energy supplied by the stretching force goes into building new string. It is fun to consider what happens when such a string, with massless quarks at the end, is not held fixed, but allowed to move. The longitudinal oscillations are interesting. They can be discussed in an elementary course by thinking about a parallel plate capacitor with massless plates! In fact, this is a simple and amusing finite system in which it is fairly easy to see how relativity works. But I don't have time to talk about this now. If people are interested, we can discuss this after my talk.

Instead I will briefly discuss what happens when you take the quarks off the end so that you have just string and you swing the other end around your head very fast. You might think that it would be impossible, with no quark at the end to carry energy or momentum, to swing the string fast enough to keep it from being slurped into your hand like a doomed piece of spaghetti. But in fact it is possible because of the Lorentz transformation law of force. If you swing it so fast that the end moves at the speed of light, then the transverse force on the end gets Lorentz contracted to zero (in fact, I use this as a mnemonic device—I can't forget that the transverse force is smaller in the frame of reference in which the object is moving). This configuration can then hold itself together.

The interesting thing about this spinning object can be seen by dimensional analysis. In units with $\hbar = c = 1$, the only parameters on which the physics of this system can depend are the length of the string, l, and the string tension, T, the energy per unit length of the relativistic string at rest. In these units, all dimensional quantities can be expressed in terms of energy. l has units of $1/E$, while T has units of E^2. Both the energy and the angular momentum of the system must be proportional to T. Then dimensional analysis tells us that the energy is proportional to Tl while the angular momentum is proportional to Tl^2. That means that the square of the energy is proportional to the product of the string tension and the angular momentum. When this was realized in the sixties, particle physicists were interested because experimenters had discovered families of short lived hadrons with exactly this property—their energy, mass, was proportional to their spin angular momentum. Thus, not surprisingly, some theorists thought that hadrons might all be made out of this stuff. The suggested string tension was about 13 tons. Many people worked very hard in the early seventies to build a theory of all the hadrons based on this idea.

This amusing story is not yet finished, but I am afraid that it will have a sad ending. String theory did not describe hadrons, but that is OK, because we have since found QCD instead. In fact, it was found that the quantum mechanical theory of the relativistic string does not make sense at all in space with the right number of dimensions, but the mathematics of the string theory was so much fun that it refused to die. Some theorists have resurrected the idea and found that a related idea may make sense in a ten dimensional space-time. Furthermore, the theory has many very remarkable mathematical properties. Even though ten is not equal to four, many theoretical physicists are studying the idea that this theory may describe the world, with the extra six dimensions curled up into a little balls with size of order 10^{-35} meters. This time, the string tension is more like 10^{38} tons. Personally, I suspect that this is so far removed from anything that Leon and his friends can ever measure that it should be regarded as mathematics rather than physics, but that may be just a matter of taste.

At any rate, it is clear that the idea of gauge symmetry is very important. As Chris Quigg told you yesterday, we can divide the particles that we know about into two sets—matter particles and force particles. The matter particles that carry color are the quarks and antiquarks. Those that do not carry color are *leptons*, including the electron, and their antiparticles, called antileptons, including the positron. The force particles include the photon and the gluons and a couple of objects I'll discuss in a moment, the W and Z. All of the force particles seem to be associated with gauge symmetries.

INTERLUDE

[During a pause to change the videotape, Professor Georgi answered some questions. Some of the questions and answers are included here.]

Question: I've always been told that when a neutron changes to a proton, one of the down quarks turns into an up quark. Yet as you put it on the transparency, it seems that you are saying that all the downs become ups in the transformation of a neutron into a proton. What is going on?

Professor Georgi: I think that we a talking about two different things. In the decay of a neutron into a proton, which is a process I haven't talked about yet but will soon, what happens is precisely what you have been told, one of the d quarks inside the neutron decays into a u quark plus other things (an electron and an antineutrino).

The isospin transformation, however, is different. It's a transformation that acts on all of the quarks at once. And the laws that describe the way the quarks are put together into the proton and the way protons are put together into nuclei are invariant only if you change all the quarks at the same time.

In the physical process of neutron decay, only one of the quarks changes, but in the invariance principle of isospin, we think about what happens if we change (or more correctly—rotate) all of the quarks simultaneously.

Question: Will you explain exactly, if you can, what is a string?

Professor Georgi: A relativistic string? You mean one of these things that I was just talking about?

Question: Is it the same kind of string that we've been reading about lately in the news and so on?

Professor Georgi: Almost. I'm not going to say very much about this because as I explained I am not really very sympathetic with this particular direction. The idea is to replace the concept of *particle* with the concept of—how can I say it? The concept of an elementary particle doesn't make any sense. In a quantum field theory, a particle is described as a point-like object, and anything that depends on the concept of a point is silly be cause we physicists don't know about points. Points are idealizations. Points are for mathematicians.

And in place of this concept of a point, the string theorists have put a different, but equally silly concept, the concept of an infinitely thin, one dimensional object—a superstring. This has some nice features because the string has more structure to it than the point. A string, for example, can have waves in it, modes of oscillation. It is these modes of oscillation that the string enthusiasts are excited about. They are interpreted as the particles that we actually see.

The string that I was discussing is not infinitely thin. In fact, there is another way of talking about quark confinement that Vicki Weisskopf likes that goes with the name "bag" in which one can understand, maybe, a bit more clearly what the string I talked about is. One can think of the mechanism by which the color electric field between quark and antiquark gets squeezed into a tube when the separation is large as another property of the vacuum. Suppose that lowest energy state of the interacting QCD system cannot be penetrated by the color electric field lines produced by a color charge like a quark. Then any source of color field must push the vacuum state away from it to make room for the color field lines it produces. When a quark and antiquark are far apart, the field lines must go from one to the other, and the lowest energy state of such a system has the interacting vacuum pushed away only in a thin tube between them. This is the QCD string. On the other hand, if the quark and antiquark are close together, the interacting vacuum is only pushed away from a small region nearby. The region in which the color electric field is large is the bag inside which the quarks and antiquarks are confined. The spontaneous chiral symmetry breaking that I talked about earlier occurs at the surface of the bag, because it is here that the confined quarks have to turn around to remain confined.

Of course, there's another difference between the QCD string and the superstring. You don't have to pull very hard to stretch the string between a quark and an antiquark. The tension is only about 13 tons. The superstring is a little stiffer. The tension is something like 10^{38} tons. That is the whole problem. It is very difficult to get at the dynamics of a string that you have to pull on with a force of 10^{38} tons to move. That is why I am afraid that the superstring is going to remain mathematics rather than physics.

Just to complete the confusion, I should add that there is a third kind of string that has received some attention in the semipopular press recently—the so-called cosmic string. These are relativistic strings that exist in some of the grand unified theories that I will talk about later. Some people believe that such things might have been important in the formation of galaxies. Perhaps Dave Schramm will talk about this tomorrow.

Question: About half of the audience, like myself, are high school teachers, and there are two problems that we face. One is that our students are probably in their first experience in physics, and the second is that we don't have any experience with the concepts that you are discussing. After you are finished with your presentation, we are supposed to go out and meet and discuss ways to try to incorporate these ideas into the high school physics curriculum. Perhaps there are ways we can do it. I would like to hear some reactions to this.

Professor Georgi: First of all, what I should probably do is to junk the second half of my talk, because the things that I have talked about so far are the easy ones! The strong force is much simpler to explain that the weak force, because of the analogy with electromagnetism, which can help high school students get an idea of some sort of what is going on. The strong force, at least, is a force in the sense in which one normally thinks of a force; it pulls on things. The job of the strong force is to hold quarks together. Honestly, though, I don't think that many of these things should be taught in a systematic way. You certainly should not claim that you are teaching high school kids particle physics because you are not. On the other hand, I think that some of the examples that I and the other lecturers have given could be incorporated by way of enrichment. But don't forget that your primary job is to give them a good introduction to mechanics, because it is from our understanding of mechanics that we get our ability to wave our hands.

Question: You left us hanging with the W and Z. I would like you to get back to your talk.

THE STANDARD MODEL

Okay, you asked for it. The physics is very different when it is a gauge symmetry rather than a global symmetry that is spontaneously broken. Consider the stretched string with a wave moving through through it. Now imagine for a moment that the symmetry under transverse translations is a gauge symmetry. Then the wave would not really be a disturbance at all. By making a transverse translation that depends on space and time, I could transform the disturbance away. It would be completely equivalent to the state in which the system is fixed at $y=0$. Only the gauge field would be different. The Goldstone boson would not be there at all as a physical particle. It would be an artifact of the space and time dependent translations.

For definiteness, imagine that it is color that is the gauge symmetry of our string. We saw that if the color symmetry is a gauge symmetry that is not spontaneously broken (as in QCD), it doesn't make sense to distinguish between red quarks and blue quarks, for example, except when they are sitting at the same point at the same time. But when the symmetry is spontaneously broken by the position of the string (in the color space), this is no longer true. We can use the position of the string to compare the colors at distance points. If the quark color is the same as that picked out by the string at the same point, we can define it to be a red quark (say). This gives us an absolute definition of color that is good for all places and times. There is no gauge symmetry left. Furthermore, there is no longer any reason for the color force to have infinite range because the colors do not get mixed up by the dynamics at long distances. In fact, you can show that the gauge bosons get a nonzero mass when the gauge symmetry is spontaneously broken. There is an interesting sort of conservation of states going on here. We saw earlier that a massless gauge boson has only two spin states while a massive one has three. The extra state for the massive gauge boson is provided by the Goldstone boson, which is no longer a physical massless particle. In a sense, one of the spin states of the massive gauge boson is the Goldstone boson. At energies much larger than the gauge boson mass, this spin state behaves like the Goldstone boson (or its pieces, if the Goldstone boson is a composite object like the pion). This process of trading a massless gauge boson and a massless Goldstone boson for a massive gauge boson is called the Higgs mechanism, after Peter Higgs who was one of the first to describe it in a particle physics context.

The Higgs mechanism is similar to the Goldstone mechanism in that it is very general. Anytime a global symmetry is broken spontaneously, a Goldstone boson is produced. Anytime a gauge symmetry is broken spontaneously, a gauge boson becomes massive. Neither mechanism tells you much about what breaks the symmetry. In fact, the two are closely related. The properties of the massive gauge boson in the Higgs mechanism are determined entirely by the original gauge symmetry and the properties of the Goldstone boson that is absorbed to become its extra spin state. The existence of a massive gauge boson tells you that a gauge symmetry has been spontaneously broken, but it doesn't tell you what breaks it.

In fact, as I've said the color symmetry of QCD is an unbroken gauge symmetry. But there is a beautiful example of the spontaneous breaking of a gauge symmetry—the $SU(2) \otimes U(1)$ of the weak interactions.

The electromagnetic force is very familar. The strong color force, although it is more complicated because of the nonabelian nature of the gauge symmetry, is similar in many respects to electromagnetism. In particular, it is probably obvious why both of these phenomena are called forces. The primary job of the strong QCD interactions is holding the quarks together. The weak interactions, by contrast, are very unlike familiar forces. They manifest themselves primarily by causing particles to decay.

Most types of particles do not live forever. Instead, they blow up into two or more lighter particles. This process is inherently both relativistic and quantum mechanical.

Because of quantum mechanics, the decay process involves an element of chance. Given a single particle, you cannot say when it will decay. But if you have lots of identical particles, the average rate at which they decay can be used to define what is called the particle's lifetime. The lifetime is the total number of particles divided by the rate of decay. If you have one hundred thousand particles and one thousand of them decay in a year, then the lifetime is one hundred years. Since decay is a random process, any given particle would have only one chance in one hundred of decaying in any given year.

It is relativity that allows some mass to be converted into energy in a decay process. If you stop the products of a decay and weigh them, the result for their total mass is always less than the mass of the original particle. The missing mass is converted into energy that is used to set the decay products flying apart. Of course, just because relativity allows a decay kinematically doesn't mean that it will happen. It must be produced by some dynamics.

Some particles do not decay. They are *stable*. The electron is stable for an interesting reason. Electric charge is conserved. But the electron and antielectron are the lightest electrically charged particles. Any other combination of particles with the same electric charge as the electron must have a total mass greater than or equal to the electron mass. Thus the electron cannot decay. The antielectron is stable for the same reason. An electron and an antielectron together can annihilate each other into photons because the total electric charge is zero, but neither one, by itself, can decay.

Now, let us look at some particles that can decay. The π^0 is unstable. It is built out of a combination of $u\bar{u}$ and $d\bar{d}$ quark-antiquark pairs. What this means is that the π^0 state is a sum of a $u\bar{u}$ state and a $d\bar{d}$ state. The π^0 has a 50% probability of being a $u\bar{u}$ and a 50% probability of being a $d\bar{d}$. In either form, the quark and antiquark sometimes annihilate into photons through the electromagnetic interactions. This happens fairly quickly, because it is fairly easy for a particle and its antiparticle to annihilate each other. The lifetime of a π^0 is only about 10^{-16} seconds.

129

A neutron is a bit heavier that a proton. A neutron is also unstable. It decays with a lifetime of about 1000 seconds into a proton, an electron and a very light or massless particle called an antineutrino (because it is the antiparticle of a particle called the neutrino). The neutrino and antineutrino do not feel the strong forces that hold quarks inside hadrons. They also have zero electric charge, so they do not feel electric forces.

Two questions may occur to you about neutron decay. What kind of interaction causes it? And why does it take so long, compared, say, to the π^0 decay? We can give a provisional answer to the second question. Whatever the interaction is that causes neutron decay, it must be much weaker than the interaction that causes π^0 decay, so that the decay is less probable. Historically, this fact gives the force its name. It is called the *weak interaction*. Weak interactions are also seen in many other radioactive decays of nuclei, and in decays of more exotic particles.

The weak interactions have these peculiar properties because they are associated with a nonabelian gauge symmetry that is spontaneously broken. In the breaking process, the weak interactions and the electromagnetic interactions get mixed up, partially unified. We now call the combined gauge theory that describes the weak and electromagnetic interactions, the *electroweak* $SU(2) \otimes U(1)$.

The electroweak $SU(2) \otimes U(1)$ is, as the name implies, a combination of an $SU(2)$ gauge symmetry, acting on doublets, and a $U(1)$ gauge symmetry like electromagnetism. I'll first describe the $SU(2)$ component. The electroweak $SU(2)$ is a *chiral* gauge symmetry. The left handed quarks and leptons and the right handed antiquarks and antileptons are organized into doublets that rotate under the nonabelian $SU(2)$ part of the gauge symmetry while the right handed quarks and leptons and the left handed antiquarks and antileptons do not. The lepton doublets have the neutral neutrinos on the top and the charge -1 leptons on the bottom, while the antilepton doublets are the other way around as follows:

ν_e	ν_μ	ν_τ	e^+	μ^+	τ^+
e^-	μ^-	τ^-	$\overline{\nu}_e$	$\overline{\nu}_\mu$	$\overline{\nu}_\tau$

| | left handed leptons | | | right handed antileptons | |

The νs are neutrinos, very light or massless neutral particles. The top of each doublet has an electric charge that is one unit higher than the bottom, and a different mass. That is something that could never happen in an unbroken gauge symmetry, where the states that are mixed up by the symmetry are physically equivalent.

The gauge bosons associated with the rotation from the top to bottom of these doublets and back are the W^+ and its antiparticle, the W^-. These gauge bosons carry electric charge because the tops and the bottoms have different charges. Because of the spontaneous symmetry breaking, because the tops and the bottoms of these doublets are physically distinguishable objects, the field of the W can produce particle decays by simultaneously causing a transition from the top to the bottom of one doublet and from the bottom to the top of another. For example, a muon, μ^- can turn into ν_μ and simultaneously a ν_e can turn into an e^-. The ν_μ of this process can be related to a $\overline{\nu}_\mu$ going backwards in time (remember the discrete symmetry TCP). This gives the decay process μ^- goes to $e^- + \nu_e + \overline{\nu}_\mu$.

Another important thing about this is that all of the charged leptons rotate into their corresponding neutrinos in exactly the same way. The $SU(2) \otimes U(1)$ interactions do not distinguish at all between the three different pairs of lepton.

Of course, the fact that there are three different pairs of lepton is slightly peculiar. We call the different kinds of leptons *flavors,* but we use this innocuous sounding word only to hide our total ignorance of why the multiplicity should exist. It would be easier to swallow if there were some obvious difference between them. The fact that they interact identically with gauge bosons makes they seem even more mysterious.

I've already discussed the six quarks that go with the six leptons. They are paired up in a similar way, with the charged 2/3 quarks at the top and the charge -1/3 quarks on the bottom, and the antiquark doublets reversed:

u	c	t	$\overline{d'}$	$\overline{s'}$	$\overline{b'}$
d'	s'	b'	$\overline{u}$	$\overline{c}$	$\overline{t}$
	left handed			right handed	
	quarks			antiquarks	

The primes indicate that something new has been added — flavor mixing. The d' that is part of a doublet with the u quark is mostly d quark, but it has a small admixture of s quark and an even smaller admixture of the still heavier b quark. Likewise the s' is mostly s with a small admixture of d and b and the b' is mostly b with a small admixture of d and s. In fact, such mixing might be taking place in the lepton sector of the theory as well without our knowing it. The only thing that distinguishes the d, s and b quarks from one another in this theory is their masses. If the neutrinos have mass then the lepton doublets could look like the quark doublets. But the neutrino masses, if they have any at all, are so small that they have not been conclusively observed, so for now we can just call the thing in a doublet with the electron by the name *electron neutrino.*

Flavor mixing allows the $SU(2)$ rotations of the electroweak gauge symmetry to mix up all the different flavors. By going up and down in these doublets, you can get from any quark to any other. This, in turn, means that the W fields can produce flavor changing transitions between quarks that can cause heavy quarks to decay into lighter quarks.

There is one electroweak $SU(2)$ charge that does not involving the flavor changing transitions from the top to the bottom of the doublets. It is a charge (I'll call it T_3), like the electric charge, that just acts as a number on each type of particle type. It is just 1/2 for particles at the top of an electroweak doublet and -1/2 for those at the bottom. In fact, T_3 is related to the electric charge in a very simple way:

$$Q = T_3 + S,$$

where S is the charge associated with the other half of the electroweak symmetry, the electroweak $U(1)$ symmetry. The $U(1)$ charge has the same value for each member of a doublet. For leptons, S=-1/2, for quarks, S=1/6, etc. Because this combination of the $SU(2)$ and $U(1)$ charges is the electric charge, some combination of the corresponding gauge bosons must be the photon. This is the sense in which the electromagnetic interactions are contained within the electroweak $SU(2) \otimes U(1)$. The other combination is another electrically neutral gauge boson. It is called the Z.

The force mediated by the Z field does not produce particle decays. One might think that it could because of the flavor mixing in the quark doublets. But because all three quark doublets rotate in exactly the same way under the electroweak $SU(2) \otimes U(1)$, the flavor-changing transitions can only come from W exchange. The point is this. I have written the doublets with the quarks at the top unmixed. But because the doublets are identical, as far as their $SU(2) \otimes U(1)$ properties are concerned, I could just as well mix the doublets up so that the quarks on the bottom are unmixed with all of the mixing being in the quarks on the top. This implies that any possible flavor-changing effect can come only from the comparison of the top and the bottom of the doublets. It is only the Ws that cause transitions from the top to the bottom, so only the Ws cause flavor changing effects. This would not be the case if some of the quarks behaved differently under $SU(2) \otimes U(1)$. Then Zs could also change flavor.

This structure of flavor mixing was very important in unraveling the structure of the $SU(2) \otimes U(1)$ interactions. Much of what know about the weak interactions comes from studies of how heavy quarks decay into lighter ones through W exchange. But the absense of flavor changing effects in Z exchange caused enormous confusion for many years. The fact that when all the quarks are in identical doublets, the flavor changing effects come only from W exchange is called the "Glashow, Iliopoulos, Maiani" or GIM mechanism, after the wonderfully international collaboration of Shelley Glashow of the USA, John Iliopoulos, a Greek who lives and works in Paris, and Luciano Maiani, who works in Geneva but was born in San Marino. They realized the importance of this mechanism and used it to predict the existence of the c quark before it had been seen experimentally.

The $SU(2)$ and $U(1)$ gauge interactions, because they are two different gauge symmetries, have two coupling constants. One combination of these is related to the electromagnetic coupling, α. The other determines what combination of the two gauge bosons is the Z. It is usually expressed in terms of an angle, the weak mixing angle, θ_w. While the Z field does not cause decays, it does do something interesting. It is felt by the neutrinos, so unlike electromagnetism, it causes the neutrinos to scatter off of quarks or leptons without changing to charged leptons, as the W force would require them to do. Events of this kind are called "neutral current" weak interactions. They were seen at FNAL, CERN and Brookhaven in the early 70s, and were eventually found, for an appropriate value of θ_w, to have exactly the properties predicted by the electroweak theory.

The spontaneous breaking of the electroweak $SU(2) \otimes U(1)$ gauge symmetry has many important effects. We have already seen that it allows the weak force produced by the W to cause particle decays. It also allows the quarks, the charged leptons and the W and Z to have nonzero masses. The left handed quarks and leptons behave differently under the electroweak symmetry than the right handed quarks and leptons. It is a chiral symmetry. If it were unbroken, the left and right handed components of the matter fields could not be combined into massive particles. Likewise, if the symmetry were unbroken, the W and Z would be massless gauge bosons, like the photon and the gluons.

In fact, the symmetry is spontaneously broken. The quarks and charged leptons are massive. The W and Z are also heavy. This fact is crucial because it results in the fact that the weak forces, unlike the electromagnetic force, has a short range. Particles can interact weakly only when they get close together, because the W and Z fields fall off exponentially e^{-rM}, where M is the mass of the W or Z. Thus the interaction is negligible when the separation, r, is greater than $1/M$, the distance at which the quantum mechanical properties of the W and Z become important. It is this short range that is actually responsible for the apparent weakness of the weak interactions. At low energies, particles only rarely get that close together.

The theory of the $SU(2) \otimes U(1)$ gauge interactions was worked out in the early 60s by Shelly Glashow. Later, in 1967, Steve Weinberg and Abdus Salam independently constructed a simple model in which the $SU(2) \otimes U(1)$ symmetry was spontaneously broken. They predicted the W and Z masses in terms of θ_W. Finally, in 1971, Gerard 't Hooft showed that the resulting theory actually made sense at short distances. By 1976, it had become clear that the theory accounted for a vast array of data on the weak interactions, including the neutral current effects that had been first predicted by the $SU(2) \otimes U(1)$ theory. Glashow, Weinberg and Salam were awarded the Nobel Prize for their contributions to the theory in 1979, before the W and Z had even been seen directly. If nothing else, this was a dramatic demonstration of the faith that the Swedish Academy of Sciences has in theoretical physics. Of course, as I'm sure you all know, the W and the Z have since been discovered at the proton-antiproton collider at CERN, and will soon be seen, we hope, here at FNAL and at Stanford. They seem to have just the properties predicted by the Glashow, Weinberg, Salam theory.

Our present understanding of the $SU(3) \otimes SU(2) \otimes U(1)$ theory of the strong and electroweak interactions is a majestic scientific accomplishment. But it is clearly incomplete. It describes everything we have seen so far, but it leaves too many questions unanswered, too many patterns unexplained. Today, I will discuss a few of these, beginning with those that are most likely to lead to interesting physics in the near future.

The only real missing link in the theory is that we have no experimental information about what physical mechanism is breaking the $SU(2) \otimes U(1)$ electroweak gauge symmetry. Because it is a gauge symmetry that is spontaneously broken, the Higgs mechanism must be operating. That means that one of the spin states of the massive W and Z bosons that have been seen at the CERN collider is a converted Goldstone boson associated with the spontaneous symmetry breaking. One way of asking about the physics of the spontaneous symmetry breaking is to ask, "What kind of objects are these Goldstone bosons?"

In the Weinberg-Salam model, the Goldstone bosons are an entirely new kind of fundamental particle, unlike anything else that we have seen. All of the objects we have discussed so far are either spin 1/2 particles, like the quarks and leptons, or spin 1 gauge bosons, like the photon, the W and Z, and the gluons. The Goldstone bosons of the Weinberg-Salam model are fundamental spin 0 bosons. The three Goldstone bosons, associated with the Z and W^+ and W^-, are part of a quartet of fundamental spin 0 bosons that rotate into one another under the $SU(2) \otimes U(1)$ symmetry. It is the interactions of these four bosons among themselves that give rise to the spontaneous symmetry breaking. It is important to note that these self interactions are not simply the result of a gauge symmmetry. Like the objects themselves, their interactions are entirely different from anything that we have actually seen. Furthermore, the interactions themselves lack some of the nice properties of gauge theories. If the Weinberg-Salam model is part of a unified or otherwise modified theory in which there is interesting physics at energies much higher than the W and Z mass, it is particularly difficult to understand why the the Higgs boson does not break the symmetry at higher energies.

133

The fourth member of this quartet is not eaten by the Higgs mechanism. It survives as a massive spin 0 boson. This is the famous or infamous Higgs boson, predicted by the Weinberg-Salam model, but so far not observed experimentally.

BEYOND THE STANDARD MODEL

One could interpret the fact that the properties of the W and Z seem to be correctly predicted by the Weinberg-Salam model as evidence for this particular mechanism for spontaneous symmetry breaking. But the generality of the Higgs mechanism makes this unwise. The properties of the W and Z really only depend on the properties of the Goldstone bosons that are absorbed in the Higgs mechanism. The relevant properties of these Goldstone bosons are determined almost entirely by the symmetries of the theory that produces them. It turns out that the properties of the W and Z in the Weinberg-Salam model are determined primarily by an approximate isospin symmetry of the dynamics of the Higgs boson and its three Goldstone bosons. The same symmetry, and therefore the same masses and couplings for the W and Z can come out of theories in which the dynamics of spontaneous symmetry breaking is analogous to the dynamics that produces the pion in QCD. In this picture, $SU(2) \otimes U(1)$ is a chiral symmetry of some hypothetical fundamental spin $1/2$ particles that are confined by a gauge interaction much stronger than QCD that is sometimes called *technicolor*.

In a technicolor model, the Goldstone bosons that are absorbed in the Higgs mechanism are bound states of the hypothetical strongly interacting *technifermions* (as the spin $1/2$ particles are called) just as the pions are bound states of quarks and antiquarks in QCD. The technicolor force is mediated by gauge bosons, *technigluons* just as the QCD force is mediated by gluons. In fact, the weak $SU(2) \otimes U(1)$ gauge symmetry in a technicolor model is very much like the chiral isospin symmetry in QCD. In the simplest technicolor model, the left handed technifermions are a doublet that rotate into one another under the weak $SU(2) \otimes U(1)$ gauge symmetry, while the right handed technifermions do not rotate. When the technifermions are confined by the technicolor interactions, the left handed and right handed technifermions are combined into dynamically massive techniparticles and the chiral $SU(2) \otimes U(1)$ symmetry is spontaneously broken. But an analog of the ordinary isospin symmetry remains because the two techniquarks get equal dynamical masses so they can still be rotated into one another. It is this remaining isospin symmetry that guarantees that the W and Z have the same properties in this technicolor theory as in the Weinberg-Salam theory.

There is something very appealing about this mechanism. It puts the interactions that produce spontaneous breaking of the $SU(2) \otimes U(1)$ symmetry on the same footing as all the other interactions that we have seen. They are just another example of a force determined by a gauge symmetry and mediated by gauge bosons.

Technicolor is a beautiful idea that might be right, but it seems to have a couple of serious problems. The trouble starts with the fact that the quarks and leptons do not feel the technicolor force. There is nothing to put the right handed quarks and leptons together with their left handed partners into massive states, so unless there is something else going on besides technicolor, the quarks and leptons remain massless.

In the Weinberg-Salam model, the Higgs boson provides the glue that puts the left handed and right handed spin $1/2$ particles together. This is possible because the interaction between the Higgs boson and the spin $1/2$ particles makes sense even at energies very large compared to the

W and Z masses. But the interaction of the Higgs boson with the spin 1/2 particles, like its self-interactions that cause spontaneous symmetry breaking, are yet another kind of interaction, unlike anything else in the theory. They do not have any sort of gauge symmetry to tell us anything about their form. In fact, these interactions do almost nothing except give masses to the quarks and leptons. And by the same token, since they have been put into the theory solely to generate these masses, they give us no insight into why the masses are what they are.

It ought to be possible to enlarge the technicolor gauge interactions to allow for quark and lepton masses, but so far, no one has managed to do it. Two things usually go wrong. It is hard to get the peculiar masses and mixings that we actually observe from any theory. And as soon as any mixing is introduced, the GIM mechanism is destroyed and the theories predict Z mediated weak decays that have not been observed.

Under the pressure of these difficulties, the interest in technicolor has waned in recent years. Many theorists devoted some of their energy towards trying to make sense out of the Higgs boson by embedding it in a theory with a very different kind of symmetry, a so-called *supersymmetry*. What is "super" about supersymmetry is that it puts particles with spin 1/2 together with particles of spin 1 or spin 0. Since at least two of these types of particles are known to exist in the world, this seemed very exciting for a while. Equally exciting were some of the peculiar properties exhibited by the interactions between supersymmetric particles. They could have gauge interactions just like those in ordinary theories, but the other interactions, those not associated with a gauge symmetry, are more constrained than in a theory without supersymmetry.

Unfortunately, it seems to me that supersymmetry has failed to help us with any of the really interesting questions. For one thing, it has become clear that the particles that we see are not rotated into one another by supersymmetry rotations, but into an entirely new set of particles, none of which have been observed. In addition, if, as seems likely, the theory has in it objects with very large masses, supersymmetry by itself does not explain why the W and Z are light. Finally, supersymmetry gives no insight into the origin and significance of the quark and lepton masses.

My personal guess is that it may be possible to address these questions in theories in which not only the electroweak Goldstone bosons, but also the Higgs boson and quarks and leptons are composite states, built out of more fundamental matter particles interacting through another yet stronger force, again determined by a gauge symmetry. We will see.

Of course, that's the whole point. Unless we are going to learn something about the answers to these questions some day, there is not much point in speculating about them. The nice thing about the question of electroweak symmetry breaking is that we know where to look for it—in the interactions of the extra spin states of the Ws and Z, at energies well above the W and Z masses, the physics of $SU(2) \otimes U(1)$ breaking must show up, whatever it is. That is the reasoning behind the proposed SSC. We need a machine that can produce Ws and Zs at energies large compared to their masses to study this physics.

UNIFICATION

Finally, let me discuss the possibility that the $SU(3)$ and $SU(2) \otimes U(1)$ gauge interactions may be pieces of some larger, more symmetrical gauge symmetry. There are indications that there is something very special about the particular pattern of gauge symmetries and fermions.

We can see one of them by putting together information from color SU(3), weak SU(2) and electromagnetism in a special way.

First consider color. The color charges, as I have explained, are complicated. They must be represented by matrices, rather than just numbers, because their multiplication law is not commutative. However, you can find two of them that do commute with one another, so that they behave like numbers. If you plot these two numbers on the x and y axis for each of the three color states, it will probably not come as any surprise that you get the most symmetrical figure that you can make with three objects, an equilateral triangle. This is a useful way of representing the color properties of the quarks. The antiquarks have exactly the opposite charges. In the same two dimensional space, they are represented by another equilateral triangle with the opposite orientation. The leptons and antileptons have zero color charges, so they sit in the center of these triangles.

Next consider weak SU(2). I will look at the subset of the particles that sit at the top of the weak SU(2) doublets, the neutrinos, the charge +1 antileptons (like the positron), the charge 2/3 quarks and the charge +1/3 antiquarks. Plotting all of these in the color plane as just discussed, we have two oppositely oriented equilateral triangles and two points in the center.

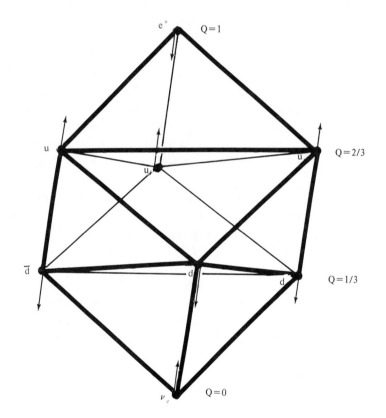

Fig. 1 A cube showing the particles from weak SU(2) doublets is projected onto a page. The z-axis is electric charge; the x- and y-axes represent color coordinates of the particles. Cube edges and diagonals nearest the front are shown by heavier lines. Note the opposite orientation of the equilateral triangles representing the charge 1/3 and charge 2/3 quarks described in the text.

Now finally consider electric charge. Let me lift each of the states out of the color plane a distance proportional to its electric charge.

The result is a cube. This suggests that there could be some other symmetries of the quarks and leptons beyond those we have discussed. In fact, the most symmetrical way of combining these particles cannot be accurately represented in three dimensional space. The $SU(3)$ and $SU(2)$ symmetries, acting on triplets and doublets, can be combined into an $SU(5)$ symmetry acting on a quintet. Remarkably, if the quintet is taken to be an antilepton doublet and the charge $-1/3$ quark color triplet, then the $SU(5)$ charges also contain the $U(1)$ charge needed to complete the description of the electroweak interactions. Furthermore, the rest of the matter particles fit neatly into other simple representations of the $SU(5)$ symmetry.

This fit of all the known forces and particles into $SU(5)$ is extremely nontrivial and suggestive. Of course, we know that the $SU(5)$ symmetry, if it is there at all, must be spontaneously broken to produce the rather different properties of the pieces that we see. When the symmetry is broken, the extra gauge bosons of the $SU(5)$ symmetry that are are not associated with either $SU(3)$, $SU(2)$ or $U(1)$ (it turns out that there are 12 of them), will get a large mass. Call it M. Only at energies larger than M or distances smaller than 1/M will the full $SU(5)$ symmetry of the physics become apparent.

Can we somehow extrapolate what we know about $SU(3) \otimes SU(2) \otimes U(1)$ to higher energies? In fact we can, as long as we assume that nothing else is going on at intermediate energies. Quantum mechanical effects cause the values of the separate coupling constants of $SU(3)$, $SU(2)$ and $U(1)$ to get closer together as the energy increases. They seem to meet, or come very close, at a large energy of nearly 10^{15} GeV, 10^{15} times the mass of the proton. The fact thay they meet at all is nontrivial. It amounts to a prediction of one of the couplings in terms of the other two. But what is more, the energy at which they meet must be about equal to M. This means that the new gauge bosons in a theory of this kind are absurdly heavy. The forces that they produce have an incredibly short range and are therefore very weak. Nevertheless, it may not be impossible to get information about these forces.

Perhaps we can learn about unification not by experimenting, but by observing the universe. If the standard big-bang cosmology is correct, the universe was once very hot and dense. If we could follow the history of the universe back far enough, the temperature might be so high that the typical energies of the particles bouncing around in the primordial fireball are of the order of 10^{15} GeV. At such ridiculously early times, the interactions of all the gauge particles in $SU(5)$ are equally important. One might then hope to find effects of the $SU(5)$ interactions that could be followed forward in time to give observable features of the universe as we see it today.

The most obvious feature that could be explained in this way is the fact that the universe seems to made much more out of matter than of antimatter. If $SU(5)$ has anything to do with the world, baryon number is not really conserved. The extra gauge fields of $SU(5)$ mix up quarks and antiquark and leptons in a way that allows processes that change baryon number. Perhaps the protons out of which our world is built were cooked into existence shortly after the big bang by the interactions in some unified theory. This interesting speculation was anticipated by the great Russian physicist Andrei Sakharov, who suggested that interactions that can create and destroy protons were needed to account for the observed assymetry between matter and antimatter in the universe. In the modern context of unified theories, this speculation was first made by Motohiko Yoshimura of KEK and was subsequently elaborated by many physicists. They have shown that to produce an excess of protons over antiprotons requires that the quark number changing interactions look different when they are run backwards in time. This condition is satisfied in realistic unified theories, however, it is an open question whether any of these theories can account quantitatively for what we see.

It might be possible to observe a different sort of remnant of the big bang that would shed light on unification. The SU(5) vacuum can twist up into configurations that act like magnetic monopoles. Such a twist in the vacuum is called a *soliton*. A simple example of a soliton is a twist in an infinite weighted ribbon, fixed on one edge. It is only at distances at which all of the SU(5) fields are important that unification prevents the unwinding of these configurations. Thus the monopoles are very small and very heavy. They are expected to be even heavier than the heavy gauge particles. These objects are far too heavy to be produced in accelerators, but some of them might have been produced in the big bang. If they were produced, and the north and south magnetic poles separated from each other, they may still be around today, because a single monopole cannot decay. Its magnetic charge makes it stable (although if it can find a pole of the opposite kind, it can annihilate). If such monopoles exist, it should be possible to recognize them, by their magnetic charges, and perhaps by other peculiar properties associated with their large mass.

Many experimental physicists of have set up detectors to look for monopoles that may be floating around loose in the universe. On St. Valentine's Day, 1983, one such detector, put together by Blas Cabrera at Stanford, recorded an event that looked very much like the passage of a monopole through the apparatus. This caused great excitement at the time. Unfortunately, neither Cabrera, nor anyone else, has seen such a convincing monopole since. Most people assume that the St. Valentine's Day event was some kind a rare experimental glitch, not a real monopole. However, people are continuing to look.

But to my mind, the most spectacular effect of the new SU(5) force is that it allows the proton, the lightest particle carrying baryon number, to decay into particles with no baryon number, for example a positron and a π^0. Because the range of the force is so small, the rate at which protons decay is very small. The simplest sort of dimensional estimates in the simplest SU(5) model give a lifetime of about 10^{31} years. That is a long lifetime. The universe has only been around for 10^{10} years or so since the big bang. Still, decay is a probabilistic process. A lifetime of 10^{31} years means that any given proton has one chance in 10^{31} of decaying. That doesn't sound quite so bad. In a kiloton of matter, for example, there are nearly 10^{33} protons and neutrons. Why not just take such a large chunk of matter and watch it for a few years to see in any of the protons decay?

I have, in fact, just returned from Japan, where one of the two large experiments in the world designed to look for proton decay in this way is being done. This experiment is in a lead mine under a mountain in western Japan. The other large experiment is in a salt mine under Lake Erie, outside Cleveland. The experiments are done underground to minimize the effects of cosmic rays. Both of these detectors are large tanks of very pure water, lined with thousands of photoelectric tubes to record the tiny flash of light that would signal a proton decay in the center of the detector. The real problem is distinguishing the event you want from all of the other things that can be happening in a kiloton or so of matter.

Most of the detectors see events that could be proton decay, but all of these events are such that they might have been produced by neutrinos in cosmic rays. None of the groups see conclusive evidence for proton decay. Their results indicate that the lifetime of the proton is at least 10^{32} years. This is disappointing because it seems to indicate that the simplest $SU(5)$ theory is too simple. This conclusion is not 100% certain, because all of the estimates of the proton lifetime contain large and unknown uncertainties due to the difficulties of dealing with the color $SU(3)$ theory at large distances (like 10^{-15} meters, the radius of the proton) where the effects of quark confinement (as expressed by the coupling constant α_3) are large. But the likelihood is that the simplest theory must be modified. Unfortunately, it does not take much of a modification to raise to proton lifetime by several orders of magnitude, because the extrapolation from 10^{-18} meters to 10^{-31} meters is so huge that the numerical results are very sensitive to small changes in the parameters of the theory.

It seems to me that the future of unification is now in the hands of the experimenters. The ideas of $SU(5)$ and other unified theories are rather wild speculations, but they are solidly built on all the painstaking experimental work and clever theoretical insights that lead to the $SU(3) \otimes SU(2) \otimes U(1)$ model of strong and electroweak interaction. These ideas, in turn, lead experimenters to explore new ways of looking at the world in the large proton decay detectors. If proton decay can be seen, it will gives a new dimensional scale in physics that can help focus more theoretical work. If not, we will probably not be able to understand what is happening at such short distances for a very long time.

It would be sad, I think, if Nature denied us the chance to look through this window into the world of very short distances. But in any event, the story of $SU(5)$ and the search for proton decay is a nice example of modern particle physics. At its best, it involves an exciting dialog between theoretical and experimental work. Interesting theoretical ideas are stimulated by and stimulate experiments that push back the limits of technology.

I was in Japan last week to attend a conference on ideas and experiments relevant to unification. After the conference, most of the attendees, including Leon and me, went for a visit to the lead mine where the Japanese proton decay experiment is being done. To get to the experiment, in the heart of the mine, we took the little train that the miners use to get in and out. It takes about twenty minutes, twenty minutes of black darkness and white noise. It was a wonderful time to think about the process of physics. Some of my thoughts were of the evening over 12 years ago when I first fit together the pieces of $SU(5)$ and worried about proton decay. Some were of the meeting just past, where I learned that my experimental colleagues, far from being angry at me for sending them to hard labor in mines and tunnels, were excited about pushing the search for proton decay. The Japanese wanted my help in marshalling support for a much bigger detector. I hope that in this Conference, we will see how the teaching of some bits and pieces of modern physics can help to communicate some of this excitement to the next generation of physicists, and indeed to all who are fascinated by the mysterious workings of our world.

David N. Schramm, Louis Block Professor in the Physical Sciences at the University of Chicago, received his BS in 1967 from MIT and his PhD in Physics in 1971 from Caltech. He has written or edited nine books, and has authored over 200 technical articles as well as numerous popular pieces. In addition to doing research and writing in physics and astrophysics, Schramm enjoys outside activities and has climbed mountains throughout the world and has written about them in various outdoor magazines. He was also a 1971 National Greco-Roman Wrestling Champion.

He has received many awards, including the first Trumpler Prize of the Astronomical Society of the Pacific, the Warner Prize of the American Astronomical Society, the Richtmeyer Lectureship of the American Association of Physics Teachers, and shared the 1980 Gravity Prize with Gary Steigman. He is an active consultant and advisor to the government and has served on panels and committees for the National Academy of Sciences, the Department of Energy, the National Science Foundation, NASA, the City of Chicago, and the State of Illinois. He has held numerous research grants from the federal government. He has given over 250 invited technical and popular lectures throughout the world. He is listed in *Whos Who in the World* and other biographical sources, was cited in *Next* magazine's list of 100 people to watch in the 1980s, and his age of the universe is quoted in the *Guiness Book of World Records.* He is committed to the principle that as many people as possible should be exposed to the ideas and activity of basic scientific research; as a consequence, he teaches freshman non-science majors as well as advanced graduate students, and has been a guest on numerous radio and television shows.

The Big Bang Creation of the Universe

David Schramm
Department of Physics
University of Chicago
Chicago, IL 60637

and

Fermi National Accelerator Laboratory
P.O. Box 500
Batavia, IL 60510

We are living at a rather interesting time—a revolutionary time, I think—in physics and cosmology. In many ways one can compare it to the turn of the century when relativity and quantum mechanics were being developed. Right now we are beginning to see that the unification of the forces and the understanding of the early universe are merging together the concepts of particle physics and cosmology. We are coming to a realization that these two fields are intimately related.

The last few years have been an interesting and exciting time. This is particularly true in cosmology. That makes it very difficult to teach about this rapidly-changing field at a high school or beginning college level.

What I will try to do today is lay out an outline of what we know about the universe, particularly emphasizing those areas that have had input from particle physics as well as the interplay between particle physics and cosmology. Toward the end we will get the current speculations and the latest things published in the "great scientific journals" such as Time and Newsweek. One of the things about this field is that it seems to have really caught on with the popular science writers. There are constantly articles on the subject in the popular press. Frequently, these articles are wrong. Thus, how to evaluate what is going on is always difficult: One of the problems is that sometimes there will be an announcement about something that will get published in the popular press and then, a few weeks later, people realize there was an error (such as the annoucement on the fifth force that came out not too long ago). The popular press never bothers to print the retractions and what's wrong. So people who are not aware of the scientific literature, particularly high school students and college students, are not aware that these things end up not being as valid as they might have seemed by the front-page headlines.

What I will be discussing here is "the complete history of the universe." If we were doing this same talk, five or ten years ago, the entire talk would center on the astronomical side, the stars, galaxies and three-kelvin background radiation. Nowadays, when we talk about the universe and cosmology, we realize that to understand the stuff that you see with telescopes, you really have to go back and look at the earlier times where particle physics interactions were important. We must consider not only nuclear reactions but go back to quark confinement, QCD, and unification that have been discussed here in the other presentations. We now know that to understand the universe we need to look at the physics of the high temperature early universe—the physics of elementary particles.

At the same time we realize that to understand unification, we need cosmology. The one place where the kind of energy that one talks about in unification actually occurred was in the Big Bang. Accelerators are not going to reach energies of 10^{15} GeV. In order to get those energies, you have to look at where they did occur. Where were those energies ever found? That's the Big Bang, itself. This has meant that particle physics theorists, in worrying about unification, have been forced to become cosmologists to try to find the observational or experimental data to support their theories or to try to constrain things in one way or the other. By the same token, the astronomers and cosmologists in trying to understand the early universe, have recognized that the physics that dominates the early universe is the physics of elementary particles. Thus, one now must understand elementary particle physics. The two fields have become very symbiotic. Developments in one affect the other.

We also have the curious situation, which I will go into in this talk, that some of the best tests for cosmology are not with telescopes but with accelerators. Cosmology now predicts things that you can check with accelerator colliding beam machines. In fact, CERN, SLAC, and Fermilab are the places where there are going to be the best tests of the standard Big Bang model. This gives a better glimpse of higher temperatures and earlier times than any other investigation that has been done.

The other side of the coin is that to probe the grand unified epoch and the phase transitions at the birth of the universe, one may need large telescopes.

Toward the end of the lecture today, I will discuss some of the latest developments with regard to the distribution of galaxies in space—the so-called large-scale structure. There has been a discovery of a foaminess in space published in the New York Times in January, and more recently published in the Astrophysical Journal. This large-scale foam is related to phase transitions in the early universe. Thus it may be an observation of that primoridal phase transition and it might be studied in the same way that you might study the crystallization as a solid condenses out of a liquid. The way galaxies are laid out in the sky may be indicative of the phase transition—the grand unified phase transition, the birth of the universe. Thus one way of studying that phase transition is with telescopes.

What we have seen is perhaps not only a symbiotic relationship of the two fields but, in some cases, an exchange of the tools used in one field for the other field.

To begin with let me just discuss what we know that is cosmologically significant. What cosmological facts are there?

It is actually rather hard to think of an observation that has cosmological significance, and most of them end up being rather simple. In fact, each of these rather simple observations has rather profound consequences.

The first one is that the universe has three dimensions plus time. I attribute this to Aristotle, clearly recognizing it was probably noticed even before Aristotle. It used to be that this was considered more of a philosophy or metaphysics problem and had nothing to do with physics. Within the last year or two, the question of how the universe came to end up with three dimensions plus time has become a forefront question. I am not saying we understand the answer to it, but at least it's a question that is being attacked by physicists. It is no longer a philosophy or theology question; it's a physics question! As we will see towards the latter part of this presentation, and as Howard Georgi mentioned in his presentation, there are ideas now about 10-dimensional super-string models. The question is, if you had some higher dimensionality of space, how do we end up having only three dimensions and time? As a result, it's a non-trivial observation that the universe has three dimensions plus time.

The next observation of cosmological significance is probably that of Hubble, around 1920, that the universe is big. Now, this at first seems again trivial but the universe is really extraordinarily big. How it got to be that big is again another problem that has steered the way cosmology has gone. The observation that Hubble made was that the spiral galaxies—spiral nebulae in the sky—were external galaxies, outside of our own.

I always like to mention things about Hubble. He was a curious character. He was on the University of Chicago 1907 and 1908 Big Ten Championship Basketball Teams. He also was a Rhodes Scholar, went to Oxford and studied law, came back to Louisville, Kentucky and was going to practice law and teach school there, but he decided, according to a letter that he sent to his father, that he was "going to chuck the law for astronomy" because he would rather be a third-rate astronomer than the country's leading lawyer, since for him astronomy was much more important. I particularly like this story because my ex-wife is a lawyer.

Hubble not only talked about the universe being big, but he went on to show that it was getting bigger; that the universe was expanding; and that the galaxies, or actually clusters of galaxies, are moving away from each other.

Remember prior to Hubble, people thought the universe was just the Milky Way Galaxy, our Galaxy, and that it was an assemblage of stars that was static. Since any gravitational theory, including Einstein's theory as well as Newtonian theory, leads only to dynamical solutions, it is very hard to make something static. This was a difficult problem.

If Einstein had not introduced the cosmological constant and stuck to his guns with the way the General Relativity was originally written, he would have predicted the expanding universe. But even Einstein goofed sometimes and he did not predict the expansion.

The galaxies are moving away from each other in a special way—the velocity of recession is proportional to their distance. This is a rather special kind of expansion. It is an expansion that keeps the shapes the same.

The example I like best is a loaf of raisin bread. A loaf of raisin bread, as it rises, continues to look like a loaf as it gets bigger. The raisins move apart from each other, obeying Hubble's law. A more distant raisin will be moving apart from a nearby raisin at a higher velocity and that keeps the loaf looking like a loaf.

I think the raisin bread model is a much better one than a model of the surface of a balloon, because the surface of a balloon always leads the student to ask a question about the center of the balloon. You've got to tell the student that the surface of a balloon is the only part that's the universe and the center is not in the universe, but yet students are three-dimensional creatures and they tend to see that center of the balloon there, and they get all confused, because the center has nothing to do with the universe. There is no center to the universe, and so the balloon naturally gets the student thinking the wrong way about the curvature of space and the expansion. That, I think, is the wrong way to approach it.

It is much better to think in terms of a loaf of raisin bread. A loaf of raisin bread has three dimensions. Now, you say, "What about the center?" Well, all you have to do is just move the pan walls out to infinity and then you've got a good model for the universe. So, it's a rising loaf of raisin bread that happens to be infinite in all directions. But at least, if you are viewing it that way, then the expansion actually is easy to relate, because that's just the dough rising; the raisins stay the same size; the dough rises. So, you can understand again a relationship with the standard cosmological model because the galaxies stay the same size. It's the distance between the clusters of galaxies that increases. It's space that's expanding; it's not the galaxies flying out through emptiness. It's that there is space being added.

This is something that gets confused even in some of the elementary textbooks where there is a token chapter at the end on cosmology. Frequently the people writing the book do not understand that a cosmological expansion is not galaxies flying out from some center. It is really a homogeneous expansion with space being what's expanding, space expanding everywhere, and galaxies existing everywhere.

With those three observations, the stage was set for modern cosmology.

In the 1940s and 1950s, there was a big debate between two ways of explaining those cosmological observations that had been done up until then—the Hubble cosmology.

There was the Big Bang, which is in many ways the simple way of doing it. It's saying that the density evolves with time, and so, as time goes on, you are getting more space between the raisins or between the galaxies; so, the density drops. That's all it's saying. Things are homogeneous in space but, with time, the density evolves.

The Steady State model was invented by Fred Hoyle, Herman Bondi and Tommy Gold, after going to see a science fiction movie in Cambridge, England one evening. After the movie, they went to the local pub for a few pints of bitter and out came the Steady State theory. The idea in the Steady State theory is that, in addition to the universe being homogeneous in space, it is homogeneous in time, as well. So, as the space is added between the galaxies, new matter is also added with the space, so that the density stays the same for all time.

In many ways, the Steady State model looks prettier than the Big Bang. It's keeping everything the same for all time as well as in all directions of space. Unfortunately, the universe has turned out not to be as clever as Fred Hoyle and decided not to operate in this way but rather by the Big Bang. Nowadays people do not really talk about the Steady State theory other than to give it as an example of how you can do a proof in cosmology to eliminate models. However, it is a very nice demonstration of how a model is testable and shows that cosmology is no longer just an intellectual game but instead is a real science that has experimental consequences and can function in the same way as other branches of physics. Thus the Steady State theory played a very valuable role.

Let us now go through how we know with such confidence that we live in a Big Bang Universe. In the Big Bang model, as you extrapolate back to earlier and earlier times, the density got higher and higher. Of course, that means eventually the galaxies would overlap. In fact, eventually, the universe would be continuous matter or continuous plasma for early times. Whereas in the Steady State model, the universe is always the same density as today. As a result, the galaxies never overlapped, and there was never a continuous plasma everywhere.

Of course, whenever you have a continuous body, it radiates blackbody radiation. Since the blackbody radiation cannot escape the universe even though the universe is no longer a continuous body, the radiation left over from when it had been would still be bouncing around the universe. So that means in the Big Bang model that there should be some background radiation; in the Steady State model there would not be.

We all know the story of how Arno Penzias and Bob Wilson discovered this background radiation in the late 1960s, and it was confirmed by work done in the early '70s. So, by the mid-1970s, people were confident that we lived in some kind of hot, Big Bang cosmology.

As you know Penzias and Wilson discovered the background radiation by accident. They were given the antenna to track the Echo I satellite. Some other people had used the antenna before. There was some background noise in it. The other people just reset the zero and just looked for the bumps above the background noise, but Penzias and Wilson, being a little more persistent, said, "Hold it. That background noise shouldn't be so loud, we should get rid of it. The antenna is better than that." And they tried all sorts of things to get rid of it, such as scrubbing away the pigeon droppings on the antenna. Even after scrubbing away the pigeon droppings it still didn't go away. There's a great picture of them on their hands and knees, scrubbing away the pigeon droppings. It is particularly good because now Arno Penzias is Vice President of AT&T, the Director of Bell Labs, and frequently drives around in a chauffer-driven limousine. So it shows that hard work does pay off.

I think one thing that not everybody does appreciate is that at the same time Penzias and Wilson were doing their measurements, other people had used this antenna and published results showing no background noise because they just reset to zero. Those people didn't get a Nobel Prize.

It is interesting to note that there was a paper written by Zel'dovich, in the Soviet Union. (Zel'dovich and Sakharov were the co-inventors of the Soviet hydrogen bomb.) Zel'dovich is a very, very bright cosmologist, and was aware that this antenna existed in New Jersey. He was aware that papers had been published with it, showing no background noise, and so he concluded, on the basis of these publications, knowing about Gamov's earlier work in the 1940s predicting the three-kelvin background radiation (actually, Gamov, via his colleague Alpherod Herman, predicted 5 kelvin, but that is pretty close). Zel'dovich then combined all of this, and said, "This shows we do not live in a hot Big Bang universe." He was not aware of the fact that some American scientists just reset the zeros on their experiment. So, as a result, Zel'dovich missed tying all of this together, although he did his work at the same time that they were in the process of making their measurements.

It is also interesting because Zel'dovich in Moscow was aware of the Bell antenna that, at the same time, there was a group at Princeton, which is in New Jersey rather than in Moscow, that decided to try to build an antenna themselves to do this measurement. They must have been aware that the Bell Lab's antenna existed and they could have gone over and collaborated and done the work. Then Dicke at Princeton would have shared in the Nobel Prize, but he didn't so he didn't. All of this makes for interesting scientific sociology and politics.

In addition to the three-kelvin radiation that Penzias and Wilson discovered, there's also the ^{4}He observation. I will come back to that in more detail later on, because I think in many ways that's actually more important in establishing the details of the early universe and the physics of the early universe than the three-kelvin radiation. It should also be remembered that, on its face, the prediction that George Gamov and his coworkers made in the 1940s about the background radiation was based on saying there were nuclear reactions in the universe. So, in fact, these two observations are tightly connected with each other.

The universal helium abundance comes from the fact that, when the universe was very dense, much denser than just touching, but back when it was at a density and a temperature sufficient for nuclear reactions to take place everywhere, it would have automatically converted one-quarter of the entire mass of the universe into ^{4}He. All the other heavy elements combined—carbon, oxygen, iron, uranium—make up less than 2 percent of the mass of the universe. Yet, one-quarter of the mass is in one single isotope, ^{4}He. The other 73 percent is hydrogen. Thus, to get 25 percent ^{4}He means you cannot do it in a normal, stellar processing way. Supernova and nucleosynthesis in stars doesn't yield 25 percent; it only yields on the order of 2 percent, because that isn't the way to make the other heavy elements. To get a quarter of the mass of the universe converted to one single isotope takes something very special. It is interesting that the Big Bang automatically does it, and, in fact, if the observations did not come out around 25 percent, it would prove the Big Bang wrong.

The amazing thing is that every place you look, roughly 25 percent of the mass of any astronomical object, other than little things such as planets, is ^{4}He. So there's a tremendous confirmation in a quantitative way.

This work on ^{4}He was beginning to be established in the 1960s. But in the 1970s, in addition to the ^{4}He, we recognized that the other light elements—lithium, heavy hydrogen, deuterium, the other isotope of helium, ^{3}He, which are also made in the Big Bang but in trace amounts, have good agreement, in fact, exact agreement between the observation and the theory of what comes out of the Big Bang. And that actually turns out to be a much tighter constraint on the understanding of the early universe. I will come back to that in more detail, but this seems to confirm that we really understand the universe back to times when the temperature was an MeV, 10^{10} K, back at a time of the order of one second after the creation event.

These observations are what led us to know that the universe was hot and dense. We will come back to this in more detail later. Thus we understand what was going on, in detail, at a time of one second by direct observation. It is not just waving our hands and extrapolating our theories; these are direct predictions that are measured and found to agree quantitatively with the calculations. So, again, it is making cosmology into an experimental science, not just abstract thought.

Another observation about the universe is that the universe is old. Now, in fact, if you look up in the Guinness Book of World Records, you will see that the age is listed at about 15 billion years, determined by a variety of techniques. Because my name is attached to the number in the Guinness Book of World Records, I was asked by the ACLU at a trial in Arkansas why I got 15 billion rather than 6,000 years; that took a while to explain. Of course, one could always say that it was made 6,000 years ago to look as if it's 15 billion, and I had trouble arguing with that, although that seemed a little pathological.

From a scientific point of view, though, it is not so much that the universe is not 6,000 years old with which we are concerned; it is the fact that it is not 10^{-43} seconds.

You might say, "Why is that such a worry?" That turns out to be a very serious worry, if we look at the dynamics of the universe. If we take the standard expanding Big Bang model, we know we are expanding; space is increasing; we used to be hot and dense. We have three different options for the universe. If the density is high so that eventually the matter is stopped by its gravitational field, the universe falls back into what we call the "Big Crunch." It is similar to shooting a rocket up from earth. Initially the rocket is going up, the expansion phase that we are in now, but if the Earth's gravity is great enough, it will eventually pull the rocket down as an intercontinental missile. This is the same kind of effect that you would have for a closed universe.

Another option is the "Big Chill," in which the universe would just keep expanding forever. If the density is not high enough, the expansion would just continue. If you shoot the rocket up from Earth with a high enough velocity, it goes off, leaves the Earth's gravitational pull, and becomes an interplanetary probe, like Pioneer 10. This is the model for the universe where the density is low.

We use the parameter Ω to characterize the density. A value of Ω greater than one is above the critical density, and characterizes a universe that falls back in. A value of Ω less than one, below the critical density, characterizes a universe that continues to expand forever.

If you go to the boundary between the two, $\Omega = 1$, it is sort of analogous to what would happen if a rocket were shot up from the Earth and tuned just right, so that the rocket would go into orbit and become a satellite. While the universe doesn't exactly go into orbit around itself, it continues expanding forever marginally slowly. It will never fall back in, and also would end in a "Big Chill," which is something that, again, people forget. They think of a critical universe as a closed universe. A critical universe is actually an open one that expands forever but just marginally so.

Those are the options we have for the universe. In all of them, the universe is changing with time. Notice that if Ω is greater than one, eventually Ω is going to get much greater than one because the density will increase. Density is a dynamic quantity. As you crush the universe together, the density goes up to infinity. So, in a closed universe, an $\Omega > 1$ universe, the "Big Crunch," the density goes to infinity. Ω evolves with time. Similarly, if $\Omega < 1$, Ω goes to zero. In consequence, the density would go to zero. If Ω is any value other than unity, it will either go to infinity or zero on a time scale that's the time scale of the expansion rate. The gravitational time scale is roughly related to the age of the universe. The universe is expanding on a gravitational time scale today. You can work out the expansion rate and relate it to the density.

That means that we can estimate the rate at which Ω should change. Right now the universe is about 15 billion years old. We don't know from direct measurements whether Ω is 0.1 or 3 by direct measurements, but we certainly know it is not zero or infinity. So, that means that Ω had to be close to one for a very long time, and it couldn't have been evolving away from one.

In the current epoch of the universe, Ω is changing on time scales of the age of the universe, 15 billion years. However, we can extrapolate our physics back with great confidence, certainly back to the time of nucleosynthesis. We know we can measure the properties of the universe then and we know that the universe was only about one second old then. That means that densities were changing on time scales of one second. Thus, to get from then to now, 15 billion years later, meant that Ω had to be equal to one to within the ratio of the present age; age measured in units of one second, that's 10^{17}. The universe had to be tuned to an Ω of one at the time of Big Bang nucleosynthesis to one part in 10^{17}. That's an amazing fine tuning.

However, we can do even better than that. We can extrapolate our laws of physics (while we don't know all the details) to very early times. We don't really give up until we get back to times the order of the Planck time, when quantum gravity should come in. At the Planck time — 10^{-43} seconds — that's when gravity should become quantized. Basically, we can extrapolate physics in a rough way back to that order before our ideas of space and time get screwed up.

That means that at the Planck time, the universe was evolving on a time scale of 10^{-43} seconds, 15 billion years is 10^{60} Planck times. That means that, at the Planck time Ω was one to 60 decimal places—one of the greatest achievements in fine-tuning ever. You know, if you tell an engineer to make something smooth to 60 decimal places, they would have a great deal of difficulty. So, obviously, the engineer who designed the universe had tremendous skills at being able to do this fine tuning.

What we would like to do is understand how the universe was able to get to be 15 billion years old. That tells us that we've got a problem. We've got to understand how Ω, in the beginning, got to be so close to unity in order for us to be here. That doesn't even worry about the problem of whether it is greater than one now; it is just saying that for us to be here and to exist required this tremendous fine-tuning. What mechanism produced that fine-tuning? We would like to avoid appealing to arbitrary initial conditions.

While going through the cosmological fact list, let me mention another fact we know about the universe. The universe has matter in it. It's not empty. How much? What kind? These are the questions. We know that in this room there is good old baryonic matter—regular stuff that you've been hearing about at this meeting; but is that the only kind of matter in the universe? What I am going to argue in the later part of the talk is that we think that the baryonic matter is only a small fraction of the matter of the universe, probably about 10 percent, and that the other 90 percent of the universe is probably something non-baryonic. This immediately ties you to elementary particle physics, because the only place you study non-baryonic matter is in elementary particle machines. So, there is again an interaction between particle physics and cosmology.

The bulk of the matter of the universe is probably something that's different than the normal, everyday matter we deal with, and yet it is very important. The fact is the universe does have matter in it. The baryonic matter is distributed in galaxies, in stars, and so on. But there does seem to be a number or arguments that point toward the need for something else, some other kind of stuff. We will come back to that.

Now, there are two more arguments I wanted to mention about the universe—two more facts—cosmological facts. One is that the universe is smooth, and the other is that the universe is bumpy.

Now let me explain what I mean. On the large scale, the universe is very smooth. If you look out toward one direction that three-kelvin radiation that we discussed is found to be the same temperature to parts in a thousand as when you look in exactly the opposite direction. Each of those photons that you see today have been traveling for 15 billion years. Those two regions that have emitted those photons that are traveling for 15 billion years are 30 billion light years apart from each other. Thus, there is no way that a signal from one side of the universe has gotten over to the other side of the universe to be at the same temperature. How did they get to be the same temperature? The problem gets worse if you go back in time, because at the time the photons were actually emitted those two regions were a hundred times the age of the universe apart from each other. So, in fact, they were a hundred times causally disconnected as opposed to now when they are only a factor of two causally disconnected.

We had a real problem. How did the universe get to be such a uniform temperature, if the regions that we are seeing now are causally disconnected from each other?

While the universe is very smooth on the large scale, on the small scale, galaxies, clusters of galaxies, stars, people, we see bumps. A beautiful example of a bump is in this room. The typical average density of a person is one gram per cubic centimeter. The average density of the universe is about 10^{-30} grams per cubic centimeter. We've got bumps in this room of 30 orders of magnitude. Where did those bumps come from? Well, most of the bumps in this room come from chemical reactions, say 30, 40, 50 years ago. But the environment to enable those chemical reactions to occur had to have some sort of gravitational ensemble that enabled the appropriate chemicals to be near each other. So, we had to have the Earth, the sun, the galaxy, and so on.

What perturbations in this very smooth, large-scale background enabled small-scale clumps to form? While you are doing the smoothing on large scales, you've got to be able to have bumpiness on small scales and allow those bumps to grow to big bumps non-linearly. Since initial growth is linear in density, that means you've got to be able to induce perturbations in that very smooth, large-scale structure. That is a problem.

These are our cosmological facts—what we know about the universe as a whole. How do we understand them? As I mentioned before, we had established by the mid-1970s without any serious doubt that the early universe was hot and dense, and once we'd done that, that tells us that the physics of the early universe is the physics of elementary particles.

Throughout this conference, you've had quite a few discussions about elementary particle physics, and quarks, and so on. I just wanted to review a few key points that I am going to utilize a little bit later on.

Remember in another lecture, Howard Georgi talked about the unification of the forces. There are four forces—strong, weak, electromagnetic and gravitational. As you go to higher energy one may argue that the symmetry is restored. We certainly know it is restored by about a hundred GeV for the weak and electromagnetic interactions. Once you get above about a hundred GeV, the weak and electromagnetic interactions are one and the same thing, even though at the low energies we live, that symmetry is broken. The idea of grand unification is that by the time the energy is around 10^{14} or 10^{15} GeV, strong, weak and electromagnetic interactions get unified and that symmetry is restored. By the time you get to Planck energies (10^{19} GeV), gravity should be brought in as well, and all four interactions would hopefully have their symmetry restored. There would be only one super grand unified interaction which we refer to as the "theory of everything"—TOE.

How do we get the "theory of everything?" It seems we have to go to higher and higher energies; that the symmetries that we are talking about don't come into play until we are at extremely high energies. Accelerators at CERN just got to the electroweak symmetry energy of 100 GeV a couple of years ago. Fermilab is now a little beyond it. Next year, when Fermilab is operating the Tevatron, it will get to about 2 TeV. That's only a little bit beyond the electroweak energy. 10^{14} GeV is a lot higher.

If you just scale up Fermilab, and say, "How big of an accelerator do you have to build to get to 10^{14} GeV?" Well, you would have to build one from here to Alpha Centauri. Now, with Gramm-Rudman funding, I don't think we are going to be able to do that. So, we've got to confine ourselves to small-scale things like SSC, which is only 100 kilometers around. We just can't get up to these kind of energies.

If you want to study the physics at those energies, where did that physics occur? It occurred in the Big Bang. And so this has been a reason why so many particle theorists have become very interested in the early universe. It is the one experimental area that you have to test out these theories—where that physics dominated the situation. Of course, to study that means studying the universe as a whole or the debris of that experiment.

151

If you want to go all the way up to bringing in gravity, you've got to build the accelerator to the center of the galaxy. One is forced even further towards thinking in terms of the early universe.

Remember, energy and temperature are roughly equivalent. Temperature is just a statistical way of measuring the energy, and the time scale in the early universe that it corresponds to.

The kind of energies that one has in nuclear physics, MeV, were encountered at times the order of one second. That's the reason why when I talked about Big Bang nucleosynthesis for which relevant energies were about an MeV, we were exploring times in the universe one second after the Planck time. At a GeV, you are getting back up to about a microsecond. A hundred GeV — now, that's the energy at which CERN, LEP, and the Stanford Linear Collider, SLC, will be operating (SLC is the new version of SLAC). At a hundred GeV the machines are probing back to about 10^{-10} seconds. Fermilab, with the Tevatron, will be getting up to the order of a TeV; that's 10^{-12} seconds. SSC will be back at about 10^{-15} seconds, probing earlier still.

But to really get back to the GUT time scales, we are talking about 10^{-34} seconds. You've got to get up to 10^{11} TeV. For the Planck scale, the energy would have to be 10^{16} TeV. That's beyond the reach of even the most optimistic accelerator designers.

Let's also look at these elementary particles that we have been hearing about. We have heard about fermions (the quarks and leptons), and the bosons that carry the interactions. At the grand unified scale, you can rotate the quarks into leptons, leptons into quarks; all fermions can then rotate into each other. That symmetry is restored once you are at grand unification, because the only difference between quarks and leptons is that quarks have color and leptons don't. So, once you've got up to an energy for which all forces, the strong, weak and electromagnetic, are unified, then these can rotate into each other. They are the same. The bosons that do that rotation are called the "X" and "Y" bosons.

Notice that there's a very nice match-up; that every family consists of two flavors of quark and two leptons. There's a nice match-up and pairing, and you can easily see a nice rotation between the things.

However, if we want to bring gravity in and unify all four interactions, not just three, we've got a problem. Because if we want to bring gravity in, gravity interacts with everything, not just with fermions but also with bosons. Gravity doesn't pay any attention to whether it's a fermion or a boson; it just looks at mass-energy.

If we want to unify gravity, then we must rotate bosons into fermions and fermions into bosons. If gravity is really unified with everything else, it can't tell the difference. So, when all the forces are unified, we should be able to have fermions and bosons interchanged. That's called supersymmetry, if you can change fermions into bosons and bosons into fermions, but there's not a nice pairing anymore. You don't have the nice match-up that you have here between the leptons and quarks. The family is not complete. This meant, when people started talking about supersymmetry, which they felt they needed to have to bring gravity in and unify the whole picture, they needed to have new particles, because you needed to have a match-up. You needed to have boson partners for all the fermions and fermion partners for all the bosons.

I'm sure most of you have heard these terms before. You needed to have "squarks" and "sleptons," (supersymmetric quarks, supersymmetric leptons). And you needed to have fermion partners for all the bosons (photinos, gravitinos, gluinos, winos, and zinos), so that every particle would have a nice supersymmetric match-up. And these whole ensembles are called the supersymmetric particles or "sparticles."

That's one of the motivations behind building SSC and maybe some of them might even be found with the Tevatron. You seem to have masses of all the normal particles up to about a hundred GeV, and if we go beyond the mass of the "W" and the "Z," maybe we'll start to enter the world of the "sparticle." So we might discover a whole new kind of matter, once we get above the energies of normal particles.

Now, you might say, "Well, maybe you just find more and more families of quarks as you go up." I will try to argue that you don't; that if we believe the cosmological arguments, that it should end; that there should be, at most, one more family and probably no more families of quarks. So, if you go up to higher energy, you shouldn't find any more quarks and leptons. What's left is to find the "sparticles."

This again is an example of cosmology telling something about fundamental physics. I will come back to that because I think that's a very important thing to recognize, that cosmology can constrain the fundamental physics arguments. It's the first time that that's occurred in modern times, since Newton, when he used observations of the moon to figure out something about fundamental physics; now, we are starting to use cosmological observations about the Big Bang to constrain fundamental physics.

We might also mention this unification — the unification of strong, weak, and electromagnetic interactions. At last the electroweak interaction has now been proven by Rubbia's UA1 group at CERN. When you put it into historical perspective, it again reiterates that we are living in a very exciting time.

Remember, Einstein spent most of his life, from 1916 until he died in 1955, trying to find a unified field theory — unification of the forces — and he failed. He didn't make any real progress on it. And yet now, we have at least a unified theory of the weak and electromagnetic interactions, and there are a lot of indications about directions to go for the unification of the other forces. We are now making progress that Einstein spent his life on and could not succeed at. We really are at a time of revolution in physics.

Why did Einstein fail and people succeed now? That's because the success is coming from quantum field theory; that is, the recognition that understanding these forces involves "gauge" theory and, in particular, the exchange of quantum; that you can describe forces in terms of quantum exchange — the boson exchange particles — gauge particles. Einstein didn't like quantum mechanics. He stayed away from quantum fields, and he tried to do it all with geometry. That turns out to be the hard way, although I will show you at the end that some of the current ideas on supergrand unification come back again to geometry but in a very different way, in 10 dimensions rather than three.

Howard Georgi discussed proton decay. Proton decay is a very nice test of these grand unified theories because in that unification, quarks and leptons interchange; protons are no longer stable, and at today's temperatures, protons would then decay in about 10^{32} years. Various experiments at a number of places are looking for proton decay and so far have not seen the simplest mode for proton decay, $e\pi$. They may have events that could be supersymmetric proton decay, but there's not enough information to really confirm or deny that, as yet.

David Schramm

There are some events they can't explain. Some people say that is due to one of the super-symmetric modes, μ K. It certainly is not due to eπ; you can rule that out. Supersymmetry might work but simple grand unification without supersymmetry doesn't seem to work. There are problems right now, but supersymmetry is still hanging in there; simple grand unification is not.

The inverse of proton decay is something that is very important. As we get up to the kind of energies at which quarks and leptons interchange, we can have the origin of matter. In 1967, the Soviet physicist Sakharov showed that, if you have a theory that violates baryon number, has CP violation in it, and functions in the early universe (that is, all the things that grand unified theories have), you can create matter out of radiation. Thus, you can understand where our matter came from; why we have a net excess of quarks over antiquarks in the universe; how we got an excess of matter over anitmatter.

I mention this because this shows an example again of how developments in particle physics have been able to answer cosmological questions. Where did the matter come from? We now understand grand unified theory can give us an answer to that.

While we are talking about proton decay, let me show you a calculation done by one of the earlier workers in the field, the Buddha. The Buddha was one time asked how long the universe would live, and he replied in typical Buddha-ese, that it would live as long as it would take, if a very old man goes to a very big mountain and takes one stroke of a silk handkerchief every 100 years, to grind that mountain down to zero.

We can calculate that! Assume 100 microjoules per stroke. Here's the big uncertainty in the calculation, the efficiency of silk against rock — let's say about 10^{-3} is probably a good estimate, but it has an uncertainty in the exponent. Assume 0.5 eV for the binding energy per atom, and we find 10^{12} atoms per stroke. For the size of a big mountain, let us use a 10 kilometer-high Everest-class mountain; base about a 1,000 square kilometers, with the known density of rock and average atomic mass of rock, you get about 10^{42} atoms. Thus, the age as calculated by Buddha must be 10^{42} divided by 10^{12} times 100 years, which is 10^{32} years. Clearly the Buddha was thinking of proton decay.

In the first part of the talk, we discussed the cosmological facts and the current state of particle physics and unification. Now I want to go into how these two fields interact and give a few examples.

In particular, an example that I alluded to earlier has to do with Big Bang nucleosynthesis. I mentioned that we know, in the Big Bang, that a quarter of the mass of the universe is converted to ^{4}He. In the Big Bang, no matter what you do with the baryon density, the ^{4}He abundance is relatively insensitive. It stays almost constant. The other abundances, however, have large variations. For example, ^{7}Li is way down to the level of 10^{-10} relative to the ^{4}He abundance of about 25 percent.

The observations of the abundances of ^{7}Li, deuterium, ^{3}He and ^{4}He all agree with the Big Bang for a very narrow range in baryon density. That is, Big Bang nucleosynthesis agrees with the observations over a dynamic range of almost 10 orders of magnitude — this is one of the most profound agreements I think that you will find in physics — a huge, dynamic range of consistency.

The original calculations done in the mid-60s really concentrated on the ^{4}He. Although they calculated these other nuclei, too, it was not considered relevant at that time because these other abundances were thought to be due to nuclei made in stars. We then showed, in work over the 70s, that deuterium could not be made any place but the Big Bang. The observed deuterium abundance had to come from the Big Bang! Then in subsequent work we have done similar kinds of things, although the arguments are a little more complex, for the ^{3}He and the ^{7}Li.

Thus we have been able to show how you can use measurements today to relate to the Big Bang abundances that came out, and that has really tied down this agreement over 10 orders of magnitude, telling us that the density of baryons in the universe converted to Ω in baryons must be in the range from about .03 to .12. Thus, the universe cannot be closed with baryons.

From these measurements and from the consistency with the Big Bang, we not only understand the early universe back to times of one second but we also know that the universe is open with regard to the baryons in it. So, if the universe is going to have an Ω value of one, that additional material—that other 90 percent—has to come from something other than stuff that enters into nuclear reactions.

With this great agreement we've got in nucleosynthesis, we pushed a little further. In particular, let us look at ^{4}He in tremendous detail. From the bound on the baryon density from deuterium, ^{3}He, ^{7}Li, we know we have a minimum bound. From observations of the helium abundance, we know that the helium abundance in objects that have no heavy elements is slightly less than 25%. The helium abundance that came out of the Big Bang had no additions due to stel-

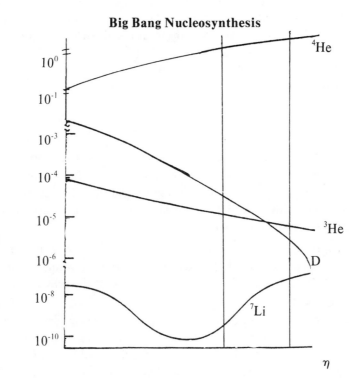

Fig. 1. Predicted abundances of light elements in terms of the baryon-to-photon ratio η.

lar evolution. This is because stellar evolution makes heavy elements, and when it makes heavy elements it also makes a little bit of helium; so it adds to the helium abundance, but if we look at just the primordial helium—the helium with no heavy elements associated with it, we obtain the above mentioned limit. The helium from the Big Bang is affected by the number of types of low-mass neutrinos. If you increase the number of neutrinos, you slightly increase the amount of helium you get out due to the fact that you change the expansion rate a little bit which then changes neutron-to-proton ratio which then affects the ^{4}He abundance. The interesting point here is that, for three neutrinos, electron, muon and tauon, you get a good fit to observation.

Remember, each neutrino is in a family with two quark flavors and one lepton. So, if you are limiting the number of neutrino types, you also are limiting the number of families and thus the number of quark flavors. Three neutrinos are allowed. If you add a fourth, you might marginally make it, but it's pretty hard, because these are rather extreme values that were chosen. It's unlikely that you could squeeze a fourth one in, although we certainly wouldn't throw away all of Big Bang nucleosynthesis if there were a fourth family. From this, we would argue that there really are only three families (at most four). As I said, this is the first time that cosmology has actually predicted something about fundamental physics in the last few hundred years.

In addition, this is testable in an accelerator, because the Z^0 boson can decay to neutrino pairs as well as lepton and quark pairs. Thus, the more types of neutrinos and quarks there are, the faster the Z^0 decays. Its quantum mechanical width is proportional then to the number of neutrinos. If you work it out, it turns out that the width is about 3 GeV for three generations, and each additional neutrino will add about 0.2 GeV, if it is just adding the neutrino forgetting about the lepton and quarks that would go with it. As a result, you can measure the width of the Z and tell how many neutrinos there are, since all the neutrinos would couple to the Z. That is being done.

From a particle physics point of view, there is no reason to limit the number of generations. They could go arbitrarily on as you go to higher and higher energy. Cosmology tells us, though, that there should only be 3 or 4. I should mention that a few years ago, prior to the CERN data on finding the Z, the limit from experiment could have allowed thousands. However, now that the Z has been discovered at CERN, Carlo Rubbia just told me a couple of weeks ago that the current best limit on the width says that the number of neutrinos is less than seven. So, it is coming down toward the cosmological value.

It is unlikely, with the current CERN machine, that they are not going to be able to get all the way down to the 3 plus or minus 1, the kind of number that we would like to really prove the cosmological answer. It is, in fact, going in that direction with the only real force theoretically pushing that way being the cosmological argument.

This is the first time that an accelerator experiment is verifying cosmology and our understanding of the universe back to one second after the creation event. There are experiments that are being done over the next couple of years at SLC, the Stanford Linear Collider, where, once they get SLC operating, they will be making Z^0s. There they should be able to get the width down to the equivalent of less than one neutrino. Fermilab's Tevatron, plans on getting down to the order of—being able to count a single neutrino. Eventually LEP at CERN, when it operates in 1989, should be able to count the equivalent of fractions of neutrinos. By the end of this decade, we should have a very definitive number, and I hope it comes out in agreement with cosmology, or we've got a real problem.

We are finally seeing where experiments are being done to check comology. In fact, I like to tease people at SLAC that, now that the "W" and the "Z" have been discovered at CERN, the main purpose of SLC is to do cosmology. It is a big telescope. It's long and skinny, too, so it looks the part.

It is interesting that we first did these calculations almost 10 years ago. This shows the length of time between theory and experiment in these areas. (It is also rather nice when a paper doesn't go out-of-date in 10 years, which is rare in this field.)

The next thing I wanted to turn to is *inflation*. In fact, it is rather nicely described in a Scientific American article by Alan Guth.[1] Inflation is a very nice idea that enables us to understand a lot of the facts that I describe about the universe. In particular, it can solve problems: Why is the universe so smooth? Why is the universe so bumpy? How did the universe get to be so old?

It also solves a fourth, relatively new, problem that I haven't mentioned to you. It's not an old cosmological problem. It's a problem that's induced by grand unification. With grand unification, when baryons are made, there can be an excess of quarks made in the early universe without making antiquarks. That's very nice because it finally explains where the baryons come from in the first place, and why the universe just has matter without antimatter. What I did not mention when I told you how nice grand unification theories look to solve that problem is, that at the same time they make excess quarks, they also make — magnetic monopoles. They actually make monopoles just slightly earlier at slightly higher temperature, because a monopole has a mass about a hundred times the GUT scale. So, it makes them at 10^{16} instead of 10^{14} GeV.

It is rather bad to make magnetic monopoles because we don't see any. You might have remembered a few years ago they thought they saw one monopole out in Palo Alto, and the current explanation is that somebody just bumped into the detector because they haven't been able to find any more and they've done more sensitive experiments and it doesn't seem to be there. But nobody has found a confirmed monopole yet. In these GUT models, in the standard Big Bang, you would end up having the number of monopoles roughly equal the number of protons in the universe. Now there might have been one monopole in Palo Alto, but there are not as many monopoles in Palo Alto as there are protons. That would be a real problem. Somehow, you've got to get rid of these monopoles.

In fact, that problem was the motivation that led Guth into his thoughts about the early universe. He was actually not paying any attention to these other problems at first. He then realized that, in his solution to the monopole problem, it happened to solve a bunch of other problems as well. This again shows how, when people do science, their initial motivations are frequently not what lead to their fame-producing discovery.

To understand inflation, let us recall again the expansion of the universe versus time in the standard Friedmann cosmology (standard Big Bang). Remember that the scale factor gets bigger with time but rate of expansion is gradually decreasing. This is because the rate of expansion goes like the square root of the density, and the density is a power of the temperature. As the universe expands, it cools. So the energy density drops, and thus the rate of expansion drops. How much it falls off is another matter, depending on the ratio of the density relative to the critical density, but it certainly falls off to some degree if there is any energy in the universe at all.

Way back in the 1920s, de Sitter showed that if you made up a model that at that time seemed to be totally ridiculous, that had a constant density (the density just stayed the same for

all time and was not related to temperature and did not cool as the universe expanded), then the universe would expand exponentially. This occurs because, as you are adding more space to the universe, you are adding that equivalent density, so you just drive the expansion faster and faster.

Guth recognized a rather interesting point that these grand unified theories give us. They have an energy to the vacuum, a so-called *Higgs Field.* This field gives the energy that creates the mass of the gauge bosons that breaks the symmetry. These fields give X bosons their 10^{14} GeV mass and the W and Z boson their 100 GeV mass.

When you are at the time of unification at temperatures above the unification scale, that energy, that *Higgs Field,* is in the vacuum. So, every place you look there's that vacuum energy. Saying that the vacuum having energy is like saying there is a cosmological constant, or an energy associated with space independent of the matter. That is, there is an energy to the vacuum when symmetry is restored. However, when you break the symmetry, then that vacuum energy goes into producing masses for the gauge bosons, which are what break the symmetry and give us the universe that we know and love today. The energy of the vacuum disappears.

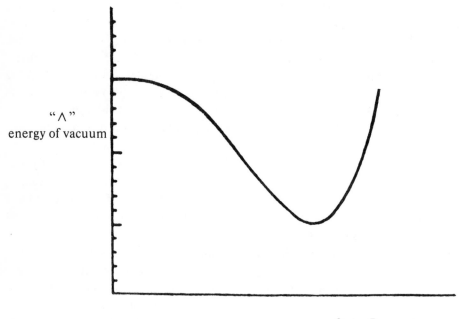

"Λ"
energy of vacuum

Order Parameter

Fig. 2. The vacuum potential energy in terms of a parameter related to the "Higgs field" which describes the breaking of the symmetry. Inflation results from spontaneous symmetry breaking. At the curve's minimum, the theory describing the universe is SU(3)⊗SU(2)⊗U(1).

Why the vacuum energy happens to go exactly to zero as opposed to some finite value is a problem that has not yet been solved, but there clearly is some drop between a higher energy vacuum and a lower energy vacuum. That difference means that, at early times, the energy of the vacuum will drive a de Sitter kind of expansion. Whenever the universe cools down—as the universe cools down to where its thermal energy gets below the vacuum energy scale—then the energy density of the universe will no longer be dropping with expansion, but will stay constant. Once the universe gets down to this scale, it can't keep dropping in density because the vacuum energy will dominate. The vacuum energy means space, itself, has an energy, and so there is no further drop in the energy with expansion. Then the expansion turns into an exponential expansion until you get over to this state where you have some phase transition between the universe in its symmetric state and in its broken-symmetric state. At that point, that expansion will revert back again to the normal cosmological expansion.

The idea that Guth had is that you have an initial expansion, and then you go through this de Sitter phase, "inflation," and then you have some phase transition where the symmetry is broken and you revert back again to our normal expansion.

Basically, all this is doing is saying that the initial conditions of the universe for the standard Big Bang are set by inflation. This is contrary to the way it has been described in some popular articles in which it seems as if the inflationary universe replaces the Big Bang. It does not replace it at all. All it does is set the initial conditions. Instead of being arbitrary, they are set by the inflation. We will come back to that in a moment. But the standard Big Bang then proceeds as normal. You can extrapolate back until you get to the end of the inflation, and that's a scale of about 10^{-32} seconds. The normal Big Bang is fine back to scales of 10^{-32} seconds. Then you've got this inflationary phase before you go back even earlier to the Planck scale, where nobody knows what's going on.

Now, what does that do for us if you've got that inflationary phase in there? Well, it solves a number of our long-standing problems.

For example, it solves the *horizon problem,* or how the universe got so smooth on large scales. During the de Sitter phase, with its exponential growth, the scale of the early universe gets very, very big. It gets bigger than the entire horizon of the universe is today, appropriately normalized. As a result, the entire universe today is expanding in a region that was causally connected back at that very early pre-inflation time, because the inflation just took a little, tiny region that was in thermal equilibrium, and made it very, very big. Thus the entire universe just came out of that little, tiny region of space in which the whole universe was in thermal equilibrium. So you can understand why the temprature is so uniform.

Another problem solved is the *flatness problem,* or how the universe got to be so old. In this inflation, the universe bacame very big. If you make it very big, you make the geometry very flat. You know, if you blow up a balloon and look at a small scale of the balloon, it looks pretty flat, so that's all you are doing with the universe. You are making it so big that the curvature of space, whatever it might have been before inflation, ends up negligible on the scales that we look at. A flat universe is an $\Omega = 1$ universe. As a result, we made the universe exactly at the critical density. Regardless of what the initial conditions were, inflation converted it into a universe that has an Ω of one, flat; and so, we can understand how the universe could be here after 15 billion years. It's right at the critical value. So, Ω doesn't go to infinity or zero. Inflation puts it at the critical value.

The puzzle that the grand unified theories introduced and was the original motivation that Guth had when he inflated was the *monopole problem* and it is also trivially solved, because each horizon size back then contained about one monopole. Monopoles were formed slightly before the GUT phase transition. So, each horizon contained about one monopole. You make one horizon now bigger than the entire universe today; so, the entire universe today would only have one monopole in it. (Maybe it was in Palo Alto...) Thus, we end up with only one monopole or less in the universe.

There is a new problem though. This was a problem that Guth noted. It is a problem that governments always encounter when they deal with inflation. How do you stop it? And Guth had this difficulty, because while this model looks very nice, it's hard to stop an inflating phase transition. Let me relate this to phase transitions you are more familiar with.

What if we convert water to ice? As we freeze a bucket of water, little ice crystals form on the water and then gradually, if we keep the water cold, the ice crystals grow and interconnect until the entire bucket is solid ice. When that happens, since the little nucleation sites for the ice crystal growth have been in a number of different places in the bucket, there will be irregularities on a small scale. The different crystals will bump into each other in different ways, and so you get little white lines, and so on, in your ice bucket, not one single, solid crystal.

At first that sounded great, because if you did that in inflation at phase transition, with different nucleation sites all intermixed, you could end up with bumpiness on a small scale, even though, it's homogeneous on a large scale. It has the density of ice, or whatever, and so it is uniform on large scales, while on the local scales you would have this irregularity, an origin for the bumpiness; great! No, unfortunately it doesn't work that way.

The little nucleation sites—the bubble of the new phase, unlike in the bucket of ice, where they are just sitting there and growing, in the inflationary universe are moving apart from each other, inflating away exponentially, whereas the bubbles themselves are only growing at slow speeds. Thus, they can't intersect each other, because space is being added arbitrarily fast. There's nothing moving faster than the speed of light; it is just that you are adding space at a faster rate, and you can add space at any rate you want. Nothing is moving; things are just sitting there; there's just more and more space between everything. As a result, the bubbles are growing on a slow scale while space is being added on a fast scale. The bubbles never intersect each other. The phase transition never goes to completion. It would be as if the bucket of ice were exploding, and so the individual crystals never ran into each other. That was the problem; you couldn't stop inflation.

The solution to that problem called the *new inflationary model,* was proposed independently by Linde[2] in the Soviet Union, and Albrecht and Steinhardt[3] in the U.S.: we live in a single bubble. Forget about the other bubbles. Who cares what they're doing? We just have to have one bubble. As long as one bubble had a nucleation site, it doesn't matter that the others are inflating away from us in some way. You just have to have that one bubble be rather well behaved, and you can have our whole universe be a single bubble, and there are a bunch of other bubbles out there, but who cares? They can't affect us. They are causally disconnected from us. They are not within our horizon. They are irrelevant. It's like talking about angels on the head of a pin.

That was the solution to that problem. However, once you live in a single bubble, then you've got the problem of where the bumps come from on a small scale because no longer do the different granules bump into each other. That problem seemed to be solved because within the bubble there are quantum oscillations. We're looking at a little, tiny bubble that just got very big. Little, tiny things have quantum oscillations in them. So, there will be quantum oscillations. Great; we've generated fluctuations! But it turned out that the scale of the quantum oscillations in the standard GUTs models ended up being such that all the matter instantaneously became black holes!

Since we have galaxies and people, and we don't just have black holes around, we have the problem of how to get the bumps small. That took a lot of fine-tuning. You can take an inflation model and can fine-tune it a tremendous amount and somehow get the bubbles small. The natural models tend to give these black holes, but if you fine-tune you might get the bumps small enough to give us galaxies, and so on.

It turned out that the kinds of grand unified models that were able to be fine-tuned most effectively were the superymmetric models. In fact, there was a paper written by John Ellis, Keith Olive, and Dimitri Nanopolis at CERN, entitled "Inflation Cries Out For Supersymmetry (SUSY)." Maybe in order to get inflation to work we need something like supersymmetry, and that's rather nice, because then it is saying that we need to bring in gravity into this whole scheme. We can't just talk about grand unified theory without gravity. So, maybe we need those kinds of models. At present there is a leaning in that direction; however, the final word on inflation has certainly not been written. There's a lot more that is being done. There are many uncertainties in detailed models.

I think everybody would agree that you need something like inflation. Let us recall here that inflation requires that Ω is one. It also severely restricts the acceptable GUT, as we mentioned, and it also tells us that perhaps each inflating bubble is a separate universe. Thus it gives a lot of implications. However, rather than go into details of all the different fine-tuned grand unified models, I want to go through one quick argument of why we are pretty sure Ω is one.

Inflation gives us a way to make Ω one, but we are pretty sure Ω is one due to the time scale argument I mentioned earlier. Ω varies on a time scale of the age of the universe. Ω had to be one at the Planck scale to 60 decimal places, and Ω had to be one to 17 orders of magnitude at the scale of nucleosynthesis. The only way you can avoid a tremendous fine-tuning is just have it be one, and inflation gives you that. It gives you a physical mechanism to get Ω of one. So, I think we are pretty confident, on theoretical grounds, that Ω is probably one. If it has any value other than zero, one, or infinity it is a short-lived transient value. Thus if Ω is not unity, we live at a very special time which is now Copernican.

Observationally, however, we are not so sure, and this leads me to a discussion of the dark matter problem. If we look at galaxies rotating in space, you can compute their mass just from normal Newtonian mechanics, and you get a typical mass for a galaxy like ours of 10^{11} solar masses, or so.

Now, you take that mass, and you find that if every galaxy in the universe has that much mass associated with it, then you get an Ω of about 0.006. So, all the shining regions in space give you negligible mass density. If the universe only has that much mass in it, the stuff that you see shining, it's a very open universe.

In fact, we already know that that's wrong. Remember I said that, from Big Bang nucleosynthesis, we know that the density of baryonic matter cannot be below about 0.03. That's about five times the value of the shining matter. As a result, we know that the universe must have a lot of matter, just regular old baryonic matter that is not shining. So, that's one dark matter problem.

The next dark matter problem arises when you look at the same galaxy interacting with another galaxy, like our galaxy and Andromeda, M31, or other galaxies in binary pairs. They are interacting with each other and going around each other. You can again do normal Newtonian mechanics; solve the same equations, nothing exotic; and the typical mass you get is about 10 times the mass you get from looking at the light.

This tells us something. It says that the mass associated with the galaxy is a lot bigger than the shining region. It says there are dark halos, because the mass you are measuring in this inter- action is all the mass interior to the orbit of the other galaxy; whereas, the mass shining is just the mass interior to the orbit of the stars. As a result, you know there have to be dark halos.

Some people used to call this problem the "missing mass problem." There's no mass that's missing here. The mass is there. You know it's there. You've measured it. It's just the light that's missing. The "missing light problem" is probably a better way of describing it. What we call it now is the *dark matter problem.* There has to be dark matter there. It's not missing mass; it's dark matter. We just don't know what the matter is that's dark.

Notice that the Ω that you get, even including the dark halos, is still very low. A lot of people in the popular literature, even in the scientific literature, seem to become confused by this problem. They talk about how, if you have dark halos, it makes the universe closed. That is false reasoning. You compute the masses of those dark halos and the universe is very open, indeed. In fact, this number is roughly the same as the number we got out of Big Bang nu- cleosynthesis for baryons.

So, we are not forced from the dark halos to have non-baryonic matter. We are forced to have some baryons be dark but it doesn't have to be non-baryonic. There's nothing that forces us to be non-baryonic at this stage.

Even when we go to large clusters of galaxies (a thousand galaxies whirling around each other such as the Virgo cluster or Coma cluster) and compute statistically the mass per galaxy, we come up with a number that's maybe a little bit bigger than the number from the binary inter- actions, although the lower bound agrees with it. The higher number goes up a little higher, and the typical Ω you might get is a few tenths—still not a closed universe, not even up to critical. The measurements are statistically significantly different from the critical density. No direct ob- servation of clusters gives you an Ω of one.

Now, there are some hints that Ω is one on a very large scale that have just been discussed recently, within this past year. Some geometric arguments and looking at scales of hundreds of megaparsecs, scales much bigger than superclusters, imply that $\Omega \simeq 1$, but those are still very preliminary. On scales of clusters of galaxies, scales of five or so megaparsecs across, the uni- verse still looks like it has a low Ω .

And Big Bang nucleosynthesis gives us a value for the density of baryons that is completely consistent with the dark halos and the matter associated with clusters. You don't need to have non-baryonic matter there. That doesn't mean you don't have it, but you are not forced to, to ex- plain the astronomical observations.

We can review our situation: the shining regions are very low; nucleosynthesis is higher than that, telling you that you have some dark matter that is baryonic. The bulk of the baryons are not shining. Inflation, though, tells you Ω is one. We've got a problem, because Ω in baryons cannot be one. That means if we have inflation, we have these arguments that I have just given you that Ω is probably one from theoretical grounds. Then the bulk of the matter of the universe is not made out of good, old-fashioned quarks; it's made out of something else. And it may be one of the SUSY particles, for example.

It also tells us something that is even more difficult to understand, and that is that whatever the non-baryonic stuff is, you do not want the stuff to cluster with the galaxies, because the dynamics of the clusters measure everything that gets gravitationaly clumped with the galaxies. Even stuff in big, loose, diffused halos will be measured when you measure the dynamics of clusters out to five megaparsecs. And out to five megaparsecs, or so, even out to 10 or 20 megaparsecs, like the distance between us and Virgo, Ω is small.

That means if you want to have an Ω of one, you've got to have the stuff not clumped with galaxies. It's got to be a more uniform background, and that's a puzzle.

Initially, people talked about having neutrinos with small masses which are now called "hot, dark matter." If you give each neutrino a little mass, about 20 electron volts, then because there are so many neutrinos in the universe left over from the Big Bang, you end up having enough density in the neutrinos to get an Ω of one; very nice.

In fact, neutrinos are very nice because they only cluster on very, very big scales like superclusters. The characteristic scale for the neutrinos to clump is about 40 megaparsecs. (A megaparsec is about three million light years.) This much larger than the scales of single galaxies or clusters. This is very nice, big scales give a relatively smooth background. You could understand why Ω might be one on very large scales, even though on small scales it is never measured to be one. That was great.

The problem with it is that with the neutrinos smearing everything out on scales out to 40 megaparsecs you don't make galaxies. You end up just with great big neutrino pancakes and big structures on scales of 40 megaparsecs and no galaxies. That doesn't look like our universe. You can make galaxies but they don't end up getting made until very late, almost the present epoch, and, yet, we know that galaxies and quasars existed at very high red shifts (the order of three). Thus we've got a problem with neutrinos.

The alternative (very popular about two years ago) is to choose, instead of hot matter that has very high velocity as the neutrinos would, stuff called cold matter. And there were various examples, axions, little black holes, GeV-*inos* such as photinos, gravitinos, and so on, that might have masses of GeV, shadow matter—all sorts of weird stuff coming out of particle physics. Because it is heavy in mass, like a proton, it would end up clustering on small scales. Thus, it ends up making galaxies very well. People who were worried about making galaxies, really liked the cold matter stuff. People who worried about getting an Ω of one would like the hot matter. No single thing seemed to solve both problems, because the cold matter would then all clump on small scales and then you would measure an Ω of one on small scales. The hot matter would only clump on large scales, so you wouldn't make galaxies. So either way you lost.

People then tried hybrids. It turns out hybrids ended up having the worst aspects of both rather than the best of both. Some sort of hybrid model with some hot, some cold, failed. The large scales smeared out the small scales, so the cold matter didn't work as well on the small scales and, yet, there was no clumpiness on the large scales—a total mess.

People also tried warm particles, something that was tuned right in between the hot and cold matter. That also failed.

One thing that you can make work but is very ugly is the so-called decaying scenario, where you start with a heavy thing, a cold matter particle, and you have it decay and turn into a hot, light matter particle. That works but it's really ugly. You've got to fine-tune all sorts of parameters. So that didn't seem very good either. Another way out was to have more clumps of cold matter and baryons that did not shine, thus making most of the universe clumpy yet invisible. This would make observational astronomy almost useless.

How we've got some new observations that have come along that are changing the whole thing just within the last few months. One is the large-scale velocity field. People are seeing the whole local 40-megaparsec region moving coherently at about 600 kilometers a second. That's a big hunk of stuff all moving coherently. If it was cold matter, you couldn't do that because the cold matter would all clump on small scales and you would have no large coherent effect. So this certainly seems to argue toward having these big neutrino pancakes—going right back in that direction—big scales.

Another thing that has come along is the Harvard Redshift Survey of Gellar, Huchra and Lapparent.[4] They looked at the distances to all galaxies in our region out to about 100 megaparsecs. They plot the number of galaxies at those distances, and they find that the galaxies, when you look at them three-dimensionally, instead of being just in random clumps distributed on the sky, seem to form a pattern of foam with holes about 20 to 40 megaparsecs. The galaxies and clusters are distributed on the walls of the foam holes. (This was published, as I mentioned, in the New York Times and Time Magazine and now recently in the Astrophysical Journal.)

We have known that there were holes in the sky for some time, actually. You might have remembered about five years ago there was talk about the big void in Böotes. A big hole with no galaxies was found in that direction, but many ignored it saying that was just somebody doing a pencil-beam observation in that particular direction and accidently finding the one place with a big gap in redshift distribution where there were no galaxies along one direction in space. You could statistically have one region or maybe a couple of regions in the whole horizon of the universe that had big holes like that. But this new work is saying they are ubiquitous. There are these big holes all over the place.

You might ask, "Well, didn't the astronomers start to map out space a long time ago in distance?" The answer has to do with sociology and we won't go into that. It is hard to measure redshifts but it is also that astronomers tend to like to look at weird objects and clusters but they won't map everything. And so they ended up not mapping all of the galaxies. They only mapped certain peculiar things. Only recently have people done very systematic work, and that's what this tells us, that there is this large-scale structure.

Another piece of work is by Koo and Kron.[5] What they did is use the pencil-beam approach but they did it in more detail to much greater depth than people had done before. They looked in different directions in space to a really great depth and counted the number of galaxies versus redshift. They found that instead of being a smooth distribution, that they are in clumps, with big gaps. Many different directions in space show this sort of clumpy behavior.

Again, it looks as if you are looking through sheets of galaxy or walls of foam. But it is saying not just, as the Harvard survey showed, that the nearby region is like this, but the Koo and

Kron work shows that this kind of structure persists all the way out to redshifts the order of unity, not just the local 100 megaparsecs. This same kind of structure persists throughout all space.

So, that seems to be telling us we may need those neutrinos to give us some sort of 40 megaparsec scale. But if we've got the neutrinos to give us that scale, how do we make galaxies?

It turns out there are two ways you might be able to save the neutrino picture, as far as making galaxies. One of these involves strings, *cosmic strings* left over from the grand unified transition, and the other involves explosions. To get explosions to work in the neutrino pancakes to move things about, you also needed some seeds left over from early times. Strings of the quark hadron phase transition things might give you that. Something from very early times could give you the explosions, but you needed something coming out of very early times to enable the neutrino model to work. So neutrinos, which two years ago were very out-of-favor, are now coming back into favor. I like to say for the neutrinos, that the new large scale structure results can bring back Mark Twain's famous quote that "rumors of my death are greatly exaggerated."

Now, between the strings and explosion hypothesis, which one do I currently favor? This is all stuff being done over the last couple of months. The field is changing rapidly. I thought I would just, in the last bit, mention why I favor the string model.

One has to do with the correlation function—the relationship of galaxies and clusters of galaxies to each other. Jim Peebles showed that you can analyze how stuff is distributed in space in the same way you analyze phase transitions in condensed matter physics. You look at the correlation function, the excess probability over random that you find two objects separated by a certain distance, R. If you plot the correlation function for galaxies versus the distance, you get a function which goes like $R^{-1.8}$. What this is saying is that on small scales (out to five megaparsecs), galaxies are much more likely to be near each other than random; that it obeys a power law; and that as you get out to large scales they are not as likely.

If you do the same thing for clusters of galaxies, depending on how you choose the various richness class of clusters and superclusters, you also get a relationship going roughly like $R^{-1.8}$, but much more strongly correlated.

One problem that you have with this is that the clusters are even more strongly correlated than the galaxies by factors of 20 or 50 for the clusters and superclusters. What's going on here? Because if gravity is what's doing it, the galaxies are much closer to each other and have a higher average density, so, they are going to have a shorter time scale and they are going to get more clustered than the cluster which are very far away from each other, 50 or 100 megaparsecs apart. How are they going to get clustered by gravity? Gravity is not going to move them. And yet they are more strongly correlated than the galaxies are. It seems very peculiar.

Now, we came up with an explanation for this, "we" being Alex Szaloy, who is a Hungarian physicist who has been at Fermilab for the last two years and just went back to Hungary, and I. (He also had worked with Zel'dovich in the Soviet Union; Alex is our "exchange particle.") We did what we call a *poor man's renormalization group*. We just looked at this correlation function in scale-free units. We removed the dimensions, which is always a good way to attack a physics problem, if you can determine dimensionless parameters. That is, consider the fundamental parameters rather than things that have dimensions associated with them.

And so, we reanalyzed this, just did a little algebra and, said, "Okay. Let's write this in terms of dimensionless units." And then we looked at the correlation function in those dimensionless units, just measuring everything in units of the average spacing of the galaxies in the sample. Then we looked at how this dimensionless correlation function plot, and we find that it's a straight line for the superclusters and the different kinds of clusters, with galaxies slightly higher by about a factor of three.

So, in dimensionless units, the galaxies are more stongly correlated and everything else is the same. Now, what does that tell you? That says that there is some sort of scale-free structure going on here. It's telling you, in effect, that galaxies and clusters of galaxies are initially laid out on the sky in a pattern that has nothing to do with gravity, and that that pattern has its correlation function independent of scale. No matter how far apart they are, they are laid out in the same pattern.

What do we call things that are patterns independent of scale? They are called *fractals*. There is a beautiful book by Mandelbrot on *The Fractal Symmetry of Nature*.[6] Galaxies and clusters of galaxies are laid out on the sky in a fractal pattern. You might have naturally expected the galaxies to be a little bit enhanced over the fractal pattern because, in addition to where they were laid out, they would have clustered gravitationally. This explains why they are more strongly correlated by a factor of 3 and everything else is on the same pattern because everything else hasn't been moved around by gravity.

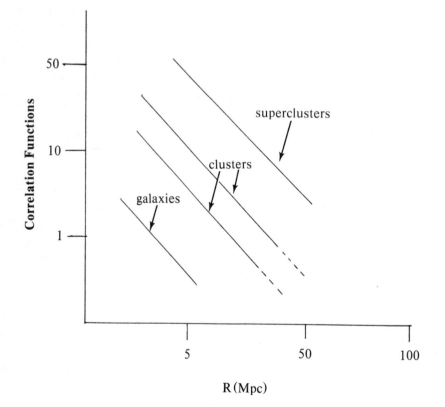

Fig. 3. The correlation function in terms of R.

What could give us such a fractal pattern? First let us find the fractal dimension of this pattern. A fractal dimension, D, is just a way of talking about the filling of space. If you talk about the effective mass, it scales with distance, R, as some power D. For example, a string is just a linear thing. The mass of the string is proportional to R. So D is one. If you are talking about a disk, the mass of the disk goes like R^2, so $D = 2$. If you fill all of space, then the fractal dimension is three.

In general, you can talk about fractional dimensions between these values, and they don't have to be topological at all. For example, in Brownian motion, which is the linear path on a surface of water of some randomly moving particles, the topological path is one dimensional, but because the Brownian motion will cover the entire surface, the fractal dimension is two. So, you can have fractal dimensions differing from topological dimensions.

An example of a fractional fractal dimension which is given in Mandelbrot's book is a coastline. In particular, I think he uses a coast of England. On the coastline you get a longer and longer coastline the smaller your ruler is, because you see more and more of the wiggles. As you go to smaller and smaller scale you continue to pick up more and more wiggles. The wiggles have the same kind of peninsular shape and everything on small scales as on big scales, so you end up adding more and more of them in, and so the coastline lengthens depending on what scale you use to measure. It turns out the fractal dimension of the coast of England is 1.2, which turns out to be the fractal dimension of the universe, as well.

If we want to then work out why the fractal dimension of the universe is 1.2, you just solve for the extra mass. That's just integrating the correlation function, which is the excess probability over random, that you are within distance R, which we mentioned goes like $R^{-1.8}$. Thus the excess mass is going like $R^{3-1.8}$. So, D equals 1.2. Thus the universe appears to be having its galaxies and clusters laid out with a fractal that's got a dimension of 1.2. Now, 1.2 is awfully close to the linear dimension of a string. So, this is the reason why I favor the string. The explosions would have a great deal of difficulty in naturally giving you that factal dimension.

The current in-vogue model as of a couple of months ago is a combination of neutrinos and strings, the neutrinos giving you Ω of unity and large scale velocities, the strings giving you the fractal dimension and the small-scale structure and correlation functions. Neutrinos and explosions can also give you this structure but it takes much more fine-tuning. So, I would argue that, while the strings are the easier way to go, they lead to a rather bizarre model.

What do I mean by strings? There are different ways a phase transition in the early universe can occur. We have talked about nucleation sites, grains forming, but, when you condense a solid out of a liquid, different kinds of materials will undergo different kinds of condensation. Some will end up with filamentary structures rather than little bubbles or grains. Similarly with the grand unified phase transition, it can go and yield filamentary structures, strings, rather than the bubbles that we've talked about.

If the strings are what give the fluctuations that grew galaxies and we need strings to get that dimension 1.2 fractal, then we are learning something about the phase transition of the universe from looking at the large-scale structure, looking at how the galaxies are distributed on the sky. It's a tie between observational astronomy and this fundamental thing, the phase transition at the birth of the universe. Naturally, we are very excited about that possibility.

These strings are just remnants of the supersymmetric GUT phase. We are talking about regular old three-dimensional strings, not superstrings. And the interior of the string has a scale like the GUT scales. The cross-section is the GUT scale. So it's very tiny, but inside of the string you've got grand unified symmetry restored. (Protons instantaneously decay and everything like that.) There's an energy density in there and where those loops of strings are would be the seeds that the baryons would fall onto to make the galaxies and clusters of galaxies.

When that model was first discussed a few years ago, it seemed very ridiculous and ad hoc and nobody paid much attention to it, except Zel'dovich.[7] He actually talked about using strings to make galaxies and then a few others, Neil Turock,[8] in England, looked at it, and Alex Vilenkin.[9] Now, strings sound like one of those examples of the famous T.H. White principle, "Everything not strictly forbidden is mandatory," because at first it appeared to be totally ludicrous. But then, as other models were ruled out, such as cold dark matter, hot dark matter, by themselves, it's interesting that the only thing that seems to be emerging and surviving all these things is the model of the strings being the seeds to make galaxies.

In addition, they have gotten a bit of a boost in that the current in-vogue theory of everything—the superstring model with 10 dimensional strings (not the three-dimensional strings that make galaxies) does seem to yield three-dimensional strings as it goes through the GUT phase transition. So it gives you things that you want. No longer are the strings just a possibility in grand unified models but they are perhaps the consequence of the current in-vogue model.

Why do I say the current in-vogue model has 10 dimensions? Remember we have been talking about SUSY and why we seem to like SUSY; that she is needed to unify all four dimensions. She seems to make inflation work well and doesn't violate the proton decay experiments. However, the best SUSY with no infinities, no anomalies, et cetera—the theory of everything—seems to require 10 dimensions.

The reason for this is that in three dimensions SUSY is rather ugly. She's got all sorts of bells and whistles. We've got to add all these funny particles and we've got to tune them in all sorts of special ways. And each of the particles are point-like, so they have self-induced infinities. But if, instead, we make the fundamental entity not a point but a loop of string, then we have removed the singularities. But if we make it a loop of string, we've got to make it a loop and we've got to get everything nice and symmetric. It turns out if we make it a loop of 10 dimensional string, all the bells and whistles that we've added to the theory, all are merged in a very nice, simple, 10 dimensional loop of string, and that 10 dimensional loop of string contains all of physics. Everything is included. It's just that it's got to be 10 dimensions. So, SUSY is very pretty in 10 dimensions.

Two years ago it was 11 dimensions. Now, it's 10; 10 seems to be much prettier. It doesn't have any infinities. It also enables you to have low mass fermions which you can include within the 10 dimensional theory but you could include in 11. Current thinking is that we need a 10 dimensional loop of string. Then we have a problem, which I alluded to in the very beginning of this presentation. How do you get down to the three dimensions that we live in? If the best theory of the universe, the TOE, is 10 dimensions, why do we live in three-plus-time? One of the solutions may be that only three-tenths of the original dimensions are inflated. In fact, there are people up here at Fermilab who spend large numbers of hours trying to get three dimensions plus time to inflate, while leaving six dimensions compactified.

What are these other six dimensions that are left over? Well, that means every point in space has a little coil of six dimensional space attached to it. Of course, we are three-dimensions beings. So, we don't see these little coils of Planck-scale dimensions all coiled up, little dough-nuts at every point in space. What do we see? What we see are that different points in space have symmetry properties. Where did they get those symmetry properties? What is symmetry? That involves orientations of different regions of space. Maybe what's going on is that the things we call forces—the symmetry properties that we give names to as different forces are be-cause every point in space has these six dimensions coiled up on it. All we're seeing is the projec-tion in our three-dimensional space of these other six dimensions and, because we can't see the other six dimensions but we are only three-dimensional beings, we say, "Ah, that's a force." And we give it the name *force,* but it's really just the geometry of these other six dimensions. In some sense, we are going back to the old geometric interpretation of things but by a quantum field theory.

This leads to my final way of viewing the universe which I call the *Extreme Copernican Princi-ple.* If we look at the way cosmology has developed over the ages, we see the following trend. Copernicus, 500 years ago, told us that the Earth is not at the center of the solar system; the sun is.

Then Shapley, at the beginning of this century, showed us that the sun is not at the center of the galaxy, it's off in the boonies, out near the edge of the galaxy, 10 kiloparsecs away.

Then Hubble went on with his expanding universe and description of space to say, "This question people worried about for 500 years was totally irrelevant. There is no center. So, why were you worring about that, anyway? All points are equivalent. Everything is expanding uni-formly, well-behaved. There is no center. It is just space expanding." So, the whole question of a center was irrelevant.

If we have inflation with the bubble universes, we are not even the only universe. To worry about being the center of something that is not even the only one was really a silly thing.

Of course, now we realize too that we are not even made out of the right stuff. The bulk of the matter of the universe is something else. 90 percent of the universe is made out of something other than baryons.

Finally, when we study grand unified theories, theories of everything, we don't even seem to have the right number of dimensions.

References:

1. A.H. Guth and P.J. Steinhardt, Sci. Am. **250** (5), 116 (1984).
2. A.D. Linde, Phys. Lett. **108B**, 389 (1982).
3. A. Albrecht and P.J. Steinhardt, Phys. Rev. Lett. **48**, 1220 (1982).
4. V. de Lapparent, M.J. Geller, and J.P. Huchra, Astrophys. J. **302**, L1 (1986).
5. D.C. Koo and R.G. Kron, Astron. Astrophys. **105**, 107 (1982).
6. A. Mandelbrot, *The Fractal Symmetry of Nature,* Freeman, San Francisco, 1977.
7. Y.B. Zel'dovich, Mon. Not. R. Astr. Soc. **192**, 663 (1980).
8. See, for example, D. Olive and N. Turok, Phys. Lett. **117B**, 193 (1982).
9. A. Vilenkin, Phys Rev. Lett **46**, 1169, 1496(E) (1981); A. Vilenkin, Nucl. Phys. **B196**, 240 (1982); A. Vilenkin and A.E. Everett, Phys. Rev. Lett, **48**, 1867 (1982).

$$\left. \begin{array}{c} \end{array} \right. gy \right) = \frac{dE}{dt} = -\frac{1}{5}$$

171

Clifford Will is Professor of Physics, and member of the McDonnell Center for the Space Sciences at Washington University in St. Louis. Born in Hamilton, Canada in 1946, he received a Ph.D. in Physics from the California Institute of Technology in 1971. He was an Enrico Fermi Fellow at the University of Chicago (1972–74), and an Assistant Professor of Physics at Stanford University (1974–81), before joining Washington University as Associate Professor. From 1975 to 1979, he was an Alfred P. Sloan Foundation Fellow. He was recently selected by the American Association of Physics Teachers to be the 1986 Richtmeyer Memorial lecturer. He has published 75 scientific articles or abstracts, including 8 major review articles, 5 popular or semi-popular articles, and two books, including *Theory and Experiment in Gravitational Physics* (Cambridge University Press, 1981), and a popular book, *Was Einstein Right?* (Basic Books, 1986). His research interests include tests of general relativity, gravitational radiation, black holes, cosmology, and the physics of curved spacetime.

Was Einstein Right?
A Topic in Modern Physics for the High School and Introductory College Physics Curricula*

Clifford M. Will
McDonnell Center for the Space Sciences,
Department of Physics
Washington University,
St. Louis, MO 63130

I. INTRODUCTION

When confronted with the subject of putting general relativity into the high school and freshman college physics curriculum, one immediately asks, Why? Why try to talk about general relativity in high schools?

I would like to offer three reasons. First, even though general relativity is an old subject—Einstein invented the theory in the late fall of 1915—it experienced a remarkable renaissance beginning around 1960 that has made it one of the most active branches of physics. For that reason, it is really "modern," and fits the theme of this conference.

The second reason why one can talk about general relativity in elementary physics classes, is that it's fun, it's neat, it's fascinating. I want to focus on one special aspect of general relativity, the quest for experimental verification of the theory: in other words, for the answer to the question "Was Einstein right?" This particular topic is especially suitable because it covers a wide variety of issues which can be understood at the elementary level. There is a little bit of theory, a taste of geometry, a dash of relativity, a helping of gravity. Furthermore, one can do some simple laboratory experiments in this field. One can talk about other experiments that clearly can't be done in the laboratory, but that involve spacecraft and the space program, a subject that always interests students. Finally, there is some astronomy in this subject, and astronomy is very interesting to young people.

Thirdly, I think most importantly, the subject is a classic example of the scientific method. This is a subject that I think is lacking in high school physics and elementary college physics. What is the scientific method? How do we determine whether or not a theory of nature is correct or viable? In high school, one is typically handed Newton's laws as if they were brought down with Moses along with the tablets. One never really discusses whether or not Newtonian theory is correct; how one can confirm it; or how one can compare its predictions with those of some competing theory. In the case of testing general relativity, one can see how a theory is produced, how it is tested, how one judges whether or not it is viable. For these reasons, I think that this is an appropriate subject for this level of physics student.

*Supported in part by the National Science Foundation [PHY 85-13953]

II. THE RENAISSANCE OF GENERAL RELATIVITY

General relativity experienced a renaissance that began during the academic year 1959 to 1960. During that remarkable year, five events occurred that, in retrospect, tell us that a new day was about to dawn for general relativity.

First, on September 14, 1959, scientists at the Millstone Hill radar antenna, about 20 miles northwest of Boston, sent a radar beam to the planet Venus, and looked for a radar echo. It turned out the echo was not initially detected in the data because there were some problems with the size of the astronomical unit, but when the September 14 data were reanalyzed a few years later, a very faint echo was found. This was the first time anyone had ever sent a signal to a planet and received an echo back. This was September, 1959.

In April of 1960, *The Physical Review* published what is now a classic paper by Robert Pound and Glen Rebka of Harvard University, called "Apparent Weight of Photons." This experiment measured, for the first time, the gravitational red shift or frequency shift of light falling in a gravitational field. This was Einstein's first famous prediction, yet it wasn't confirmed until 1960.

In June of 1960, the British mathematical physicist, Roger Penrose, published a famous paper that presented a new approach to doing general relativity. This was an approach based on spinors, rather than tensors, but it was very useful for doing certain kinds of mathematical calculations in general relativity.

Later that summer of 1960, Carl Brans, who was a graduate student at Princeton University, working under the eminent experimental physicist Robert Dicke, began to put the finishing touches on his PhD thesis. Even though Dicke was an experimentalist, this thesis by Brans was devoted to theory, and in fact it was devoted to developing an alternative theory to general relativity. Dicke and Brans called it the scalar-tensor theory, but ultimately the most popular name for this theory was the Brans-Dicke theory.

Finally, on September 26th, 1960, just over a year after the Venus radar echo, Allan Sandage and Thomas Matthews, astronomers at Caltech, went up to the Mount Palomar telescope to take a photograph of the radio source 3C48. What they expected to find was a cluster of galaxies, but instead, they found an object that looked just like a star, as far as anyone could tell from the photograph, except that it had a very bizarre spectrum that they couldn't understand and its luminosity varied widely over very short time scales, as short as 15 minutes. Ultimately, this object was called a quasar, and was the first quasar discovered.

These five events foretold a renaissance for general relativity. To explain why this was a renaissance, let me first give a bit of what I call the prehistory of general relativity, pre-1960.

After Einstein invented general relativity, he became a great celebrity. When the measurements of the deflection of light, made during a solar eclipse in 1919, confirmed the theory, he appeared in the *New York Times* and the *London Times,* and his picture was on the cover of the *Berliner Illustrierte.* He was a famous man.

But by the middle 1920s, Einstein had turned most of his attention away from general relativity and had embarked upon his futile search for a unified field theory. During the rest of his career he only occasionally dabbled in general relativity.

In fact, most research in general relativity from about the middle 1920s until the end of the 1950s was dominated by very abstract mathematics, devoted to formal, esoteric questions having little to do with physics. It became a very stagnant and sterile subject. Most people felt that the theory had very little to do with the real world, and no one really thought the theory was useful.

Furthermore, the theory was thought to be extremely difficult to comprehend. We have all heard the phrase "there must be only three people in the world who understand the theory of relativity." In fact, there are several stories as to where this kind of phrase came from. My favorite was told to me by the astrophysicist Subrahmanyan Chandrasekhar. In November 1919, Sir Arthur Eddington, the famous British astronomer who headed the expeditions to measure the bending of light, reported his results at the Royal Society of London, results that confirmed Einstein's theory. After he made this report, one of his colleagues came up to him afterwards and said, "Why, Eddington, you must be one of three people in the world who understand the general theory of relativity," to which Eddington demurred. The colleague said, "Oh, don't be modest, Eddington." Eddington responded, "On the contrary, I am trying to think who the third person might be."

Chandrasekhar believes that this kind of myth was a real impediment to progress in general relativity. People did not go into the field, or did not think of the field as being relevant for physics, and so it stagnated.

But after 1960, things began to change. Let me now explain why these five events tell us, at least in retrospect, that things would change.

Begin with quasars, the last event of that academic year 1959–60. Quasars told us that general relativity now had something to do with the real world. Because these bizarre objects emitted enormous amounts of energy that no one could explain by conventional means, people began to think in terms of highly collapsed objects that could only be described by general relativity. During subsequent years, the discoveries of pulsars, the 3K cosmic background radiation, black hole candidates, and so on, told us that general relativity would play an important role in astronomy. Here was motivation for studying and doing general relativity.

The second great event, the paper by Roger Penrose, was characteristic of a new breed of general relativists. These were relativists who knew abstract and sophisticated mathematics, but whose goal was to use this mathematics as a tool to answer observable questions. They wanted to calculate in the most streamlined way possible, what general relativity predicts, and what can be observed. This new breed of relativist wanted to make contact with the real world of physics.

Yet both these issues, using general relativity in astrophysics, and using general relativity in research, raised the question: is general relativity the right theory? Should we be using this theory, or should we use some other theory?

That question, of course, took on importance with the publication of the Brans-Dicke theory, because this theory was truly a viable alternative to general relativity. It had all of general relativity's good features, plus some other features that some people actually preferred to general relativity. The Brans-Dicke theory provided a strong competitor that motivated the idea that we really must test Einstein's theory with the best precision possible.

The remaining two events, the radar echo from Venus and the Pound-Rebka experiment, gave the signs that the tools were here to do the job. The Pound-Rebka experiment confirmed the famous prediction of the gravitational red shift, but it also demonstrated the use of the new technology that was going to explode during the '60s and '70s in these kinds of gravitational experiments. This experiment used the Mössbauer effect to reduce the width of the gamma-ray line whose frequency shift was being measured. Other developments, such as semiconductors, superconductivity, lasers, masers, atomic clocks, and so on, high-tech devices that came along in the '60s and '70s, could then be used in testing general relativity.

The radar echo from Venus gave us a new laboratory for doing experiments. The rapid development of the space program in the 1960s provided new ways to test general relativity, using radar and radio waves for tracking quasars, planets and satellites, using laser beams for tracking the Moon, and so on.

So we see during this one academic year the signs of a renaissance for general relativity. After this, the pace of research in general relativity and in a new field called "relativistic astrophysics" began to accelerate. New advances, both theoretical and observational, came at an ever increasing rate, including the discovery of the microwave background, the theoretical analysis of the synthesis of helium in the big bang; observations of pulsars and of black-hole candidates; the development of the theory of relativistic stars and black holes; the theoretical study of gravitational radiation and the beginning of an experimental program to detect it; improved versions of old tests of general relativity, and brand new tests, discovered after 1959; the discovery of the binary pulsar; the analysis of quantum effects outside black holes and of black hole evaporation; the discovery of a gravitational lens; and the beginnings of a unification of gravitation theory with the other interactions and with quantum mechanics. General relativity became an important and active subject at the forefront of physics, and it remains so to this day.

What I want to discuss is why we believe general relativity is the correct theory. What kind of experimental evidence do we have for this theory? The discussion begins at the most fundamental level, the idea that spacetime is curved.

III. THE STRAIGHT ROAD TO CURVED SPACETIME

The simplest way to introduce and motivate the idea of curved spacetime is to talk not about spacetime, but about space, and to look, for example at the surface of a sphere on which we can plot a variety of curves (Fig. 1). The "straight" lines, the lines that we would get by laying down straight rulers along the surface of the sphere, such as the equator or the lines of longitude, are

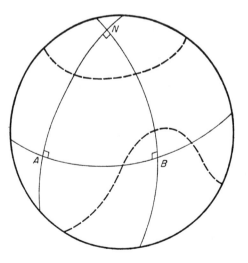

Figure 1. Curves on a sphere. Solid curves are examples of great circles or geodesics, such as the equator or lines of longitude; in fact, any curve formed by the intersection of a plane through the center of the sphere with its surface is a geodesic. Dashed curves are not geodesics. The triangle NAB has a total interior angle of 270 degrees.

of course the "great circles" on the sphere. These are called geodesics and are the straightest lines you can put on the surface of the sphere. So a geodesic is a "straight line," or straightest line, in a curved space; in this case, a two-dimensional space. The dashed curves, such as a line of latitude other than the equator, would not be a geodesic.

Imagine now a land associated with such a sphere, a land called "Sphereland," analogous to the "Flatland" of a famous 19th century book by E.A. Abbott.

Imagine these Spherelanders studying their world by laying down some triangles using aluminum rulers. When they add up the interior angles of one of the triangles, they find that the sum of the interior angles is greater than 180°. This is a phenomenon we all understand on the surface of a sphere. But since the Spherelanders are two-dimensional beings, they can't step off the sphere to look down upon it, and so they are surprised by this observation, because somehow, they have found out about Euclid who says the sum should be 180°, whereas they get more than 180°. One Spherelander postulates that there is some strange force acting that bends aluminum rulers in just such a way as to make the angles sum to greater than 180°.

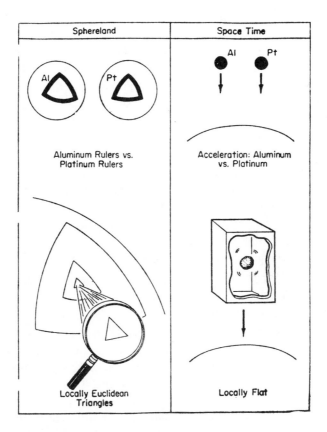

Figure 2. Curvature of Sphereland and curved spacetime. Properties of triangles on Sphereland are independent of material making up rulers, so must be a property of underlying geometry. Acceleration of bodies in spacetime is independent of material making up bodies. Is it a consequence of underlying spacetime geometry? In small enough regions of Sphereland, triangles obey Euclidean postulates of flat space. In small enough, freely falling laboratories, bodies experience no gravity, as in flat spacetime of special relativity. Conclusion: spacetime is curved.

To check this they try a different experiment, using platinum rulers instead of aluminum. For the same size triangle, they find exactly the same sum of angles as they found for aluminum. Whatever is making this sum greater than 180° is universal. It has nothing to do with the nature of the rulers they put down. Somehow, it has to do with the underlying space in which they live. Furthermore, they find that, if they take a sequence of triangles, ever smaller, and measure the interior angles as the triangles get smaller and smaller, the sum of the angles gets closer and closer to 180°.

One imaginative Spherelander concludes from this that they must be living on a curved space, which explains why they violate Euclid's postulates, but it is a space that is locally flat. If they go to a small enough region, everything looks flat as if it were Euclidean space (Fig. 2).

I now want to make a similar leap of imagination, much like the one that Einstein made, in considering spacetime instead of Sphereland.

Instead of triangles, we now talk about the fall of bodies. Let's consider two bodies, one made of aluminum and one made of platinum, falling in some external field, such as that of Earth. We find that both fall with the same acceleration. (I'll talk later about the experimental evidence for this.) Whatever it is that is making these balls fall, it doesn't have anything to do with the balls themselves, it has something to do with the underlying space in which these balls are moving, or rather space-time, because they also move through time.

We also know that, if we put ourselves in a freely-falling frame and float freely, gravitation seems to disappear. When gravity is absent, we say that one should use the laws of special relativity to discuss physics. Special relativity replaced the Newtonian concepts of space and of a separate absolute time with a single geometrical framework of spacetime, in which the spatial degrees of freedom and the time are treated on a more equal footing, and can be interrelated. For example, the rate of flow of time depends on the state of motion of the observer: a clock in a moving laboratory appears to tick more slowly than a set of identical clocks distributed throughout a reference laboratory. As another example, two events or occurrences at two different locations can be seen to be simultaneous by one observer, but will be seen to be not simultaneous by a moving observer. It was Hermann Minkowski who developed the idea that a "spacetime continuum" was the underlying geometry behind the time and space relationships proposed by special relativity. The spacetime of special relativity is flat: for example, for a given observer, the spatial part of his world has the properties of a normal Euclidean space.

So in this local freely-falling frame where gravity is absent, you think in terms of the ordinary flat space-time of special relativity.

But everything I've said parallels what the Spherelanders did. We have something that is associated with the background geometry. Locally, it looks like the flat space-time of special relativity. And so with a leap of imagination, just like the leap that Einstein took, we postulate that space-time must be curved, and it is curved space-time that makes these balls fall. The acceleration does not depend on the structure of the balls, because it has to do with the background space-time.

So the conclusion from this idea that bodies fall with the same acceleration is that space-time must be curved.

Let me review briefly the experimental evidence for this equivalence of free-fall acceleration, known as the principle of equivalence. It actually has a very old history, dating back to 500 A.D., to Ioannes Philiponos, who stated that bodies should fall with the same acceleration, des-

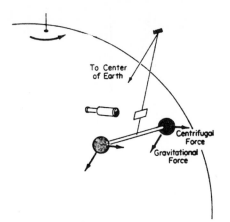

Figure 3. Eötvös' experiment. Fiber supporting rod does not hang exactly vertical because of centrifugal force from Earth's rotation, so downward gravitational force on balls is not parallel to fiber. If gravity pulls one material more strongly than the other, rod will rotate about the fiber axis. If entire apparatus is rotated so that the two balls are interchanged, the resulting rotation will be in the opposite sense.

pite the prevailing Aristotelian theory that this should not be the case. Many people since then have discussed equality of free-fall, including Benedetti, Stevin, and Galileo in the 16th century. In the 17th century, Isaac Newton proposed equality of acceleration as a fundamental principle of mechanics, and performed experiments using pendula to test it.

The first of the truly high-precision tests of the equality of free-fall of bodies was performed around the turn of the century by the Hungarian Baron Roland von Eötvös. In Eötvös' experiments, schematically, two balls are attached to the ends of a rod, and the rod is suspended by a fiber. Because of the rotation of the Earth, the centrifugal force pushes the balls away from the Earth's rotation axis, so that the fiber is no longer vertical (Fig. 3). The gravitational force pulls the balls toward the center of the Earth. If bodies fall differently in a gravitational field depending on their composition, relative to how they respond to the inertial centrifugal force, there will be a torque about the fiber axis. The torque is then measured by looking, say, at light reflected off a little mirror suspended on the fiber. Eötvös found no evidence for any effect in this experiment, and claimed a confirmation of the equality of free-fall of different materials to a part in 10^9.

Another version of this experiment was done by Dicke at Princeton, and ten years later by Vladimir Braginsky in Moscow, in which one asks whether the balls fall toward the Sun with the same acceleration. As the Earth rotates, the orientation on the balls relative to the Sun's direction will change. So if one ball falls with larger acceleration, there will be a torque in one direction, giving a positive reading. When the Sun is parallel to the axis of the rod, there will be no effect. When the Sun is in the opposite direction 12 hours later, the rotation will be in the negative sense. The idea then, is to look for a torque in such a pendulum that oscillates with a 24-hour period. No evidence for such a torque was found down to the experimental accuracy, and these experiments confirm the equality of acceleration of different materials such as platinum and aluminum to a part in 10^{11} in the Dicke experiment, and to a part in 10^{12} in the Braginsky experiment. These are extremely high-precision confirmations of the idea that bodies fall at the same acceleration, making us confident that space-time is curved.

What do I mean by curved spacetime, and how is it related to the gross, everyday effects of gravity, such as the tossing of a piece of chalk across a lecture room on Earth, or the orbit of a

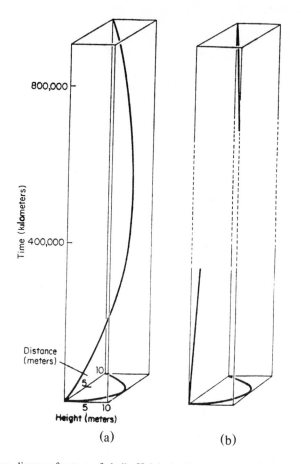

Figure 4. Spacetime diagram for toss of chalk. Height is plotted to the right, distance along the ground into the page, and time upwards. In (a), scale of time axis has been severely compressed from 900,000 km to the size of the page. Spacetime path of chalk runs from lower left corner to upper back corner. Projection of the path in space (a parabola) is shown at the base. If time axis is stretched toward its true length, spacetime curve appears almost straight, as in (b).

planet or spacecraft. I stated before that Einstein's proposal was that gravitation and curved spacetime are in some sense one and the same thing. How on Earth (or anywhere else for that matter) is the trajectory of a piece of chalk related to curved spacetime? Einstein postulated that the motion of a freely falling body, such as a thrown object or an orbiting planet was along a geodesic, a "straight line" of the curved spacetime. How can we reconcile that with what we observe to be the motion of a piece of chalk or planet, which certainly is not along anything approximating straight lines?

The reconciliation is really quite easy, once we have learned to distinguish spacetime from space. The best way to do that is to draw a picture that includes some of the spatial dimensions, as well as the time dimension. We obviously can't draw all four spacetime dimensions on a two-dimensional page, but if we ignore one of the spatial dimensions, then we can use perspective to represent the two spatial dimensions, plus time. These so-called "spacetime diagrams" are a popular tool for talking about special relativity.

Consider the problem of a piece of chalk, thrown to a height of ten meters, and coming down a distance ten meters away. The trajectory of the piece of chalk is a parabola, certainly nothing like a straight line. However, let us consider the motion of the ball in a spacetime diagram (Fig. 4). The base of the diagram shows a line that records the height of the ball (parallel to the page), and perpendicular to that (into the page) a line drawn between the starting and ending points of the ball to record its ground track. The third spatial dimension is not shown. The vertical line in the spacetime diagram is the time direction. Now, because we know that we must treat space and time on an equal footing, we must therefore give them the same units. But how do we do this if spatial distances are measured in meters, say, and time is measured in seconds? The way to do it is through the speed of light, which according to special relativity is a universal constant, the same value when measured in any local freely falling frame. Therefore, if we take a time interval of one second and multiply it by the speed of light, approximately 3×10^8 meters per second, then we have something with units of length. This length is just the distance traveled by light in one second (300,000 km, or 186,000 miles). If we now look at time using units of distance, we would say that "one meter of time" corresponds to the time it takes light to travel one meter, or about 3.3 nanoseconds. We are told to treat time and space equally, so if we mark off distances on the spatial lines of our spacetime diagram in intervals of one meter, we must also, on the same scale, mark off intervals of one meter on the time line. Now draw the trajectory of the piece of chalk on the spacetime diagram, each point being determined by the horizontal distance, the height, and the corresponding time. Right away we have trouble. The time taken for the chalk to reach its highest point is only 1.4 seconds, but on the time line, this corresponds to 430,000 km, or a bit farther than the distance to the Moon. And that is only one half of the trajectory of the chalk! Clearly, our spacetime diagram will not fit on the blackboard. Nevertheless, we begin to see the sense in which the trajectory of the piece of chalk is a geodesic or "straight line." The trajectory begins at the starting point, then as we move up the time line, it moves into the page along the ground track line as well as to the right as it ascends. Continuing up the time line (we are now past the distance to the Moon), we find that the point stops moving to the right (the high point of the chalk's motion) and begins to move to the left, while continuing along the ground track line. Finally, at the top of the spacetime diagram, the point reaches the end of the trajectory, on the original line. It is clear that the trajectory has curved, but because it has been stretched in the time direction to more than the distance to the Moon, we would be hard pressed to look at any piece of the curve and to say that it was anything but a straight line. The point is that, when we look at the trajectory of the chalk in space, it describes a parabola, but when we look at it in spacetime it is almost, though not quite, a straight line. The fact that it is nearly straight in spacetime is a consequence of the smallness of spacetime curvature on the Earth.

In summary, we have this basic idea that space-time is curved, but in a freely-falling frame it is locally flat, and all the laws of physics take on the forms specified by special relativity. These two statements are all we need to understand curved space-time at this level. What then is general relativity?

General relativity is not the same as curved space-time. The principle of equivalence tells us that space-time must be curved, but it doesn't tell us by how much. General relativity, then, takes curved space-time and provides us with field equations that can be used to calculate *how much* space-time is curved by the Earth, by the Sun, by a neutron star, and so on. The Brans-Dicke theory provides a different set of field equations to determine how much curvature is produced by matter. Both theories, however, have curved space-time as their basic foundation.

181

IV. THE GRAVITATIONAL RED SHIFT OF LIGHT AND CLOCKS

The gravitational red shift was the first of Einstein's three great predictions, in fact, he made it in 1907, eight years before he had the full general theory of relativity. But, ironically, it was the last to be verified experimentally.

Although it is difficult to measure, the gravitational red shift is easy to derive, once one understands the principle of equivalence.

Imagine a tower, at the top of which is an emitter of light with some well-defined frequency, and at the bottom of which is an identical receiver that can receive this light. Imagine a freely-falling laboratory that at some earlier time was blasted off from the center of the Earth. After some interval, the rockets were turned off, and the laboratory is now in free-fall. It reaches the top of its trajectory at the moment the emitter emits a wave packet or a photon. We won't worry about how to achieve such an incredible feat; this is a *gedanken* or thought experiment, so we can achieve any feat we want (Fig. 5).

An observer inside the laboratory measures the frequency of the emitted packet of light. Since he is in free fall, and since he is well versed in the principle of equivalence, he realizes that gravity is absent, from his point of view, and so the emitted frequency can be calculated using

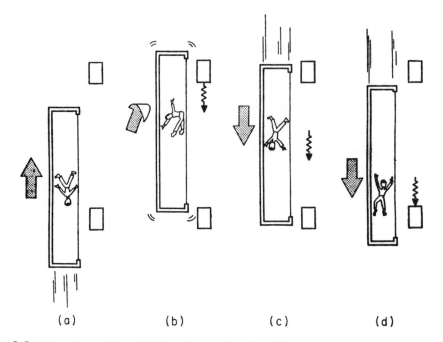

(a)   (b)   (c)   (d)

Figure 5. Sequence of events in redshift thought experiment. In (a), the laboratory is still moving upwards, nearing the top of its trajectory. In (b), the laboratory is momentarily at rest with respect to the emitter, at the moment it sends out the packet of light. The frequency of the packet measured by the observer in the laboratory is the standard value. In (c), the laboratory has started to fall downward, but since the observer inside senses no gravity, he sees the packet of light propagating with the same frequency as before. In (d), the laboratory is falling faster, and the observer sees the receiver coming up toward him. Since the light packet still has the same frequency as seen by him, the ascending receiver will see a higher frequency (a blueshift), because of the Doppler shift. The velocity that determines the amount of the shift is just the speed that the laboratory picked up in the time it took for the light packet to go from the emitter to the receiver.

the laws of special relativity. This of course, is why it was important to make his frame freely fall-ing. Also, since the laboratory is at rest with respect to the emitter, if only for a moment, the emitted frequency is the "rest" frequency, unaffected by any slowing down of moving clocks that is predicted by special relativity. The measured frequency is therefore the standard value for that emitter, and could be looked up, say, in standard tables of physical constants, or calculated using the standard laws of atomic or nuclear physics. The observer then follows the progress of the wave packet as it travels toward the bottom of his laboratory. Since he is still in free fall, the motion of the wave continues to obey the laws of special relativity, including moving at the con-stant speed of light with an unchanging frequency. But as the wave packet moves downward, the observer's laboratory also begins to fall. It was at its apogee for only a moment, then the gravita-tional pull of the Earth began to pull it back down. Nevertheless, the frequency of the wave packet remains unchanged as seen by him. After a moment, the observer notices the receiver coming up toward him, since he is falling, while the receiver is at rest on the surface of the Earth. Thus, when the onrushing receiver receives the packet of light, it is going to detect a higher frequency than that measured in the freely falling laboratory, because of the Doppler effect. The emitter and receiver are still at rest with respect to each other; the important point is that from the point of view of the observer in the freely falling laboratory, in which the frequency has its standard value, the receiver is moving toward him. The velocity of the frame relative to the receiver is the same as the velocity which the freely falling frame has picked up in the time taken for the wave packet to travel the distance between the emitter and the receiver. This is given by the acceleration of gravity (9.80 meters per second per second) times the time, which is the separation between the emitter and the receiver divided by the speed of light. From the Dop-pler formula, the fractional shift in frequency is then given by this velocity divided by the speed of light. For example, for a difference in height of 100 meters, the shift would be only one part in 10^{14}. If the emitter and receiver are at the same height, but separated in the horizontal direction, there is no frequency shift.

In this thought experiment, the observed shift was toward higher frequencies, or toward the blue end of the visible spectrum, because the freely falling frame was heading toward the receiv-er. If the emitter had been at the bottom and the receiver at the top, the shift would have been toward lower frequencies, or toward the red, because by the time the wave packet reached the top, the freely falling frame would be falling away from the receiver. Even though the result can be either a redshift or a blueshift, depending on the experiment, the generic name for this effect is the gravitational redshift. It is called a "gravitational" shift because it occurs only in the pre-sence of a mass that exerts a gravitational force.

Since 1960, there have been several experiments to test this gravitational red shift. The 1960 Pound-Rebka experiment checked it to about ten percent precision, and the 1965 Pound-Snider version of the experiment improved that to one percent, by looking at how gamma ray photons emitted by iron-57 would shift their frequency as they descended or ascended the Jeffer-son Physics Laboratory tower at Harvard University.

Another test of the gravitational red shift is called the "jet-lagged clocks" experiment, in which atomic clocks were flown on jet aircraft around the world during October 1971, to see how much time difference there would be when they returned. This experiment measured not only the gravitational red shift effect, but also the time dilation of special relativity in combination.

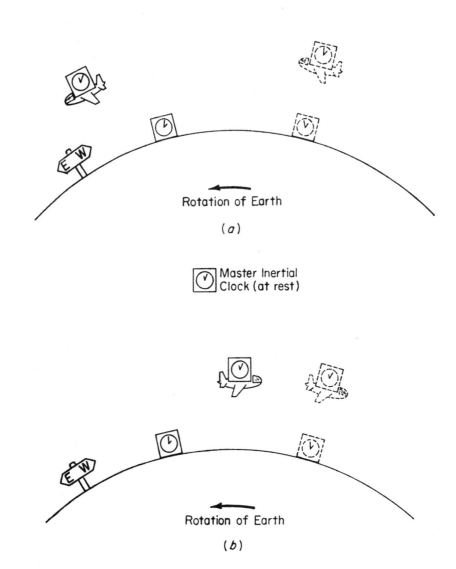

Figure 6. Jet-lagged clocks. (a) Eastbound flight. At the start, the flying clock is directly above the ground clock, after some time it is to the East. The flying clock travels more quickly relative to a stationary fictitious master clock than does the ground clock. Thus the flying clock ticks more slowly relative to it than does the ground by the time dilation of special relativity, and so the flying clock ticks more slowly than the ground clock. On the other hand, the gravitational blueshift makes the flying clock tick more quickly than the ground clock. The two effects can offset each other, so whether the net effect is a gain or loss of time by the flying clock depends on the speed and altitude of the flight. (b) Westbound flight. At the start, the clock is directly above the ground clock, after some time it is to the West. But because of the Earth's rotation, it has moved to the East relative to the inertial clock, with a lower velocity than has the ground clock. Thus the ground clock ticks more slowly relative to the master clock than does the flying clock, so the flying clock ticks more quickly than the ground clock, as a result of time dilation. The gravitational blueshift also causes the flying clock to tick more quickly, so the two effects augment each other.

Consider for simplicity a clock on Earth at the equator, and an identical clock on a jet plane flying overhead to the east at some altitude (Fig. 6). Because of the gravitational blueshift, the flying clock will tick faster than the ground clock. What about the time dilation? Here we must be a bit careful, because both clocks are moving in circles around the Earth rather than on straight lines. Now, according to special relativity, the rate of a moving clock must always be compared to a set of clocks that are in an inertial frame, in other words, clocks that are at rest or moving in straight lines at constant velocity. Therefore we can't simply compare the flying clock directly with the ground clock. Let us instead compare the rates of both clocks to a set of fictitious clocks that are at rest with respect to the center of the Earth. The ground clock is moving at a speed determined by the rotation rate of the Earth, and thus ticks more slowly than the fictitious inertial clocks; the flying clock is moving even more quickly relative to the inertial clocks, so it is ticking even more slowly. Thus time dilation makes the flying clock run slowly relative to the ground clock. The two effects, gravitational blueshift and time dilation, tend to offset one another, and whether the net effect is that the flying clock ticks more quickly or ticks more slowly will depend on the height of the flight, which determines the amount of gravitational blueshift or speed-up, and the ground speed of the flight, which determines the amount of time dilation slow-down. Consider now a westward flying clock at the same altitude. The gravitational blueshift is the same, but now the flying clock is traveling more slowly relative to the inertial clocks than is the ground clock, and therefore it is the ground clock that ticks more slowly relative to the inertial clocks than does the flying clock. So the flying clock ticks more quickly than the ground clock. In this case, both the gravitational and time dilation effects add together, to cause the flying clock to tick more quickly. Therefore, if we were to start with three identical, synchronized clocks, and were to leave one at home while sending one around the world to the east and the other around the world to the west, we would expect the westward clock to return having gained time, or aged more quickly, while the eastward clock would have gained or lost time depending on the altitude and speed of the flight.

The actual experiment, coordinated by J.C. Hafele, then of Washington University in St. Louis and Richard Keating of the US Naval Observatory, used cesium-beam atomic clocks. During the course of the experiment, the air speeds, altitudes, latitudes and flight directions all varied. The flights included numerous stop-overs. But by keeping careful logs of the flight data, they could calculate the expected time differences for each flight. The eastward trip took place between October 4 and 7 and included 41 hours in flight, while the westward trip took place between October 13 and 17, and included 49 hours in flight. For the westward flight the predicted gain in the flying clock was 275 nanoseconds (billionths of a second), of which two-thirds was due to gravitational blueshift; the observed gain was 273 nanoseconds. For the eastward flight, the time dilation was predicted to give a loss larger than the gain due to the gravitational blueshift, the net being a loss of 40 nanoseconds; the observed loss was 59 nanoseconds. Within the experimental errors of ± 20 nanoseconds, attributed to inaccuracies in the flight data and intrinsic variations in the rates of the cesium clocks, the observations agreed with the predictions.

But the best gravitational red shift test ever performed was called "GPA," (which stands for "Gravity Probe A"), or the Gravitational Red Shift Rocket Experiment. It was done in the summer of 1976, by flying a hydrogen maser clock, the most stable kind of clock we know on Earth, on the top of a Scout-D rocket, to an altitude of about 10,000 kilometers. An identical hydrogen maser clock was on the ground, and the rates of the two clocks were compared by telemetry. A signal was sent from the rocket clock to the Earth, and the frequencies were compared.

Consider what happens during ascent of the rocket, say, when the rocket clock emits its signal, and the signal is received at the ground and compared with the frequency of the ground clock. The received frequency differs from the ground clock frequency because of the gravitational blueshift, and because of the time dilation, due to rapid movement of the rocket. However, the received frequency is also shifted toward the red because of the usual Doppler shift caused by the rocket's motion away from the ground clock (upon descent of the rocket, this effect would be a blueshift), and this Doppler shift is 100,000 times larger than the gravitational redshift, for a typical Scout D velocity of several kilometers per second. We would like to eliminate this huge effect somehow, in order to see the much smaller effects of interest. This was done in a very elegant way as follows: suppose a signal is emitted from the ground clock toward the rocket clock. When received by the rocket clock, the received frequency differs from that of the rocket clock by the Doppler shift, and by the gravitational redshift and time dilation. Incorporated into the rocket payload is a transponder, a device that takes a received signal and sends it right back with the same frequency (and with a little more power, to make up for any losses during upward transmission). When the transponded signal is received back on Earth, its frequency is further redshifted by the Doppler effect, since the transponder is receding from the Earth, but it is now gravitationally blueshifted by an amount that exactly cancels the gravitational redshift experienced by the signal on the uplink, and is changed by time dilation in a way that also cancels that experienced on the uplink. Therefore, when received back at the ground, this two-way signal has had its frequency changed twice by the Doppler shift, and that's all. The one-way signal sent by the rocket clock and received at the ground has been changed by only one factor of the Doppler shift and by the gravitational blueshift and the time dilation. All one has to do then is take the frequency change on the two-way signal, divide by two, and subtract it from the one-way frequency change, and presto: no Doppler effect. This "Doppler cancellation" scheme in fact was incorporated directly into the electronics that gathered the data from the two radio links, and so it disappeared from the experiment altogether.

Shortly after launch in the morning of June 18, 1976, the payload containing the clock separated from the fourth stage of the rocket, and was in free fall thereafter. At this point data could be taken, since the rocket clock was no longer affected by the high accelerations and vibrations of launch. For about three minutes, the one-way "downlink" frequency from the rocket clock (with the Doppler piece cancelled automatically, remember) was lower than that of the ground clock, because the high velocity of the rocket caused a time-dilation redshift to lower frequencies, while the altitude was not yet large enough to produce a gravitational blueshift. Shortly after that, the frequencies of rocket and ground clock were exactly the same: the gravitational blueshift cancelled the time dilation redshift. After that, as the altitude increased and the speed of the rocket decreased, the gravitational blueshift dominated more and more. At the peak of the orbit, the shift was predominantly the gravitational blue shift, amounting to almost one Hz out of 1420 megahertz, or 4 parts in 10 billion. Since both the rocket clock (after separation from the fourth stage of the Scout) and the ground clock maintained their intrinsic frequencies stably to a part in a million billion, these changes in frequency could be measured to very high accuracy. Data-taking continued during descent, with the cancellation between gravitational blueshift and time dilation occurring again, shortly before the rocket landed in the Atlantic Ocean, some 900 miles east of Bermuda.

This two-hour flight produced more than two years of data analysis, but when all was said and done, the predicted frequency shifts (gravitational plus time dilation) based on the speed and location of the rocket known at all times through tracking, agreed with the observed shifts to a precision of 70 parts per million, or to 7/1000 of one percent.

By now we should have great confidence that curved space time is right. We next want to ask ourselves which theory of curved space-time is the correct theory? Is it Einstein's, or is it some other? This brings us to experiments that compare the predictions of different theories.

V. THE DEFLECTION OF LIGHT

The first test, one that made Einstein's name a household word in the 1920s, is the deflection of light. It turns out that the question of whether gravity affects light actually has a long history.

We believe the first person to really think about the idea that gravity might affect light was a British amateur astronomer named John Michell, around 1783. If you have a body of some uniform density, you can calculate the escape velocity from its surface using standard Newtonian gravitation theory, and it is given by the standard formula. Michell asked the question: for that same density, how big does the body have to be so that the escape velocity just exceeds that of light? In such a case, the light reaches some maximum altitude, and then turns around and falls back down.

Michell did not know special relativity, which demands a constant speed of light. He did this calculation based on standard Newtonian theory, plus the corpuscular theory of light, which was in vogue at that time. For a body with the density of the Sun, he found that the radius needed is 500 times the solar radius. The mass of such an object would be 1.25×10^8 solar masses. Today we recognize such an object as a supermassive black hole. A similar calculation was done about 15 years later by Laplace, the famous French mathematician.

A few years later, around 1803, a German astronomer named Johann Georg von Soldner asked, not whether light would be stopped, but whether light's trajectory would be bent by gravity. Von Soldner also assumed standard Newtonian gravity and the corpuscular theory of light.

The idea is to calculate the orbit of a body of velocity v in an unbound orbit, and to let the velocity go to the speed of light. The result is that for a ray that just grazes the surface of the Sun, the deflection is 0.875 arcseconds. This was a hundred years before Einstein.

In 1911, armed with his understanding of the principle of equivalence, Einstein derived the deflection of light, and got the same answer. It is quite straightforward to derive this deflection of light using just the principle of equivalence.

Imagine a pair of laboratories with glass sides, each containing an observer well versed in the equivalence principle. The laboratories are perfectly constructed with parallel faces and right angles at all corners (Fig. 7). They are connected by a frictionless trolley that keeps them parallel to each other, but allows them to move independently and freely in the parallel direction. With a blast of rockets, the laboratories take off from the Sun in a radial outward direction on their way to intercept a light signal that is about to pass by the Sun. After the rockets turn off, the laboratories are in free fall, and begin to slow down under the gravitational pull of the Sun. Their initial velocities have been cunningly chosen (remember, this is a thought experiment!) so that, at the moment the light ray crosses the left pane of Laboratory #1, that laboratory happens to be at rest with respect to the Sun, just about to begin its backward fall. The second laboratory's velocity has also been cleverly chosen so that at the moment the light ray leaves laboratory #1 and crosses the left pane of #2, the latter is at rest with respect to the Sun and about to fall back. Now, in laboratory #1, the observer notes that the light ray enters his window at a certain location and at a certain angle. The ray then crosses the laboratory and exits through the right pane at exactly the same angle. This is completely natural according to observer #1, since he is in a freely falling frame, in which gravity is apparently absent. Light must travel in a straight line in such a frame, and so must subtend equal angles on both sides of the laboratory. Observer #2 comes to a similar conclusion. She observes the ray enter her laboratory at some angle and leave at the same angle, and she too acknowledges that this is in accord with the equivalence principle. However, when the observers return to base and compare their data, they find that they disagree on the angle the ray subtended in each of their laboratories. In laboratory #2, the observed angle was more in a downward direction (toward the Sun) than in #1. After a little thought, they understand why. Although each laboratory was at rest with respect to the Sun at the moment the light ray entered it, the two moments were not the same. By the time the ray crossed #1 and entered #2, laboratory #1 had begun to fall, and had picked up a velocity given by the acceleration of gravity times the time taken for light to cross it. Thus observer #2 saw light enter her laboratory from a laboratory that was moving downward relative to her. Thus, the angle of entry was deflected downward, by the phenomenon of "aberration." The upshot is that observer #2 claims that the light ray has been deflected slightly towards the Sun.

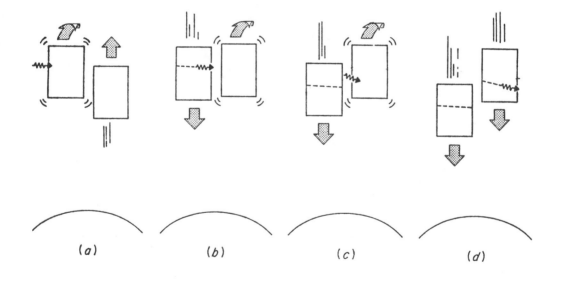

Figure 7. Equivalence principle and the deflection of light. A pair of laboratories is shot out from the center of the Sun, the rockets are turned off, and they are in free fall. (a) Laboratory #1 comes to rest momentarily at the top of its trajectory at the moment the light packet enters it. Laboratory #2 is still ascending. (b) Since laboratory #1 is in free fall, light moves on a straight line because of the equivalence principle, and leaves it at the same angle with which it entered it. By that time, laboratory #1 has started to fall, while laboratory #2 has reached the top of its trajectory. (c) The light packet enters laboratory #2, but since laboratory #1 is falling relative to #2, the observed angle of entry is "downward" relative to that measured in #1, because of aberration. (d) Light moves on a straight line through laboratory #2 because it is in free fall. When these incremental deflections are added up for a sequence of laboratories all along the light path, the result is 0.875 arcseconds for a grazing ray.

By considering a sequence of such laboratory experiments performed all along the trajectory of the light ray and adding up all the tiny deflections, the observers conclude that the net deflection of a ray that just grazes the Sun would be 0.875 arcseconds. So, just from the principle of equivalence and a simple calculation, or from Newtonian gravity and the corpuscular theory, you get the same deflection of light.

But in 1915, Einstein doubled the deflection, making it 1.75 arcseconds for a grazing ray. The doubling comes from the general theory of relativity; it cannot be obtained from the principle of equivalence alone.

Where does the rest of the effect come from? It comes from the curvature of space. As we have already seen, the principle of equivalence tells us that spacetime must be curved, but it does not tell us in detail by how much it is curved, and in particular, it says nothing about the curvature of space itself. The curvature of space modifies the result in the following way. Imagine a large number of small, perfectly straight rulers with ends cut at perfect right angles. Take about a third of these rulers and line them up end to end, each parallel to its neighbor, stretching across a large region of empty space far from the Sun. This is the line BO in Fig. 8. Now take another third of the rulers, and line them up end to end starting from point O again staying far from the Sun, but in such a direction that the two free ends of the lines of rulers end up on opposite sides of the Sun. Take most of the remaining rulers, and line them up end to end in the same way, starting at B, so that they go just past the surface of the Sun, and on to point A to complete a large triangle. Two sides of the triangle (OA and B) are far from the Sun, while the third side (AB) grazes its surface. With the remaining rulers, start at a point B′ setting off at the same angle as you set off at B, and complete the triangle OA′B′.

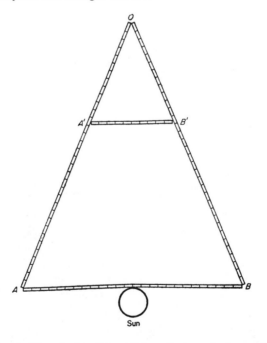

Figure 8. Solar system triangle. Line of rulers BO and OA are far from the Sun, while line of rulers AB grazes the Sun. The line of rulers A′B′ is far from the Sun, and the angle OB′A′ is the same as the angle OBA. The sum of the interior angles of OA′B′ is 180 degrees. The sum of the interior angles of OAB is 179 degrees, 59 minutes, 59.125 seconds. Thus angle OAB is 0.875 arcseconds smaller than angle OA′B′.

189

General relativity predicts that space is curved near gravitating bodies, the curvature being greater the closer one gets to the body, and negligible at large distances. Since the triangle OA'B' is far from the Sun, the sum of its interior angles is 180 degrees, just as in ordinary flat space. But a detailed calculation using the equations of general relativity reveals that the sum of the interior angles of the triangle OAB is no longer 180 degrees (remember the Spherelanders) but is 179 degrees, 59 minutes, 59.125 seconds, or 180 degrees less 0.875 arcseconds!

There is a 0.875 arcsecond difference in the sum of these interior angles. That is where the other half of Einstein's deflection came from. The deflection we calculated before using the principle of equivalence was determined relative to a string of parallel-aligned laboratories, that is, relative to a string of rulers. We obtained 0.875 arcseconds. But the rulers themselves are bent, relative to rulers at infinity, by an additional 0.875 arcseconds. Thus the total deflection of light relative to a straight line near infinity is 1.75 arcseconds.

Indeed, this part, due to the effect of curvature on rulers, is the part that differs from one theory to another. For instance, the Brans-Dicke theory gives the same equivalence-principle part but predicts a different amount of curvature of rulers, so the net deflection is slightly different.

The first measurements of the deflection of light were made during the total solar eclipse of May 29, 1919. Two teams of British astronomers made the observations, one headed by Sir Arthur Stanley Eddington, located on the island of Principe, off the coast of Spanish Guinea, the other headed by Andrew Crommelin, located in the city of Sobral, in northern Brazil. During a total eclipse, the stars in the field of the Sun are revealed in the darkened sky, and photographic plates can be exposed. The positions of the stars on these plates can then be compared with plates of the same star field taken at another time of the year, when the Sun is not in that part of the sky.

The idea of course is that since the deflection depends inversely on the distance from the Sun, stars far from the Sun will hardly be deflected at all. Stars close to the Sun will be deflected. By comparing the relative locations, you can measure the deflection.

The results of these expeditions in 1919 confirmed general relativity, with quoted errors of around ten percent, and helped make Einstein a famous person. During the next 40 years, attempts were made to repeat these measurements, during eclipses in the 1920s, in 1936, 1947, 1952, and even one as late as 1973, but there were only minor improvements in accuracy. The best accuracy achieved was about 10 percent.

By the middle 1960s, however, a new technique came along to measure the deflection of light much more accurately than could be done by optical measurements during solar eclipses. This was the technique of radio interferometry, in which an array of radio telescopes measures the signal from some distant radio source, looks at the interference between the signals received at the different telescopes, and uses that to determine the direction to the source, with very high precision.

The radio sources that were used in these experiments were exactly the sources whose discovery help bring about the relativity revival: the quasars.

Furthermore, there is a heavenly coincidence: every October, two quasars, 3C273 and 3C279, pass very close to the Sun, as seen from Earth; in fact, 3C279 goes behind the Sun. The power of radio interferometry is that it is most adept at measuring the angle between two such radio sources, with even more accuracy than it can measure their absolute positions on the sky.

Over a period of ten days centered on October 8, the pair of quasars passes from one side of the Sun to the other, and each day the angle between the two can be measured on the radio interferometer. What would be the expected output of the instrument? Initially, when both quasars are far from the Sun, the angle measured would be the true, undeflected angle between them (Fig. 9). As days pass, the pair approaches the Sun, with 3C279 much closer than 3C273. Since the amount of deflection varies inversely as the distance of the light path from the Sun, the radio

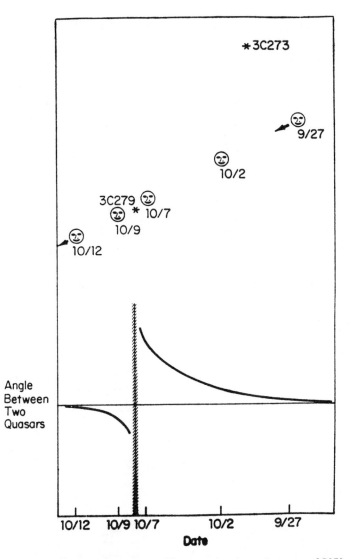

Figure 9. Quasar light deflection measurements. Upper portion shows the quasars 3C273 and 3C279 and the apparent path of the Sun between September 27 and October 12 each year. The Sun actually passes in front of 3C279. Lower portion shows the apparent angle between them that would be measured by a radio interferometer. September 27: very little change from the normal undeflected angle, since the Sun is not close to either quasar. October 2: small increase in angle, since both apparent quasar positions are deflected slightly away from the Sun. October 7: large increase in angle since the 3C279 position is deflected a large amount away from the Sun. October 8: no data because 3C279 is behind the Sun. October 9: modest decrease in angle since the 3C279 position is deflected away from the Sun, i.e., toward 3C273. October 12: return to nominal undeflected angle.

waves from 3C279 are deflected more than those from 3C273, and so the angle between the two will appear to increase. As 3C279 just grazes the Sun, its light is defected by about 1.75 arcseconds, while 3C273, 9 degrees away and therefore 35 times farther from the Sun, is deflected only 0.05 arcseconds. As days continue to pass, the pair moves further from the Sun and their separation returns to the original value. By careful analysis of the behavior of the separation angle with time, the radio astronomers can determine the deflection of light as a function of distance from the Sun and translate that into a deflection of a grazing ray. One of the advantages of this technique is that it can be done every year, as opposed to eclipse measurements, which must occur sporadically and in inhospitable locales. Here, each October, as in some ancient harvest-time Druid ritual, the radio astronomers could be imagined marching out to the interferometer to measure the bending of light.

Between 1969 and 1975, almost annually, a series of experiments was done to measure the deflection of radio waves from these quasars, and, in some later experiments, from a trio of quasars that pass by the Sun in April. Their results confirmed the general relativistic deflection of light to about 1 1/2 percent. So modern technological advances in radio astronomy, plus the existence of quasars, gave a way to measure the bending of light much more accurately than the old eclipse expeditions.

VI. THE PERIHELION SHIFT OF MERCURY: TRIUMPH OR TROUBLE?

The Newtonian theory of gravity is one of the triumphs of human thought. The Newtonian gravitational theory explained the gross motions of the planets and of the Moon. Furthermore, one could take into account the perturbations of the planets on each other, and understand the detailed motion of the planets in perfect agreement with observation, as early as the late 18th century, and well into the 19th century. But there was one problem.

By the mid-19th century, it was discovered by LeVerrier that there was a difficulty with Mercury. If you look at Mercury's motion, you find that it does not move on a closed ellipse as ordinary Newtonian two-body theory would predict: instead, the ellipse precesses at a rate of 574 arcseconds per century (Fig. 10). It was well known that the perturbations of Mercury's motion due to the other planets would lead to a perihelion advance. Venus, since it is the closest to Mercury, contributes the most to this effect, 277 arcseconds. Earth contributes 90. Jupiter, fur-

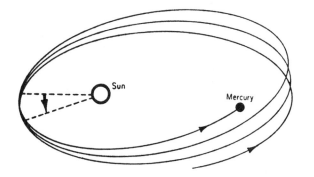

Figure 10. Perihelion shift of Mercury. Elliptical orbit rotates in the plane, so that the point of closest approach, the perihelion, rotates.

ther away, but more massive, contributes 153, and all the other planets combined contribute about 10. When you add all these up, you find only 531 arcseconds. The discrepancy is about 43 arcseconds per century, and as late as the turn of the 20th century, there was no explanation for it. Try as they might, celestial mechanicians could not account for this discrepancy. The modern value for this discrepancy is 43.11 ± 0.21 arcseconds per century, based on improved measurements of Mercury's motion plus our improved understanding of the motions of all the planets, mainly coming from the highly accurate radar tracking that evolved from those early Venus radar measurements.

With the equations of general relativity in hand, one of the first things Einstein calculated after the deflection of light was the perihelion shift of Mercury. The prediction was 43 arcseconds, in complete agreement with the observations. This was considered a great triumph for general relativity (the accurate, modern value of the prediction is 42.98 arcseconds).

However, around the middle 1960s, questions were raised about the perihelion shift of Mercury, because of the possibility, based on observations by Dicke and his student, H. Mark Goldenberg, that the Sun might be oblate or flattened. Now if the Sun is oblate, the flattening of the mass distribution of the Sun will produce a $1/r^3$ gravitational potential, in addition to the standard $1/r$ potential, that you get for a spherical Sun. This additional $1/r^3$ potential will generate its own contribution to the perihelion shift, and if you believe Dicke's observations for the amount of flattening, the contribution would be 3 arcseconds per century. That would clearly knock down general relativity: There is no room for a 3 arcsecond difference. At the time, coincidentally, this observation agreed with the prediction of the Brans-Dicke theory, which at that time predicted a relativistic contribution of 40 arcseconds.

Ironically, the solar oblateness is an unresolved problem to this day. Later measurements of the solar oblateness, using Dicke's idea of making optical observations of the shape of the Sun led to smaller values of the solar oblateness. Other values for the flattening of the Sun, based on how the Sun oscillates, were obtained in the early 1980s. These led to conflicting values for the solar oblateness, some large and dangerous for general relativity, some very small and consistent with general relativity. At the moment this is an open question. No one really understands the structure of the Sun well enough to know conclusively whether there is this extra oblateness in an amount to be dangerous for general relativity.

Now, we expect the Sun to have a small oblateness because it is rotating and it should be centrifugally flattened, but with a standard solar model, the oblateness is sufficiently small to be no problem.

One way to settle this question would be through a space mission, that once was under consideration by NASA, called Starprobe. Starprobe is a spacecraft that would take off from the Earth and go very close to the Sun, within four solar radii. By exploring the gravitational field close to the Sun, it could measure directly the $1/r^3$ part of the solar gravitational potential contributed by an oblateness, and thereby measure directly how much oblateness there is. Despite its importance however, this mission is not under serious consideration at the moment.

VII. THE TIME DELAY OF LIGHT: BETTER LATE THAN NEVER

The next test that I want to describe is one that Einstein never invented. It was discovered in 1964 by the radio astronomer Irwin Shapiro, and it is called the time delay of light. It is a retardation of light, an apparent slowing down of light as it passes by a gravitating body.

I will have to make precise what I mean by "slowing down" because we know from the equivalence principle and special relativity that the speed of light is always the same in a local inertial frame. Just as in the case of the bending of light, to illustrate how this effect works we have to do two separate calculations. One part we can get right from the principle of equivalence; the other part will come from the curvature of space.

Consider a light ray sent from one body at a particular moment of time toward a target on the far side of the solar system, in the orientation called superior conjunction. Assume the ray grazes the Sun. Again, consider a set of freely falling laboratories, all of the same width, as measured by their occupants, all cunningly shot out from the center of the Sun in sequence, so that each one reaches the top of its trajectory just at the moment the light ray enters it. Each observer has two flash cubes and an atomic clock. One flash cube is triggered to go off at the moment the ray enters the laboratory, the other at the moment the ray leaves the laboratory, each one powerful enough that a distant observer, far from the solar system, can pick up both flashes. Each observer uses his atomic clock to determine how long it took for the light ray to pass through his laboratory, and using his measured value for the width of the laboratory, he calculates the speed of light. Every observer obtains exactly the same value for this speed. This is the principle of equivalence in action. On the other hand, the distant observer receives the pair of light flashes from each laboratory, and using an atomic clock identical to the ones used in the laboratories, he measures the time interval between each flash. For those observers in laboratories which encounter the light ray far from the Sun, the interval between flashes measured by the distant observer agrees with the interval of time measured by the laboratory observers (we will ignore the fact that the laboratories fall a short distance during the passage of the light signal through them; although that effect was important for the deflection of light, it is a negligible effect here). However as the distant observer receives flashes from laboratories that are closer and closer to the Sun, he notices that his measured interval between flashes is slightly longer than that recorded in each laboratory. By now we know why. It is simply the effect of the gravitational redshift; the interval between flashes in the elevators appears longer as received at great distances than it does locally. By simply adding up all the measured time intervals from all the observers along the light path from the emitter to the target and back to the emitter, the observer can determine the round-trip travel time. The result is a delay of 125 microseconds for a round trip to Mars at superior conjunction. This delay is a consequence of the equivalence principle and hence of the gravitational redshift. The trouble is, that is only half the total effect! What went wrong?

Just as in the case of the light deflection, nothing went wrong. We're just not done. Just as in the light deflection, we have to take space curvature into account.

To do this, we imagine the following experiment. An observer equipped with a large number of rulers of equal length and with perfectly square corners sets out to build a solar-system sized rectangle, in the following manner. Starting from the emitter of the radar signal (the Earth, say), he sets rulers end to end, one set along the path of the radar signal heading for the target (the side ET in Fig. 11), and one set perpendicular to the radar path (EA). The first set of rulers is extended until it reaches the target. The second, perpendicular set is extended a large but basically arbitrary distance until its end A is far from the Sun. A third set of rulers AB is laid out perpendicular to this set in such a way that it parallels the set of rulers sent along the light path. Finally, a fourth set of rulers is sent out from the end of the third set, perpendicular to it, and parallel to the second set of rulers. The fourth set of rulers is extended until it meets the target, forming a gigantic rectangle EABT. (Practically speaking, of course, this may require a few trials, because the third side AB may initially be too long or too short to allow the fourth set of rulers to meet the target. The advantage of thought experiments is that such practical problems are never insurmountable!) The observer notes two things. First, since sides EA, AB, and BT of the rectangle were all constructed far from the Sun, where spacetime is essentially flat, or at least flat enough for the accuracy required, all his notions of "parallel," "perpendicular," "straight" are valid in the usual Euclidean sense. Second, he is well aware that the radar signal path ET is deflected slightly as it passes the Sun, so that the distance from Earth to target might be a bit different along a deflected path than along a straight path. However, a simple calculation convinces him that this effect is completely negligible. His goal now is to compare the number of rulers used in side ET with the number used in side AB, in other words to compare the actual length of the two sides. According to Euclidean geometry, in a plane figure such as EABT, with right angles at each of the corners, the opposite sides should be equal in length. But instead, he finds that the side that passed by the Sun is slightly longer (took more rulers) than the side that stayed far from the Sun. For a rectangle whose side ET extends from Earth to Mars and just

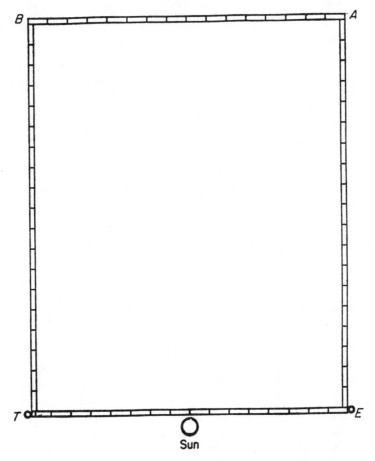

Figure 11. Solar system rectangle. Line of rulers ET follows radar path from Earth to target, grazing the Sun. Line of rulers EA is perpendicular to ET, line AB is perpendicular to EA, and BT is perpendicular to AB. Lines EA, AB and BT are all far from the Sun. Deflection of line ET can be ignored. The number of rulers in line ET is found to be more than those in AB.

grazes the Sun, he finds that the extra length of the side is 19 km. For a radar signal that covers this extra distance twice (once out and once on the return) the added delay in the round trip travel time would be 125 microseconds. This is the other half of the predicted delay! It is a direct result of the curvature of space near the Sun. It is not due to the bending of the light path, which causes a negligible delay of less than a hundredth of a microsecond.

One useful way to visualize the delay caused by space curvature is to take the two-dimensional plane formed by the light path and the Sun, and to embed it in a fictitious three-dimensional space in such a way that the added distance measured by rulers near the Sun compared to those far from the Sun can be seen using Euclidean intuition (Fig. 12). This is done by stretching the plane into the fictitious third dimension in the vicinity of the Sun. The picture that emerges is that of a rubber sheet with a heavy ball in the center, causing a depression. Earth and the target sit on this sheet (they too cause depressions, but they are too small to worry about). The path of the light ray is also confined to this sheet. When the Earth and the target are positioned so that the light path never passes close to the Sun, the time taken for the round trip is just what you would expect using the Euclidean distance and the speed of light. However, when the Earth and the target are at superior conjunction, and the light ray passes near the Sun, it must follow the contour of the sheet, and in moving into the depression and back out, it covers a greater distance, and is therefore delayed. Again, this picture gives only the curvature part of the time delay.

Just as in the deflection of light, this contribution of space curvature is the part that can vary from one gravitation theory to another. Any theory of gravity that is compatible with the equivalence principle predicts the first 125 microsecond part for an Earth-Mars experiment. The second part comes from space curvature, and it is purely a coincidence that general relativity predicts the same contribution from the two phenomena. The Brans-Dicke theory predicts slightly less curvature than general relativity, and so a slightly smaller time delay.

How can this delay be detected in practice, in other words how can we take a radar measurement to a planet at superior conjunction and tell what part of it is Shapiro time delay and what part of it is due simply to an unforeseen change in true distance between the planet and Earth?

The answer is simple in principle, though complicated in practice. Even though the radar signal may go near the Sun, the planet itself never does. Its orbit is well away from the Sun, on the order of 300 million km for Mars, for instance. Because of this, it always moves through a region of low spacetime curvature, and maintains a relatively low velocity. Therefore, the relativistic effects on its orbit are small. To the accuracy desired for a time delay measurement, its orbit can be described quite adequately by standard Newtonian gravitational theory. Therefore, even though the planet moves during the experiment, its motion can be predicted accurately. Because of this circumstance, the time delay can be measured in four steps. Step 1: by ranging to the planet for a period of time when the signal stays far from the Sun, determine the parameters that describes its orbit at that time. Step 2: using the orbit equations of Newtonian theory, including the perturbations of all the other planets, make a prediction of its future orbit and that of the Earth, including especially the period of superior conjunction. Step 3: using the predicted orbit, calculate the "naive" round trip travel times of signals to the planet assuming no Shapiro time delay. Step 4: compare the predicted "naive" round trip travel times with those actually observed during superior conjunction, attribute the difference to the Shapiro time delay, and see how well it agrees with the prediction of general relativity.

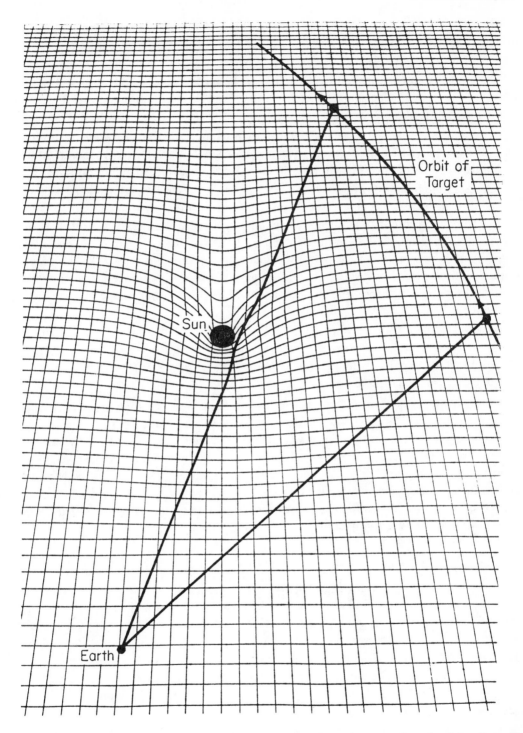

Figure 12. "Rubber sheet" picture of space curvature. Surface of the sheet is the two-dimensional plane formed by the light path and the Sun. Added distance near the Sun caused by curvature can be mocked up by stretching the sheet in a fictitious third dimension. Light path far from the Sun from Earth to target traverses "flat" part of sheet, so Euclidean intuition holds. When light path passes near the Sun, it must "dip into" the depression, and thus has more distance to cover.

197

The first measurement of this delay was in fact done by Irwin Shapiro. Soon after he discovered the time delay prediction, Shapiro and his colleagues set out to measure it experimentally using the Haystack Radar Antenna in Western Massachusetts.

These measurements, made in 1966 and 1967, bounced signals off Mercury and Venus and looked for the excess delay as the planets swung behind the Sun and found that it agreed with general relativity to about 20 percent.

Since that time there have been significant improvements in the measurement of this time-delay effect, achieved using spacecraft. The first spacecraft put to this purpose were Mariner 6 and Mariner 7, which visited Mars in 1969 and then went into orbit around the Sun.

Between March and May of 1970, the two spacecraft passed through superior conjunction, and the time delay measurements were found to agree with the general relativity prediction to about 2 percent. Another time-delay experiment was done using the Mariner 9 Mars orbiter, and validated general relativity to the one percent level. The best experiment to date was done using Viking, a mission that is usually associated with scooping dirt out of the Martian soil, or taking beautiful pictures of Mars, not with relativity. But in fact, range measurements were made to the two orbiters and the two landers as Mars went through superior conjunction in November of 1976.

Because of the very precise tracking of Viking, aided by the addition of high-frequency X-band to the standard S-band radar to help account for the effects of the solar corona on the propagation of the signals, the time-delay effect was measured by Viking to a precision of 0.1 percent, or to one part in a thousand.

Viking represents the most accurate confirmation of general relativity we have to date. The Viking time delay results also dealt a telling blow to the fortunes of the Brans-Dicke theory. One of the important features of this theory was that its predictions depended on the value of an adjustible parameter called ω. The larger the value of ω, the closer the Brans-Dicke predictions became to those of general relativity, and in the formal limit as ω tends towards infinity, the two theories become the same. The Viking results placed a lower limit of about 500 on the value of ω, making the theory virtually indistinguishable from general relativity in its predictions. Because general relativity is a simpler theory, it is therefore favored.

VIII. DO THE EARTH AND THE MOON FALL THE SAME?

There is another important experimental test of general relativity, that also was not envisioned by Einstein. It was discovered in 1967 by a young theorist named Kenneth Nordtvedt, who, at the time, was an assistant professor at Montana State University. Sitting in the wilds of the northern United States, pondering the mountains, he asked the question, do the Earth and Moon fall with the same acceleration?

It seems like a trivial question. We have already talked about how balls of platinum and aluminum fall with the same acceleration. Why shouldn't the Earth and Moon fall with the same acceleration?

Now, the laboratory Eötvös experiments that I discussed previously indicate that the nuclear and electromagnetic contributions to the energy of bodies all respond to an external gravitational field in the same way, since bodies of different structure or composition always fall with the same acceleration, regardless of whether one body might contain relatively more electromagnetic energy per unit mass than another body. It was this universality, or independence of the gravitational acceleration from the nature of the body, that allowed us to interpret gravity as a manifestation of curved spacetime, rather than as a phenomenon connected with particular objects. But what about gravity itself? Could it not be that, perhaps because of some non-linear gravity-gravity interaction between the external gravitational field and the internal gravitational field, a body that was tightly bound gravitationally might fall differently in a given external field than a body that was less tightly bound? For example, a white dwarf star of the same mass as the Sun has a diameter of only a few thousand kilometers compared to the Sun's 700,000 km, and because it is so compactified by gravity, its gravitational binding energy is about a thousand times that of the Sun. If one held a white dwarf and the Sun side by side in some external gravitational field, would they fall with the same acceleration?

To try to answer this question, Nordtvedt did a calculation using a mathematical framework that encompassed a broad class of theories of gravity, including general relativity and the Brans-Dicke theory. He found that in general relativity, they should fall at the same acceleration. This is one of the remarkable products of the elegance and simplicity of general relativity. Not only do platinum and aluminum fall at the same acceleration, but so also do the Moon, the Earth, even, as it was found later, neutron stars, and black holes. In general relativity, all bodies fall with the same acceleration in some distant gravitational field, regardless of their structure or composition, whether they are gravitationally bound or not.

However, the Brans-Dicke theory, among others, predicted a slight difference in the rate of fall between the Earth and Moon because they have different fractions of their mass in the form of internal gravitational binding energy. The Earth's fraction is about 25 times that of the Moon.

What would be the observable consequence of such an effect? Suppose the Moon falls with a higher acceleration towards the Sun than the Earth does, for example. In each orbit the Moon is pulled a little closer to the Sun. But by the time the Moon goes around one orbit, the Sun has moved about 27 degrees in the sky. So the Moon now has to be pulled toward the new direction of the Sun. The result is a displacement of the Moon's orbit, that is shifted slightly toward the Sun (Fig. 13). This occurs only in theories such as the Brans-Dicke theory where there is a difference in acceleration. In general relativity this effect does not occur. In Brans-Dicke theory, the size of the displacement could be as high as 100 centimeters.

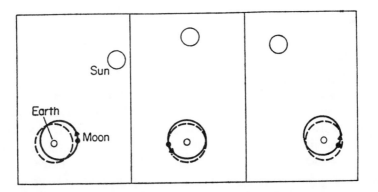

Figure 13. The Nordtvedt effect: If Moon falls with larger acceleration than Earth toward the Sun, its orbit develops a displacement or elongation which points toward the Sun. In general relativity, the elongation does not occur at all.

As minuscule as such an effect on the lunar orbit might seem, it can be checked using one of the advances in technology of the '60s and '70s, the laser, together with specially designed mirrors called retroreflectors, that were deposited on the Moon by the Apollo astronauts and by unmanned Soviet lunar landers. These reflectors take a laser signal sent from Earth, and reflect it back, exactly in the same direction from which it came. The round-trip travel time of a laser shot to these reflectors can be measured to an accuracy that translates into a distance accuracy of 150 millimeters. Of the key goals of the Lunar Laser ranging experiment was to look for this displacement of the lunar orbit, now called the Nordtvedt effect. From data taken between 1969 and 1975, the result was no evidence for this effect down to the precision of 150 millimeters. Brans-Dicke theory predicted upwards of a meter for the effect, so this experiment went against Brans-Dicke theory, and was a confirmation of general relativity's prediction that the Earth and the Moon fall with the same acceleration. Another way of stating the result is to say that the Earth and Moon fall with the same acceleration to better than 7 parts in 10^{12}.

IX. THE BINARY PULSAR: GRAVITY WAVES EXIST!

There is a remarkable test of a prediction of general relativity that really set the world of relativity on fire in the middle 1970s. This test involves a system called the binary pulsar, and has verified the existence of gravity waves.

To understand the significance of this, we must first learn something about gravitational waves, and the simplest way to do this is to compare and contrast them with electromagnetic waves. We all know what electromagnetic waves are. They are produced by moving charges or currents and propagate through space with the speed of light. When they hit an antenna, they move charges around to produce an electric current. Electromagnetic waves have a very old history; they were discovered by Hertz in the late 19th century, and of course, have countless practical uses.

There are some similarities between electromagnetic waves and gravitational waves. Gravitational waves are produced, not by moving charges, but by moving matter. A binary system will emit gravitational waves because of the moving matter. A black hole swallowing a star will emit gravitational waves. A collapse of a non-spherical object to form a black hole will emit gravitational waves. Gravitational waves also move with the speed of light. When they reach some detector, they do not make charges move, but they make masses move. For two masses on opposite ends of a spring (a simplified model of a gravity-wave detector), perpendicular to the direction of motion of the waves, a typical wave will make the masses move together and apart.

However, there is one key difference between gravitational waves and electromagnetic waves, and that is their strength. The strongest source that we can imagine at the moment for gravitational waves is a stellar collapse associated with a supernova in our galaxy, in which a black hole is formed. Such an event will emit gravitational waves whose consequence will be to change the distance between two masses separated by one meter by 1/1000 the diameter of an atomic nucleus. So the effects of gravitational waves are tiny, and extremely difficult to detect. Nevertheless, there is a world-wide effort to build gravitational-wave detectors of sufficient sensitivity to detect even weaker events. Ultimately, we believe that, once gravitational waves are detected, they will be a new tool for astronomy, but it will probably be many years before they are actually detected.

Yet we know that gravitational waves exist, not just because general relativity predicts them, but because of the binary pulsar. The binary pulsar was discovered in 1974 by Russell Hulse and Joe Taylor, who were working at the Arecibo radio telescope in Puerto Rico in a systematic search for new pulsars. The pulsar they discovered, denoted PSR 1913+16, proved to be a member of a close binary system with an as yet unseen companion. This would have been only a mild curiosity (of 300 radio pulsars, only a few are known to be in binary systems), were it not for two important properties of the system. The orbit is so close, with an average separation on the order of a solar radius, orbital velocities up to 300 km/s, and an orbital period of only eight hours, that relativistic effects such as the periastron shift (the binary-system analogue of Mercury's perihelion shift) can be significant. Futhermore, the pulsar appears to be one of the most stable clocks in the universe: Its pulse period of 59 ms drifts by only a quarter of a nanosecond, or four parts in 10^9, per day (only the millisecond pulsar PSR1937+214, discovered in 1982, is more stable).

These two circumstances made the binary pulsar an exciting new laboratory outside the solar system for studying relativistic effects. By measuring arrival times of individual groups of radio pulses at Earth to accuracies of $50\,\mu$s, the observers were able to determine the motion of the pulsar about its invisible companion (the system is a special kind of what astronomers call a "single-line spectroscopic binary"), and thereby measure many of the important orbital elements with an accuracy that boggles the mind. For instance the intrinsic pulse period (referred to September 1, 1974) is 0.059029995271 s, the rate of change of the pulse period is 0.273 ns per year, the eccentricity of the orbit is 0.617127, the orbital period is 27,906.98163 s, and so on.

This accuracy made possible the measurement of several relativistic effects. The first was the periastron shift: the measured value was 4.2263 degrees per year (compare with Mercury's 43 arcseconds per century!). According to general relativity, the predicted shift depends on known orbital parameters such as the period and the eccentricity, and on the total mass of the system, which is unknown. We are assuming here that the companion does not have a significant oblateness, in order not to generate an additional contribution to the periastron shift like that produced by an oblateness of the Sun. This assumption will be valid if, as seems likely, the companion is a neutron star, a black hole, or a slowly rotating white dwarf. Now, the observation of the periastron shift cannot be used to test general relativity because of the unknown total mass. Instead, the tables can be turned, and general relativity can be used as a tool to determine the total mass of the sysem. The result turned out to be 2.8275 solar masses.

Another relativistic effect that was observed, although with more difficulty, was the combined effect of the gravitational redshift of the pulsar signal, due to the companion's gravitational potential, and of the special relativistic time dilation due to the pulsar's orbital motion. The observed effect was a periodic variation in the arrival times of pulses, with an amplitude of 4.38 ms. By comparing this observation with the prediction, one obtains another, independent piece of information about the masses, which makes it possible to determine them separately. The results, in solar masses, were 1.40 ± 0.03 for the pulsar and 1.42 ± 0.03 for the companion. For the first time general relativity was used as a direct tool for an astrophysical measurement—the weighing of a pulsar!

What does this have to do with gravitational waves? General relativity predicts that a binary system will emit gravitational radiation. Like electromagnetic radiation, gravitational radiation carries energy away from the system, so the system should lose energy. Because the binary system loses energy, the two bodies should spiral in toward each other. The orbital period should then decrease as they get closer and closer together.

General relativity makes a definite prediction for the decrease of the orbital period due to gravitational radiation loss. Using the measured values for the orbital parameters and for the two masses, we find that the predicted decrease rate is 75 microseconds per year, or more precisely, $(2.403 \pm 0.002) \times 10^{-12}$ seconds per second. Shortly after the discovery of the binary pulsar in 1974, it was thought that measurement of such a small effect would require 10 to 15 years of data, but through brilliant efforts to improve electronic data acquisition techniques at the telescope and to refine the data analysis, Taylor and his team were able to do it in just over four years, in time to open the 1979 Einstein centenary year. Their initial result, announced in December 1978, agreed with the prediction with 20 percent uncertainties, but subsequent data to 1983 have led to the improved value of 76 ± 2 microseconds per year, or $(2.40 \pm 0.09) \times 10^{-12}$ seconds per second, in complete agreement with the prediction of general relativity. No other plausible source of orbital period decrease has been proposed which could account for all or part of the observed decrease. The recent, post-1980 data on the binary pulsar has gotten so good that it has been possible to detect even smaller relativistic effects, such as the Shapiro time delay of the pulsar signal as it passes by the companion, the first time that this effect has been seen outside the solar system.

X. THE FRONTIERS OF EXPERIMENTAL GRAVITATION

Despite the success of general relativity in confronting the experiments described in the previous sections, the subject of experimental gravitation is far from being a closed book. Work continues to improve many of the measurements, for example by further analysis of Viking radar data to improve the determinations the time delay. Other experiments are underway or are planned that will measure effects that have not been seen before.

One of these is the Stanford Relativity Gyroscope Experiment (also called GPB, for Gravity Probe B), that has been under development since Leonard Schiff in 1960 proposed a new test of general relativity using gyroscopes. The goal of the experiment is to measure the precessions of a set of orbiting gyroscopes that result from two effects, the curvature of space around the Earth (net effect $\sim 7''$ per year) and the "dragging of inertial frames" by the rotation of the Earth (net effect $\sim 0.''05$ per year). The gyroscopes are 40 mm diameter quartz spheres coated with a layer of superconducting niobium; at liquid Helium temperatures the sphere develops a magnetic moment parallel to its spin axis whose direction can then be determined by super-precise magnetometers. The precession of the gyroscope axes will be measured relative to the optical axis of a telescope fixed on a distant star (Rigel). The entire system will be in a drag-compensated satellite. Current plans call for a proving flight to test the components on a 1991 Space Shuttle mission. If all goes as planned, a science mission could follow in a few years.

Recently, there has been a renewal of interest in the principle of equivalence and the Eötvös experiment as a consequence of a reanalysis of the original Eötvös data by Ephraim Fischbach and colleagues. One of the goals of that re-analysis was to search for the effects of a hypothetical short-range (~ 100 m) force, known as the "fifth" force, that could couple to hypercharge or baryon number. Such a force would cause a difference in acceleration between bodies whose baryon number to mass ratio differed, and would yield a systematic deviation from equality of acceleration between different materials; Fischbach's group claimed to see just such a deviation. This result, they argued, was qualitatively in accord with measured deviations from the inverse square law of gravity using gravimeter data from deep mines and with anomalous energy dependences in the fundamental parameters that characterize the behavior of the unstable K-mesons, effects that would also be consequences of such a short-range fifth force. The more precise Princeton and Moscow experiments that I described above would not be sensitive to this effect because the relevant source of gravity in those cases was the Sun, and the effect of a 100-m short-range force would therefore be negligible.

A number of authors subsequently took issue with the results of this reanalysis. It was argued by some that the evidence for the deviation was much weaker than claimed by the Fischbach group because of a number of factors which were inadequately taken into account in the reanalysis, including a sign error in the interpretation of one of Eötvös conventions, uncertainties in the isotopic composition of each element and the chemical composition of each compound used in the experiments, uncertainties in the individual masses of the substances used and of the containers in which they were placed, and unknown systematic errors that might have affected the outcomes of Eötvös' experiments, given that three different methods were used by Eötvös and his colleagues. Reanalyses by others of Eötvös' data, and of the data from a series of 1935 experiments using the same apparatus by Renner, failed to support the Fischbach claim. Other authors pointed out that, even if one accepts the dependence of acceleration on baryon number claimed by Fischbach, it is virtually impossible to infer anything about the nature of a short-range fifth force, because its putative effect on Eötvös' experiments is completely sensitive to the details of the nearby mass distribution (such as the mass of the building next to Eötvös' Budapest laboratory), which are unknowable nearly a century after the fact.

At present the situation is inconclusive. However, the Eötvös reanalysis has stimulated extensive interest in new experiments to test this idea, and several are in progress.

The idea of a short-range "fifth" force, or a short-range component of gravity has also been scrutinized directly by checking the validity of the Newtonian inverse-square law at various distances. Up to the accuracy at which relativistic corrections play a role, the inverse square law of Newtonian gravity has been verified to high accuracy in planetary and lunar motion. Observations of planetary, solar, and stellar structure support the law as applied to bulk matter. Further agreement has been found for determinations of the Earth's gravitational field via comparisons of surface, rocket, and satellite measurements, and via comparison of laser ranging to the Moon and to a satellite known as LAGEOS. These observations confirm the inverse square law for distances typically from 1 km to several astronomical units.

However, at shorter distances, the inverse square law has recently come under more intense theoretical and experimental scrutiny, partly because of the possiblity of detecting "fifth" force effects. Because the range of such effects is expected to be less than a few kilometers, the experiments have been geophysical or laboratory scale, using methods that range from variants of the Cavendish experiment using various massive sources (rings, cyliners, oil tanks, etc.), to direct measurements of the divergence of $\vec{g}$, (which vanishes if the inverse square law is valid) using an array of orthogonal gravity gradiometers. However, in the range 1–1,000 m, measurements of the gravity profile in deep mines have yielded values of the gravitational constant at depth that appear to differ from the laboratory values obtained from Cavendish-type experiments. In these experiments, measurements of the local acceleration g and its gradient at different depths in mines are combined with determinations of the local matter density from core samples, in order to take into account deviations from that of an idealized Earth model. The resulting values of G are between 0.5 and 1 percent higher than the laboratory values. Other experiments are currently in progress to see if this effect is real or a result of systematic errors.

During the past 25 years, an intensive theoretical and experimental effort has gone into testing the predictions of general relativity and of other theories of gravitation in many different arenas, and to high precision. General relativity has passed every test, while numerous theories have fallen by the wayside. Although many opportunities remain for further testing of gravitational theory, we can be sufficiently secure about the empirical underpinnings of general relativity to assume now that it is right. General relativity now has taken its rightful place as both a cornerstone and a tool of physics and astronomy.

XI. STRATEGIES FOR TREATING GENERAL RELATIVITY IN ELEMENTARY PHYSICS CLASSES

General relativity may be a well-accepted part of modern physics, but what about teaching it in high school and introductory college physics classes? What kinds of things can one do with this subject in these kinds of classes? Let me mention some random ideas that occurred to me in the course of preparing this lecture.

One can give a qualitative understanding of the nature of curved spacetime using pictures, both of two-dimensional curved surfaces to get the students out of their Euclidean habits, and of spacetime diagrams (using one or two spatial dimensions), to introduce the notion of spacetime. The mathematics of curved spacetime can't be treated at this level, but pictorial analogies can be useful.

Some of the effects of relativity can be calculated quite simply, once one has developed the proper intuition concerning freely falling frames. For example, the gravitational redshift effect requires only a knowledge of the Newtonian acceleration, the speed of light and the Doppler-shift formula. The equivalence principle part of the deflection of light requires knowledge of the inverse square Newtonian acceleration, and of the formula for aberration. The net effect over a sequence of freely falling frames can be estimated, or, if the students can do simple integrals, can be integrated exactly. Alternatively, the deflection can be obtained using the limiting Newtonian hyperbolic orbit.

Because this subject involves such a broad range of allied areas of physics and astronomy, there are many opportunities for making connections with other subjects, such as celestial mechanics (Question: For Starprobe to go from the Earth to a close encounter with the Sun, why is it necessary to go to Jupiter first?), astronomy (quasars, pulsars, black holes), the space program (Apollo, Mariner, Viking), even Einstein biographies.

This subject can even be used to enliven some ordinarily deadly student laboratory experiments, such the study of pendula. Newton studied the periods of pendula of equal length, but containing weights of different materials, and found that to a part in a thousand, they were identical. This verified the principle of equivalence. I suspect that a group of high-school physics students could do at least a factor of 10 better, and could be motivated to do so by the claim that they would be verifying that spacetime is curved!

Suggestions for Further Reading

Calder, N. *Einstein's Universe* (Penguin Books, New York, 1981).

Einstein, A. *Relativity: The Special and General Theory* (Crown Publishers, New York, 1961).

Gardner, M. *The Relativity Explosion* (Vintage, New York, 1976).

Greenstein, G. *Frozen Star: Of Pulsars, Black Holes and the Fate of Stars* (Freundlich Books, New York, 1984).

Pais, A. *'Subtle is the Lord...': The Science and the Life of Albert Einstein* (Oxford University Press, New York, 1980).

Will, C.M. *Was Einstein Right?* (Basic Books, New York, 1986).

Banquet Session

Leon M. Lederman, internationally known specialist in high energy physics, is the director of Fermi National Accelerator Laboratory in Batavia, Illinois, and is the Eugene Higgins Professor at Columbia University. He has been associated with Columbia as a student and faculty member for more than 30 years, and he was director of Nevis Laboratories in Irvington, which is the Columbia physics department center for experimental research in high energy physics until June 1, 1979. With colleagues and students from Nevis he has led an intensive and wide ranging series of experiments which have provided major advances in understanding of weak interactions. Dr. Lederman has been the Director of Fermi National Accelerator Laboratory since June 1, 1979.

In 1965, Dr. Lederman was awarded the National Medal of Science by President Johnson for systematic studies of mesons, for his participation in the discovery of two kinds of neutrinos and of parity violations in the decay of mu mesons.

In 1956, working with a Columbia team at the Brookhaven Cosmotron accelerator, he discovered a long-lived neutral K-meson which had been predicted from fundamental symmetry ideas.

In 1957, in conjunction with Dr. Richard L. Garwin and Dr. Marcel Weinrich, Dr. Lederman carried out the crucial experiment on muon (or mu meson) decay at the Nevis synchrocyclotron. This work, immediately following an independent experiment in low energy nuclear physics by a team led by Columbia's Professor C. S. Wu, verified the startling prediction of the non-conservation of parity by Professors T. D. Lee and C. S. Yang. Professor Lee, the Enrico Fermi professor at Columbia, along with Professor Yang, won the Nobel prize in 1957 for this theoretical work.

Professor Lederman's fundamental experiments on the interactions of neutrinos and high energy muons have been carried out primarily as a part of the Nevis Laboratory program at the Brookhaven National Laboratory. Here, in 1961, the group discovered the second neutrino, that associated with a muon. This discovery was crucial in establishing the doublet structure of lepton currents. It also initiated the program of high energy neutrino physics which dominates programs at the major accelerators today.

Following a series of experiments at the Brookhaven 30 GeV AGS accelerator and which included a search for free quarks, the first observation of antideuterons and a study of nuclear fermi motion. In 1968 Lederman's team initiated a measurement of pairs of muon emerging from primary proton collisions. A suspicious "shoulder" in the effective mass distribution of these dimuons was clarified six years later as the J/ψ resonance. The dimuon data stimulated the "Drell-Yan" model which has been a rich source of information on the quark structure of hadrons.

Professor Lederman worked with a team of physicists from Columbia and Rockefeller University on a large experiment at the intersecting storage ring at CERN, the European Center for Nuclear Research in Geneva, Switzerland. In 1972, this group established the existence of large transverse momentum pions. This initiated a large program of research in many laboratories and has become one of the most incisive tools in the probing of the substructures of nuclear particles.

In 1973 he worked with a separate team from Columbia, Fermilab and Stony Brook in the NAL Proton Experimental area on the production of electrons and muons at high transverse momentum to study the fundamental structure of matter. The group carried out a series of experiments culminating, in 1977, in the discovery of the Upsilon particles—a series of three narrow and closely spaced levels generally interpreted as evidence for a new quark called "bottom." This interpretation was confirmed in subsequent experiments carried out at electron-positron colliders. Professor Lederman led a team which carried out one of these experiments at the Cornell Electron Storage Ring facility and is the co-discoverer of the fourth Upsilon state.

Dr. Lederman was born in New York City in 1922. He received a B.A. from City College in 1943. Columbia awarded him his M.A. in 1948 and a Ph.D. in 1951. He was appointed full professor in 1958 and was director of the Nevis Laboratories from 1962-1979. In 1979, he became director of the Fermi National Accelerator Laboratory.

Dr. Lederman is a member of the National Academy of Science; Fellow of the American Physical Society; a member of the American Academy of Arts and Sciences; and of the American Association for the Advancement of Science. He has been a member of the AEC Advisory Panel on High Energy Physics, and is a U.S. representative to the International Committee for Future Accelerators.

He is a recipient of numerous honors and awards, including fellowships from the Ford Foundation (1958-59); John Simon Guggenheim Foundation (1958-59); the European Center for Nuclear Research in Geneva (1958-59); the Ernest Kempton Adams Foundation (1961), and the National Science Foundation (1967).

In 1965, Dr. Lederman was awarded the National Medal of Science by President Lyndon Johnson; in 1973, he received the Townsend Harris Medal, City of New York; and in 1976, the Elliot Cresson Medal of the Franklin Institute. He was awarded an Honorary Doctor of Science by the College of the City of New York in 1981. He is the co-recipient (sharing with Martin Perl) of the 1982 Wolf Prize in Physics.

In 1983, both Northern Illinois University and the University of Chicago awarded him the Honorary Doctor of Science.

206

Ceremonial Session

Introductory Remarks

Leon Lederman
Director, Fermi National Accelerator Laboratory
Batavia, Illinois 60510

I want to try to avoid platitudes. We are enjoying the presence of so many enthusiastic teachers here. We are honored, privileged, to help sponsor this conference along with the AAPT, the National Science Foundation and the International Commission of Physics Education.

We have a number of interesting things happening this evening. One, there will be the awarding of a medal to Professor Viki Weisskopf by the International Commission on Physics Education. I can't resist, with Viki getting a medal, telling one or two Viki stories, even though he'll deny these things, or try to correct the accuracy.

One of the famous Viki stories that is very well known in some physics circles and that may not be known in educational circles, has to do with the time when he was young—which is a long time ago—he was in fact competing with another young man, by the name of Hans Bethe, for a prize position as an assistant to the famous Wolfgang Pauli, a great theoretical physicist. You all know about the Pauli Principle, but you don't know about Pauli's principles.

Anyway, Viki won the contest, and very proudly presented himself to the master. He said, "I'm Victor Weisskopf, I'm your new assistant." And Pauli is supposed to have looked at him, looked at him, looked at him, nodded and said, "So young, already you're unknown."

As I understand it, after— don't worry about it, this is poetic license—after carafe after carafe of wine—then after, you know, many weeks of being there, Viki worked on a paper, and he solved a problem, he thought it was very elegant. And he presented it to Pauli, who looked at the paper and said, "Ach, this isn't even wrong!" (Laughter)

From then on, every time he passed Viki in the hall, he said, "I should have taken Bethe."

But the story I would really like to tell is apocryphal. It comes from the fact that, as I already said in the beginning of this conference, Viki is noted as being a great teacher. And a teacher does many things in physics: he stands before classes and teaches. Another function of the teacher is to have graduate students. And that's a very important function and very different. It is one-on-one. After years and years you have this collection of former students and they are yours forever.

So now the story. There is a woods somewhere in the world, and in this woods there was a rabbit, and the rabbit was sitting in the woods, and along came a fox. He noticed that the rabbit was busily writing. And the fox was curious. And so he overcame his hunger pangs.

He said to the rabbit, "What are you writing?" And the rabbit said, "I'm writing my thesis." And the fox said, "Oh, that's interesting. What is your thesis?" And the rabbit looked up, and hesitated a little bit. And finally, he said, "Well," he says, "my thesis is about how rabbits eat foxes." And the fox started laughing, and he chuckled, and he said, "That's crazy. You know that foxes eat rabbits. Rabbits don't eat foxes." And the fox continued, "That's ridiculous. It just doesn't work. You'll never get anywhere with this thesis."

And the rabbit said, "Well," he says, "let's go and I'll prove it to you." So the fox said, "How are you going to prove it?" "Let's go over to that cave."

So they go up the hill to the cave, and they disappear into the cave. And a short time later, the rabbit comes out of the cave chewing on a fox bone. And he goes back and starts writing. And along comes a wolf. And the wolf says, "What are you doing, rabbit?" And the rabbit says, "I'm writing my thesis." The wolf asks, "What's your thesis?" And he says, "My thesis is how a rabbit can eat a wolf."

And the wolf says, "That's ridiculous." And they go through the same thing. They go to the cave. And pretty soon, out of the cave comes a rabbit chewing on a wolf bone.

And, you know, I can keep this up for—

(Laughter)

So after the bear also disappeared into the cave and out comes the rabbit chewing on a bear bone, the rest of the forest is alerted that something screwy is going on. And so they all collect around the rabbit and they say, "We demand an explanation for the funny things going on in this forest."

The rabbit shrugged and puts aside his thesis, and he says, "Okay, come on." And they walk to the cave, and the rabbit goes in. And out of the cave comes a ferocious, enormous lion.

And the moral of the story is, it doesn't matter what your thesis is, it's your sponsor that counts.

(Applause)

I would like now to introduce Professor Leonard Jossem, who will represent the ICPE.

Citation for the Presentation of the Medal of the International Commission on Physics Education
to
Victor Frederick Weisskopf
April 25, 1986

E. Leonard Jossem

Secretary, International Commission on Physics Education of the
International Union of Pure and Applied Physics
Department of Physics
Ohio State University
Columbus, OH 43210

It is a pleasure and a privilege for me to have the opportunity to present to Professor Weisskopf the Medal of the International Commission on Physics Education.

I would like to show you the medal, with the sculpture on the front and the inscription on the back, "Awarded to Victor F. Weisskopf by the International Commission on Physics Education of IUPAP. For long and distinguished service in physics teaching."

There is also a certificate attesting to the awarding of the medal.

And finally, I would like to read to you the citation that goes with the medal:

The Medal of the International Commission of Physics Education of the International Union of Pure and Applied Physics was established in 1979 for the purpose of recognizing contributions to physics education which are "... major in scope and impact and which have extended over a considerable period." The medal is, along with the Boltzmann Medal and the London Award, one of the three awards for excellence sponsored by the International Union of Pure and Applied Physics. It has, to date, been awarded three times: to Professor Eric Rogers in 1980, to Professor Peter Kapitza in 1981, and to Professor Jerrold R. Zacharias in 1983.

Victor Frederick Weisskopf's contributions to physics have been many. He has given us the theory of the widths of energy levels and their fundamental relations to atomic and nuclear lifetimes, the theory of the self-energy of the electron, and the "Cloudy Crystal Ball" theory of nuclear structure, to mention only three of especial importance. He has played a leading role in the task of fostering international collaboration in physics, perhaps most visibly as Director General of CERN (1961-65), but also in many other more subtle ways. He has also been, and continues to be, an eloquent spokesman concerning the social and political responsibilities of science and scientists. Over and above all this, and of special importance on this occasion, are his contributions to the teaching and exposition of physics at all levels. His books, conference lectures, review articles, and popular scientific writings all demonstrate his scope and power as an exquisite interpreter of science.

"Human existence," Professor Weisskopf has said, "is based on two pillars: compassion and knowledge. Compassion without knowledge is ineffective; knowledge without compassion is inhuman." These two qualities are also major attributes of a teacher, and they have long characterized Professor Weisskopf and his work. Those who have had the good fortune to be able to hear his talks and lectures will understand the meaning of the phrase "Education by Presence." One goes away from them having seen and heard marvelous examples of how beautifully lucid, stimulating, and inspiring really good teaching can be.

In trying to describe the attitudes and pedagogical philosophy of a teacher of the caliber of Professor Weisskopf, one can surely do no better than to allow him to speak for himself, as the following brief quotations from some of his writings will illustrate.

On the teaching of science: "The teaching of science must return to the emphasis on the universality of science and should become broader than the mere attempt to produce expert craftsmen in the specialized trade."

On the relation between teaching and research: "I discovered that the effort of explaining and clarifying a field of physics not only leads to better understanding of past work, but also produces many new ideas, explanations, and discoveries. This is yet another illustration of the close connection between teaching and research. It never seemed possible to me to do one without the other."

Victor Frederick Weisskopf has brought to many—to colleagues, to students, and to the general public—new ways of seeing phenomena that they live with daily, and by his personal example as a teacher has allowed us all to see and to understand the immensely productive uses and power of the combination of knowledge and compassion. It is especially for these reasons that the International Commission on Physics Education takes pleasure and pride in presenting him with this medal.

Response of the ICPE Medalist

Victor F. Weisskopf
Department of Physics
Massachusetts Institute of Technology
Cambridge, MA 02139

I am deeply moved receiving this medal and the acclaim from those people I value most from the community of physics teachers.

I'd like to say a few words. I cannot compete with Leon's jokes and wonderful stories. So I will not make any jokes.

For me, that medal really means a lot. Because I do believe that teaching is fundamental, a fundamental activity for any scientist. In teaching to people and not necessarily only to beginners or even to laymen, also I really believe that you haven't understood what you're doing if you cannot explain it, or at least describe the main idea to your spouse, to your girl or boy friend as the case may be assuming that he or she is not a physicist. And usually—fortunately—they are not. (Laughter) There is something else. When I was young, when I was deciding early in college, or even high school, what should I do, I had actually two things in mind: either I go into science, or into music. I was playing piano, I was conducting the orchestra in school.

And, well, I made a decision. The decision is probably the right decision. Because, if you are mediocre in science, it doesn't matter so much, you still can contribute valuable results,—but if you are mediocre in music, I guess that's not so good. Also, music you can enjoy as an amateur. And I'm not sure whether you can enjoy science as an amateur, unfortunately, because there are not the right books about, and the right spirit that leads you on. You should be able to enjoy science as an amateur, as you do music, but you cannot yet.

But when I made this decision, I found out that it isn't so—there are some similarities there, and also some things that should be more similar aren't. I'll tell you what I mean. In music, for example, here there are the composers, who are the creative people. But then, there are the performers. Now, in music the performers are usually much better paid than the composers. But how is this in physics and science? I'm afraid it's completely the other way around. And it shouldn't be so. Because what we do—now I'm identifying with you, teachers, because really I think that was my primary profession, and the things that I did in research, mostly came from efforts to teach the subject correctly.

But actually, my most important contributions to science came when I tried to teach, and then found out, in the process of teaching, that there is something there which was not fully understood. The real understanding contributed a piece of research.

Leon told us at the beginning of the conference, the story about the teacher who was asked to repeat what he taught three times, you know. And after the third time, he finally pointed out how little he understood the matter. Then he was able to make it clear to himself and perhaps to a few students.

But I would like to come back to the analogy of the teacher and the music performer. When you give a talk, or when you're giving a course, you take the same responsibility as the performer: it is honestly like playing a Beethoven sonata, for example. I mean, you have to find out what is the real essential, what is the great thing, in that piece of music. You try to bring it out in a good performance. And that is what we have to do when we teach physics: find out what is great with the Maxwell equations, or with whatever you teach. And then let it act on the students; he must be as much in awe of the deeper connections as the listener to a first rate music performer.

So there are a lot of connections here. The only trouble is that you are not paid for it as well as a good pianist is paid for it. And there is something wrong. The production, the presentation of cultural, valuable things ought to be valued by society as much as the creation itself. That is unfortunately not so in science.

Science teachers are actually very badly paid although they are actually in a very much more responsible position. Most of the researchers are sort of protected. Throughout the university, you are surrounded with people who understand that science is a great thing—I do not include these poor administrators, who have to go to Washington and fight with people who are probably even worse than high school students.

But seriously, as high school teachers and, to a great extent, also, teachers of undergraduates, you are actually facing the real world. Not like us, the research physicists who are protected by representatives in Washington who actually sympathize with what they are doing. You're facing the budget makers, you are fighting almost alone, you are out in the battlefield, with only very few helpers. Actually you should get more attention and support than the research scientist. Because it is you who help research by telling the public what physics is, by raising the importance of scientific knowledge, for our culture, for the world, for our nation. And that is your task. That is a most important task.

And it is a task that is way beyond the mere teaching of science, trying the right way to bring out the values, like the pianist—it is more than that. You will have to show what science means to us, and what science means to society.

Of course, we also worry about these things; but you are in direct contact with those who are actually skeptical, and rightly skeptical. They are saying, "What has science done, with all this abuse of technology, all this pollution, it makes more weapons, it brings us nearer to catastrophes, such as nuclear war."

So it is your task not only to show the beauty of science but also to show that the detrimental applications of science can be avoided, that we can have a world where science and the consequent technology are for the good of us. And so you are, in some ways, much deeper in this role in which we all should be. We have to tell, to discuss, what is good in science, why science is important. What are the right kinds and wrong kinds of application? Which branch of science promotes peace, and which does not? This question is often treated in pure military terms. But enviormental problems require even more science than the military. The problem of the arms race can be solved by stopping the production of arms. But the world is full of problems of non-military nature that have not such simple solutions, for example the increases of CO_2 and the destruction of the ozone layer.

You all know where I stand. I am not trying to convince you of the terrible danger of the military situation with SDI, and a proliferation of nuclear weapons, and so on. All that I want to say is, it is our task, or one of our tasks, also to discuss these problems. Because only then can you bring out what we all really want: a peaceful world, and science as the essential part of the culture of today. (Applause)

Fermilab

DR. LEDERMAN: One of the problems our friend Viki will have is transferring that medal from his jacket to his pajamas. And when you see all the other medals he has, it's a real problem. (Laughter)

Viki pointed out the difference between a high school teacher's life and a university teacher's life. With high school, of course, you have five, six, seven hours per day of facing the students. In the university, they never have meetings on Wednesdays because it kills two weekends. (Laughter)

Well, the next event on this schedule is a very special treat. We are delighted, honored to have Dr. Bassam Shakhashiri here, who is the director—I don't know the exact title, but he is the boss of science education at the National Science Foundation. And because he's in the National Science Foundation, he works in Washington, D.C.—D.C., darkness and confusion.

Washington is the only city in the world where the velocity of sounds exceeds the velocity of light. (Laughter)

Visitors to the White House are given hard hats, because so much is swept under the rugs— (Laughter)

In Washington churches, you have express confessionals for six sins or less.

I have a story about Washington that may be a little more subtle. It has to do with two female ostriches that are walking down the beach, gossiping. And suddenly, one of them says to the other, "You notice, there are two male ostriches following us." "Yes, I see them, I see them."

"They're following us. What should we do?" "Well, let's hide." So they bury their heads in the sand. And the two male ostriches look around and say, "Where did they go? Where did they go?" (Laughter)

Now, if you don't understand that story . . .

Let me say that, in introducing Bassam Shakhashiri, that I think that we are extremely fortunate to have him in charge of science education. If you remember Viki's talk with the charts, it showed the National Science Foundation as sort of starting out at 40 percent of its budget, and now down to eight percent of its budget. That eight percent isn't much, but it's powerful—it's something. And that something is in good hands, as you'll see when I introduce Dr. Shakhashiri, who will encourage us with a few words. (Applause)

Bassam Z. Shakhashiri joined the National Science Foundation in June 1984. He heads the Directorate for Science and Engineering Education which annually awards over $80 million in grants.

The focus of the Directorate is in two areas of critical concern: support of graduate education in science, mathematics and engineering; and precollege education with emphasis on improving the quality of science and mathematics teaching and development of exemplary curricular in the scientific disciplines.

Previously, Shakhashiri was at the University of Wisconsin–Madison where he was Professor of Chemistry and coordinator of the general chemistry program. In 1983 he was named the first Director of the Institute for Chemical Education. The Institute was formed to revitalize the teaching of chemical sciences at all educational levels by gaining the cooperation of academic, industrial and government chemists to ensure a high level of quality in teaching future scientists, the general student and the public at large. During his tenure he taught over 10,000 students and developed many innovative programs one of which, "The Chemical of the Week," has been adopted by the *Journal of Chemical Education*. While at the University of Wisconsin he served on various committees and as chair of the Committee on Undergraduate Education and as the first Chair of the University of Wisconsin System Undergraduate Teaching Improvement Council.

Before joining the UW faculty, Shakhashiri was a postdoctoral research associate at the University of Illinois–Urbana from 1967–68. He was visiting assistant professor from 1968–70.

In his career of teaching and research, Shakhashiri has co-authored several publications in chemical education including a laboratory testbook, General Chemistry Audio Tape Lessons and Workbook, and semi-programmed books on chemical equilibrium, chemical kinetics and organic and biochemistry. He developed CHEM TIPS, a computer-managed instructional system and has produced over 30 laboratory videotapes. He is principal author of a series of handbooks on chemical demonstrations for use by teachers at all levels. The first volume was published in May 1983 by the University of Wisconsin Press. The second volume was published in 1985.

He has conducted over 300 workshops on a variety of topics, including chemical demonstration, course content, and the use of educational technology for high school students, "Chemistry Can Be Fun," is aimed at generating interest in the chemical sciences and developing a positive attitude toward chemistry. His programs have reached almost 12,000 students. He pioneered teaching in shopping malls and his Christmas lectures have reached thousands through television broadcast over the past 14 years. He also collaborated on an interactive chemistry exhibit for the Chicago Museum of Science and Industry which opened in the fall of 1983.

PROFESSIONAL ACTIVITIES

Shakhashiri has been very active in the American Chemical Society (ACS). He served on the ACS Committee on Education. He is a past Chair of the ACS Division of Chemical Education and the ACS Wisconsin Section. He was general chair of the ACS Fourth Biennial Conference on Chemical Education in 1976. He has served on the board of editors of the *Journal of College Teaching* and on the publications board of the *Journal of Chemical Education*. He was a member of the organizing committee of the Sixth IUPAC conference on Chemical Education held in 1981 at the University of Maryland.

AWARDS

* Outstanding Lecturer of the Year in General Chemistry, University of Illinois, 1969 and 1970
* William Kiekhofer Distinguished Teaching Award, University of Wisconsin-Madison, 1977
* Visiting Scientist Award, ACS Western Connecticut Section, 1978
* Manufacturing Chemists Association College Chemistry Teaching Award for Excellence in Chemical Education, 1979
* Danforth Foundation
* Elected to Boston University Collegium of Distinguished Alumni, 1981
* Cited by the Wisconsin Society of Science Teachers for outstanding contributions to science education at the local, regional, and national levels, 1982
* Youngest recipient of the ACS James Flack Norris Award for Outstanding Achievement in the Teaching of Chemistry, 1983
* Youngest recipient of the ACS Award in Chemical Education, 1986

EDUCATION

Shakhashiri was born in 1939 in Lebanon where he completed high school and attended American University of Beirut for one year. He came to the United States with his parents and two sisters in 1957. He completed his undergraduate work at Boston University where he received a degree in chemistry. He completed work for his M.S. and Ph.D. degrees at the University of Maryland.

The Challenge of Communicating Science to Non-scientists

Bassam Z. Shakhashiri
Assistant Director for
Science and Engineering Education
National Science Foundation
Washington, DC 20550

Thank you very much. I'd like to start by making a couple of very serious comments, and then I'll get to the level—try to get to Leon's level. I know I won't succeed.

But I do want to start by telling you about the three biggest lies in the world. You know what they are. The first one is "the check is in the mail." And I see some of you recognize that one. The second one is, "of course I'll love you in the morning." A few recognize this one. And the third one is, "I'm from Washington; I'm here to help you." (Laughter)

I especially would like to dedicate the next set of statements to Professor Lederman. Because he's not letting me get to what I really want to say.

You can all see what I have in my right hand here. I have a match. If I take this match and I drop it, it falls to the floor. That's physics, Leon.

If I take the match and strike it, that's chemistry. Chemistry is more striking than physics. And physics is no match for chemistry.

DR. JACK WILSON: That really burns us up.

DR. SHAKHASHIRI: Jack has heard this before, and for six months he's been trying to think of a clever response. (Applause) I wouldn't say that was a hot retort.

PARTICIPANT: Retort is chemistry.

DR. SHAKHASHIRI: I have to confess to you that I thought very hard about using that example, especially with this audience, but some of the words of Professor Lederman prompted me to be perhaps somewhat foolish and use it.

The last time I had occasion to be on the same platform with Professor Lederman was last October, when the 1985 presidential awardees in science and mathematics were being honored at the State Department Diplomatic Functions Dining Room. And the main speaker that evening was Professor Lederman. But he was somewhat at a loss in terms of how he wanted to give his talk, because, except for the President and for the Secretary of State, slide projection equipment cannot be used in that room. And so he told me privately afterwards that he'd get back at me some day for not making better arrangements for him to get a slide projector. And so, tonight, I am being punished, because I do have some important things I'd like to tell you, and I have my slides but there's no projector.

It is really a great privilege for me to be here tonight, especially when Professor Weisskopf is being awarded the Medal of the International Commission on Physics Education. And I speak to you as a fellow teacher, although I am a chemistry teacher. And with your permission, I would like to just share with you and everyone else tonight a few important convictions that I have about good teachers. I believe *good* teachers have four important characteristics which distinguish them from all other teachers. Good teachers are:

Competent in their disciplines
Committed to their disciplines *and* to the profession of teaching
Comfortable with the methods and techniques they use
Compassionate with students (and colleagues)

The first characteristic is so obvious that I am often questioned about including it. I *insist* on including it, for it is not sufficient for a teacher to be certified as a holder of a degree in a scientific discipline or be tenured at a school or at a college or a university in order to be considered competent. All of us have to *maintain* our competence by engaging in scholarly and professional renewal activities to keep us ahead of our students and at a level of knowledge much more advanced than what is in the textbooks and manuals we use in our courses. Furthermore, if we are competent in a scientific discipline, say physics, then we must be committed to physics. But commitment to physics alone is not sufficient; we have to be committed to *teaching* physics as well. Many researchers, of course, are committed to their sub-discipline, but they cannot be characterized as good teachers unless they are committed to communicating physics to those outside their area of *sub-specialty*. They may be good at research; and if they are, that does not automatically make them good teachers.[1] In my opinion good teachers must be comfortable with the methods they use be they audiovisual aids, lecture demonstrations, computers, books and manuals, etc. As we adapt or even adopt a "new" method or technique, we often experience discomfort to a varying degree. This discomfort has to disappear; otherwise, we should abandon that particular method in order not to diminish our effectiveness as teachers. We must be careful in not becoming too comfortable and thus quickly risking becoming complacent. In addition, I believe we must be compassionate with our students—we must *care* about them and about what and how they learn. This should not be done by compromising standards; on the contrary, we should set our standards high. Since our purpose as educators is to enable students to develop fuller intellectual capabilities and emotional capacities while they are under our influence, we and others should recognize that their grades are by no means the only measure of our success as teachers or their success as students. As good teachers we must also be compassionate with all our colleagues in the educational enterprise including fellow teachers, administrators, and support staff. This will contribute to creating and maintaining an atmosphere conducive to good teaching.

I would like to share with you another strong conviction. Again, both as a teacher and as a person who has a responsibility in the nation's capital. I do firmly believe that the United States now faces a situation by far more critical and more consequential than what the country faced in the immediate post-Sputnik era. There are at least three reasons for this:

---

[1]In this connection, I am bothered when I hear about some faculty bragging that they do not have to teach (in some instances I suppose I should be happy they are not being inflicted on students); and I am saddened when I hear faculty and administrators talk about "research opportunities and teaching *loads*"—what a remarkable statement about the value system of some of our institutions of higher education!

1. We have more people living in this country now than we did 25 years ago or so. The population of the United States has increased by approximately 50 million people. To put that number in perspective, that is roughly the population of Great Britain alone. What does that mean? There are more students to teach and we need more *qualified* teachers to teach them.

2. Secondly, we need to have a good supply of scientists, engineers, and technologists coming through our educational systems in order for our society to continue to enjoy the benefits of technology. That is essentially what the National Science Foundation set out to do in the immediate post-Sputnik era. Now, we need to maintain having a good supply of scientists, engineers, and technologists for economic and national security reasons and in order to retain our international pre-eminence in science and technology.

3. The third reason, perhaps the most important and most consequential of all, is that we now live in a more advanced technological society than we did 25 years ago or so. And it is the education of the non-scientists, the non-engineers, the non-technologists in science, in engineering, and in technology that require our attention.

I submit that our greatest challenge now is to extend learning opportunities so that all individuals can continue to expand their knowledge and understanding of science. Improving science teaching is crucial, but it is not enough. Our adult population also needs to learn new science concepts. We need not only skilled scientists, engineers, and technicians, but managers and decision makers who understand the nature and implications of their fields. And we need a citizenry that can follow and weigh the progress and implications of science and technology. That is why we *must* be creative and inventive in communicating the very essence of our science and its results to *all* segments of our society .

Professor Weisskopf said it much more eloquently than I can say it tonight. It is extremely important that we, who care about science, we, who care about science education, very quickly develop effective methods of communicating our science to the non-scientist. We have to do this. We must do this. We all know that scientists are fairly good at communicating science to each other through professional meetings, scientific publications, and a variety of forums. But frankly, we do a poor job communicating science to the non-scientist. It is absolutely crucial that we—we, the people who know science, we who are competent in our disciplines, we who are committed to our disciplines and to the profession of teaching quickly develop effective ways of communicating science to the non-specialist. And the plea that Professor Weisskopf has made tonight—and not only tonight; he's been saying it over and over again—we must reach our students, both in the formal classroom setting and in the informal classroom setting. Most people most of the time learn most of what they know outside of the classroom. Our responsibility in the classroom is to enable them to appreciate the process of science, the way in which science functions, the way in which it works. So we have an awesome responsibility as teachers.

Also, we have to try to make sure that the quality of science that is taught in our pre-collegiate classes, in our undergraduate classes, is based on laboratory experience. There is a move afoot in this country to eliminate laboratory work from pre-collegiate programs and from introductory undergraduate courses. Most of the reasons that are given are based on economics. If we feel strongly about the science that we care about, we should see to it that our colleagues, the administrators who have to make decisions about facilities and about safety in the laboratories—we have to see to it that they share our conviction about the importance of laboratory work. I think the great master, Leonardo, said it much more eloquently than I can say it. He said, "There is no higher or lower knowledge, but one only, flowing out of experimentation." It is the pursuit of experimentation and it is the interplay between theory and experiment, that we must understand and we must be able to communicate to others.

I am happy to tell you that the National Science Board, exactly a month ago, accepted a report from a special task committee that dealt with the question of undergraduate science, mathematics, and engineering education. I urge you to read this report and to act on its far-reaching recommendations that are addressed to various sectors in our country including NSF. The recommendations made to the National Science Foundation call for leadership role and for leveraged program support of about $100 million dollars over the next two budget cycles—fiscal year 1988 and fiscal year 1989.

I think I did mention to a couple of people earlier tonight when I was asked about the button that I'm wearing—"science is fun," I said that, at the banquet, I will distribute the few buttons I brought with me. I have about a hundred. Also, I brought for my physics colleagues, a small gadget. And I have maybe 500 of those. It's actually a small disc. And if you listen carefully, you can hear this click. This is a bimetallic disc. And there is a message on the inside of the disc. I'll let you read the message, and then I'll also give you some instructions on a small piece of paper, in case you don't know how this works. But I do want to close by making a special presentation to Professor Weisskopf and to Professor Leon Lederman, in that order. I'd like to present you this (a green chemiluminescent light stick is presented)—for showing us the light. (Applause)

And Professor Lederman, I'd like to present you with this. The color is indicative of a message (a red chemiluminescent light stick is presented).

DR. LEDERMAN: What message?

DR. SHAKHASHIRI: That the National Science Foundation budget is in the red— (Applause)

DR. LEDERMAN: I hope this is not "stop" and "go." It's certainly not politics.

I—I don't know what to say—that doesn't work.

(Simultaneous Discussion)

Banquet Speech

Leon Lederman
Director, Fermi National Accelerator Laboratory
Batavia, Illinois 60510

I would just say that I think one of the important things that we have to look forward to with Washington is the question, which I'm sure Dr. Shakhashiri often faces, and the question is often asked, "Why should the Federal Government be involved in education today? What have they got to do with it? This is a responsibility of the states. They shouldn't be in it." There was a time not long ago when the NSF budget was essentially zero for education.

I'm really delighted to hear the good news of the intention of the Administration to improve things, as we just heard tonight. However, I am skeptical that the message has really gotten through. We read a lot about the crisis in science education but that is a long range problem and the federal deficit is immediate. In the light of actions taken it is not clear that both the Administration and the Congress are moving in the same direction.

And it reminds me of a pilot who announced to the passengers the good news and the bad news. The good news: we have a 200-mile-an-hour tailwind! The bad news: we lost our instruments, and don't know where we're going. Somehow, I feel that we also are not quite sure where we're going, but we're getting there fast.

My concern about education is derived from the fact that it's very clear that the Federal Government does have an important issue in education, in science education especially, but I think in all education.

I mean, you don't expect, say, Arkansas or Massachusetts, or even Illinois, to respond to the Japanese industrial technological challenge, or the West German challenge.

We should be concerned about a lack of a scientific manpower pool from which you can recruit, say, soldiers and sailors who can be trained to operate our trillion dollars of military technology. And we should be concerned about the flow of students into the sciences and engineering disciplines, to address crucial problems like acid rain. These issues are not confined to one state. One state rains on another state. Concerns are here too about health care, global ecological phenomena, potential shortages of materials vital to a technologically-dependent society. These kinds of problems are national and therefore it is the obligation of the Federal Government to make sure that there is this pool of scientists. If you left it up to the states, it just wouldn't work.

Indeed, it was the Federal Government, and people concerned with the Federal Government, who commissioned the various reports on science education: The Nation at Risk, and so on. These many reports that focus on the threats to the economy, defense, and the general well-being of what is perceived to be a decaying educational infrastructure.

Well, it's not only education I'm worried about. I'm also worried about my own subject. Anyway . . . you know, I do a lot of selling. My community is interested in a new machine called the superconducting supercollider. It's tremendously expensive, but enormously exciting, scientifically. We really need that machine. It's a scientific imperative.

In going around and trying to sell this thing to colleagues and other scientists I became aware of how fragile the whole scientific enterprise is in this country. Terribly fragile. I've been through this before. It's *deja vu*. After Sputnik in the 1950s there was a golden age of science. I think we're still profiting from that glorious time when anybody who had a very good idea, that could be subject to peer review, could get funded to try it. The science education budgets were blooming. There were teachers' institutes and all kinds of wonderful things that were being done. And suddenly, something happened. It didn't just level off. It didn't flatten out, it turned around and nosedived in a great slide of the 70s. And that had terrible consequences.

Perhaps the worst was: a loss of technological self-confidence. There was a decline in the graduate student population in the sciences. Actually, if you go to a graduate school nowadays, you find that it's dominated by the culturally more attuned Asian-Americans with a sharp decline in native Americans working in these fields. Other indications are a declining balance of high technology trade, decrepit equipment in the universities, and even more decrepit faculties. The average age of faculties in American universities is zooming up. That's an ominous sign.

Somewhere in the late 70s—79, 80, and 81—things started picking up. And slowly, over the last four or five years, the infrastructure began to repair, here and there. Progress was slow. It's always easy to destroy an infrastructure; it's extremely difficult to rebuild. But it began.

People started showing up in science classes a little more. Interest in science on the part of the general public increased. T.V. programs—you know, "billions and billions"—ten billion people tuned in on Carl Sagan. And a proliferation of magazines. Budgets for science started increasing. And even in education, the budget went from zero to eight percent, and hopefully it will go up even higher. Teachers started getting national attention, e.g., presidential awards. There were signs of movement in the salary levels.

Well, all of this looked like the good news. And then, suddenly, what happened in the last year is, we suddenly recognized that we have something called the national budget deficit.

Now this part of my talk should be entitled, "Einstein, Newton, and Gramm-Rudman." Because, in the infinite wisdom, Congress invented this Gramm-Rudman-Hollings doomsday machine. They buried it under the nation and they lit the fuse, and they said, "Don't worry, it'll never go off." "We'll take care of it."

But if you read the newspapers today, you realize that it will go off. It's a terrible thing. It mandates cuts across the board of all activities: the stupid activities, and the wise activities. The compassionate activities, and the cynical activities.

It looks as if this particular crisis is going to reinstitute the slide of the '70s, just when we were starting to recover. It's a terrible situation! I don't know how to fix it except to suggest that you call your congressman and complain about it. And when you do this, you will find that you get a lot of sympathy. There is a certain sensitivity in the Congress to people's complaints. Usually, their answer is, "It's not me; it's the other guys. You know, we don't trust ourselves to keep control of the deficit."

But I think if we paid enormous attention to both the requirements of science and the requirements of the federal role in science education, it would not complicate the problems of the deficit in any significant way, because of the many things this government does.

So with that I'd like to tell two teacher stories which, if you think about them deeply—you won't—

One of them, which is very deep, has to do with a teacher of English, creative writing. The teacher said to the students, "I'm going to give you an assignment for the weekend. I want you to write a short story. Now, there are four basic ingredients of all short stories that are successful. That's very important. I want you to include all four ingredients in the short story. The four ingredients are royalty, religion, sex, and mystery."

And a little girl raised her hand. Teacher: "What do you want, Mary?" And Mary said, "I'm finished." "What do you mean, you're finished? You've got a whole weekend." "No, I've finished my story." "You've finished your story?" "Yes."

"Well, would you like to read it to the class?" "Yes: 'God,' said the duchess, 'I'm pregnant, but I don't know who did it.'" (Applause)

My last story—absolute last—which also is probably well known, has to do with this racially-prejudiced teacher—nobody in this room. This teacher happened to be very racially prejudiced. And so she worked in a carefully chosen suburb, but couldn't avoid the fact that, in her new class, she noticed in the back of the room, there was an Oriental student, a Japanese kid.

Well, she decided to ignore him. And she said, "I've got to get to know the class to see what they know about American history, so I'll give them a question."

"Class," she said, "I want you to tell me who said 'Give me liberty, or give me death'." The class started at her blankly. This was an affluent suburb of Chicago.

She said, "Who said, 'Give me liberty, or give me death'?" The class was just blank.

Finally, very hesitantly, the little Japanese boy tentatively raises his hand. Well, she ignores him, but none of the other kids are doing anything—you know, blank looks. So, she says, "Yes?" And he says, "Patrick Henry, 1777." "Okay," she says, somewhat irritated, "I'll try something more modern." "Who said, 'The only thing we have to fear is fear itself'?" And the class—blank. The Japanese kid slowly raises his hand. And finally, she can't do anything, she calls on him and he says, "Franklin D. Roosevelt, in his campaign speech in 1932."

"Okay," she says, now really annoyed, "I'll try something really modern." "Who said, 'Let us not ask what our country can do for us, but what we can do for our country,' class?" The kids looked at each other—blank. The Japanese kid raises his hand, and she said "Yes!" And he says, "Kennedy." "John F. Kennedy, in his inaugural address, in 1962."

So she gets so mad that all her restraint goes away. And she says, "You damned blankety-blank Japanese!" And the kid jumps up and shouts, "Lee Iacocca, 1984." (Applause)

With that—is there anybody here who wants to make a speech? It's a tradition of Fermilab that we have a Quaker session.

If there are no other volunteers, you're going to have to go back to work.

Volunteers?

Well, in that case, if there are no volunteers—oh, yes, there's a volunteer.

State your name and institution.

MR. MANOS: Harry Manos, Schurr High School, in Southern California.

I don't want to make a speech, I want to thank the people at Fermilab for providing not only these fine meals, but for a wonderful educational experience, for outstanding speakers, and for creating a wonderful environment for us to not only learn, but to work.

And I thank you. (Applause.)

International Session

Dr. E. Leonard Jossem, Professor of Physics in The Ohio State University, was Chairman of that department from 1967 to 1980 and has long been interested and active in physics education. He was Staff Physicist and Executive Secretary of the Commission on College Physics 1963–65, and its Chairman 1966-71. He also served as a member of the Panel on Education of the NAS-NRC Physics Survey Committee 1970-72. In 1970 he received the Distinguished Service Citation of the American Association of Physics Teachers, and in successive years from 1971 to 1975 served that organization in the posts of Vice-President, President-Elect, President, and Past-President. In 1985 he received the Melba Newell Phillips Award of the AAPT. Since 1981 he has been a member of the International Commission on Physics Education of the International Union of Pure and Applied Physics, and currently (April 1986) serves as its Secretary and Acting Chairman.

Physics Around the World

E. Leonard Jossem
Secretary, International Commission on Physics Education of the
International Union of Pure and Applied Physics
Department of Physics
Ohio State University
Columbus, OH 43210

The message you see on the overhead projector screen, "We are not alone", is intended to indicate that we physics teachers are part of a world community of physics teachers. We have colleagues and friends all over the world who have the same kinds of interests and problems that we do. It is, of course, true that educational systems and languages differ from country to country, but when one talks to colleagues and looks at their textbooks, one finds that the equations they use and the diagrams they draw and the basic difficulties they encounter with students are very like our own. It is one of the very nice things about physics and physics teaching that you can go almost anywhere in the world and find people who would like to sit down and talk with you about these subjects.

What I would like to do this afternoon is to talk with you about some of the things that are going on in physics teaching in other parts of the world, and a little about some of the organizations that are involved in trying to further the cause of physics education.

On the table here is an exhibit of printed materials related to physics education to which I will refer from time to time. You are cordially invited to examine them all at the end of this session.

Let me begin with the organizations. At the international level, the principal organization for physics is the International Union of Pure and Applied Physics (IUPAP). The goals of IUPAP include the stimulation and promotion of international cooperation, the sponsoring of international meetings and conferences, the promotion of international agreements on the uses of symbols, units, and standards, the fostering of free circulation of scientists, and the encouragement of research and physics education. Some 47 countries are represented in the IUPAP governing body—the General Assembly—which meets every three years and elects the officers and members of the Executive Council and the members of the 18 Commissions in various areas of physics. Commission C 14, the International Commission on Physics Education, has a chairman, a secretary and 10 other members who serve without compensation for three year terms. The current members of the Commission come from the IUPAP member countries of Austria, Brasil, Egypt, West Germany, Hungary, India, Italy, Japan, Mexico, Sweden, the USA, and the USSR. There are also three Associate Members, one each from Scotland, The People's Republic of China, and U.N.E.S.C.O.

The ICPE acts primarily as a catalyst in trying to promote physics education at all levels on an international scale. For example, it assists persons and groups who wish to hold international conferences on some aspect of physics education in the planning of such conferences. It also assists in the process of obtaining funds for the conference from other agencies, the ICPE itself having no funds of its own for such purposes.

Since its establishment in 1960, the Commission has sponsored, or co-sponsored, about a dozen international conferences in about as many different countries of the world. Copies of the proceedings of most of these conferences are here on the exhibit table for your inspection. Two additional international conferences are scheduled for 1986. One will be at the end of August in Tokyo on trends in physics education with substantial emphasis on secondary education. The other, at the beginning of September in Nanjing, will also be on the general subject of physics education. Three conferences are scheduled for 1987: one in Cairo in mid-April on low-cost laboratory and lecture demonstration experiments, one in Hungary at the end of April on chaos and other non-linear phenomena in physics, and one in Mexico in July on networking in physics education. For 1988, so far, there is one conference scheduled. This conference is to be held in Munich on the subject of teaching modern physics: condensed matter physics. The conference will be, as indeed this one is also, one in a sequence of international conferences on various aspects of the teaching of modern physics. The goal of these conferences is to assist physics teachers in incorporating "modern physics" into their curricula by bringing them together with research physicists working at the frontiers of their respective fields. The first such conference in this sequence was at CERN in 1984, this conference at Fermi Lab is the second in the sequence.

In 1977 the ICPE inaugurated a Newsletter which appears at more or less regular intervals and which is distributed without charge to interested persons. Some copies of the Newsletter are here on the exhibit table. The Commission has also published two books. The first, "Einstein—A Centenary Volume" was edited by a former chairman of ICPE, A.P. French, and was published in 1979. The second book, "Niels Bohr—A Centenary Volume", was published in 1985 and was edited by Dr. French and a former Secretary of the Commission, Dr. P.J. Kennedy. Copies of both books are here on the table, and are obtainable from your local bookstore.

The Commission, from time to time, awards a medal for excellence to persons whose contributions to physics education transcend national boundaries and which are " .. major in scope and impact and which have extended over a considerable period." The medal has been awarded to Professor Eric Rogers in 1980, to Professor Peter Kapitza in 1981, to Professor Jerrold R. Zacharias in 1983, and is, as you know, being awarded to Professor Victor F. Weisskopf here at this conference.

There are, of course, other organizations at the international level which have concerns with physics education and with which the ICPE has at various times engaged in cooperative efforts. Let me mention in particular GIREP (Groupe International de Recherche sur l'Enseignment de la Physique); ASPEN, the Asian Association for Physics Education; ICSU, the International Council of Scientific Unions and its Committee on the Teaching of Science; ICTP, the International Center for Theoretical Physics at Trieste; and UNESCO. The relationship with UNESCO has existed since the beginning of the Commission and has been an especially close and fruitful one.

On a national scale, almost every country that teaches physics has some professional organization concerned with physics education, and the list of them is too long to detail here. Most of them also publish in their national language journals and booklets concerned with physics education, and some examples of such journals are also here on the table for your perusal. As you will see, the diagrams, graphs, and equations will give you a pretty good idea of what is being written about even if you are unable to read the language.

The contents of these journals and booklets reflect the interests and concerns of the physics teachers of the countries in which they are published, and, as I have mentioned previously, you will find that they overlap strongly with each other and with those we have in this country. There is worldwide concern with many questions of physics pedagogy. For example, how can we as teachers find out what goes on in the heads of our students, what kinds of concepts they have about the world of physics, and how they use these concepts in their thinking? If we can find out these things, how can we best use that knowledge to help our students learn? What shall we do about the new technologies—computers, video tape and disks, etc.? In what ways can they be profitably integrated into the physics curriculum? What can we do about integrating "modern physics" into the curriculum? How can we improve laboratory instruction, and how can we get excellent, inexpensive, laboratory and lecture demonstration equipment? And, by no means least, how can we motivate our students to the study of physics? These questions are earnestly discussed and debated everywhere in the world. By joining in those discussions and debates we have an opportunity to establish connections with our colleagues in other parts of the world and to exchange ideas with them. Such exchanges can often be mutually helpful in expanding horizons and in finding new ways to think about old problems. Again, let me invite you to examine the variety of physics education materials collected here on the exhibit table.

And now, finally, you may remember that I mentioned earlier that there is a conference scheduled to be held in Mexico in July 1987, the InterAmerican Conference on Physics Education. Professor Luis Ladera of the Universidad Simon Bolivar in Venezuela is a member of the steering committee for this conference. We are fortunate to have him here with us, and I would like now to ask him to talk with you about the InterAmerican Conference. It is a pleasure to introduce to you Professor Luis Ladera.

Note: Because Dr. Jossem's original presentation made heavy use of visual material, this printed version has been substantially edited and modified.

231

Celso Luis Ladera was born in Caracas (Venezuela) on April 6, 1942. He graduated (cum laude) as a teacher of physics and mathematics at the Instituto Pedagogico de Caracas and remained there, as an instructor professor, until 1972 when he got a scholarship from the Venezuelan Ministry of Education, which allowed him to go to Great Britain to do graduate work. In 1973, and 1977, he obtained the M.S., and the Ph.D., respectively, from the Department of Physics, University of Reading, for work in applications of laser light. Since 1977 he has been a faculty member of the Department of Physics of the Universidad Simon Bolivar. At present he is teaching the advanced physics laboratory for their senior/graduate students, and is involved in work on laser interferometry, non-linear optics, and mathematical physics. In 1983–84 he spent his first sabbatical in the Department of Applied Physics at Stanford University. He reports "I have always been strongly attracted by the problems of physics education, and have been active in trying to solve them, first as a high school teacher, and even later as Academic Coordinator of the Open Studies in Physics Teaching of the Universidad Simon Bolivar." In 1977 he suggested the celebration, in Venezuela, of the 1979 Maxwell-Einstein Centenary Year, and in 1979 he coordinated the celebration of a large national academic program to promote physics. In 1982–83 he was also active as a coordinator of an International Seminar on Physics Education. A member of the AAPT, and the OSA, since 1975, he is also a member of the International Committee for the Interamerican Conferences on Physics Education. He is also an Assistant to the Commision for Innovations in University Education of the Venezuelan Council for Science and Technology (Conicit). Since one year ago he has been active as a consultant in Optics for the National Service of Metrology.

The First Inter-American Conference on Physics Education

C.L. Luis Ladera
Depto. de Fisca
Universidad Simon Bolivar
Caracas, Venezuela

I wish to make an announcement that there will be a major meeting on Physics Education, in Mexico City this coming year. I will give you dates and more precise details later on, in the meantime I would just like to give you a short review on what has been going on in Latin America in recent years, in the area of Physics Education.

This year the Instituto Pedagogico of Caracas—the main national teacher training college in Venezuela—is celebrating its 50th Anniversary. This institution was created 50 years ago with the help of a mission of educators who came from Chile. This shows that the interest in Physics Education in the countries which once were Spanish colonies have been going on now for some years.

Several countries have large or medium-sized enterprising groups, Brasil and Argentina, particularly, and Mexico as well. In Venezuela we too have a group of people interested in the problems of Physics Education. This does not mean that similar groups and interests do not exist in the rest of the Latin American and Caribbean countries.

We have so far had several conferences, seminars, and regional meetings on the general area of Physics Education, and some of these have tried to solve or to attack problems, I mean specific problems in Physics Education. Since the '60s, we have had at least three major conferences and somewhere between five and 10 minor events. Two of these events are worth further mention. One was held in Brasil, in 1963, and another took place at the Universidad Simon Bolivar in Caracas in 1975.

Thanks to the work of the late Professor Robert Little, from the University of Texas at Austin, and of the AAPT Executive Board—I think Professor French was the President at that time—we received from the States an invitation to work together, for the first time, for an Interamerican Conference on Physics Education. As a matter of fact, there had been interaction taking place between physics teachers and professors from North America and the people from Latin America and the Caribbean before, in those major events I already mentioned. Now, for the first time, we are going to meet to share our common problems and to make a joint effort to try to shed some light, or to find new ways, towards the solution of those problems which are pertinent to physics education.

Thus, on the 18th of July, 1987, we will meeting in the so-called First Inter-American Conference on Physics Education in Mexico City,. This for the first time, and officially, incorporates the whole American Continent, and that meeting will run for at least nine days.

The first two days will be devoted to workshops mostly in the way or the style with which you are familiar in the States—I mean, the workshops that you usually have here run by the AAPT.

The conference in Mexico, in July '87, will be devoted to the theme of creating a Physics Education network for the Americas. We will be producing variations on that theme, and we expect to really get down to finding a solution to the problem of producing regional projects, in order to solve those problems of Physics Education which we discover we have in common in countries which are now being developed, or countries with post-industrial areas, or countries like the United States.

There is a short history which goes with this forthcoming conference. Last year, we had an event in Guatemala in which representatives from different parts of the Americas got together. We gathered in the City of Antigua, in Guatemala, as members of a Steering Committee, and proceeded to create the so-called International Council for the Inter-American Conferences on Physics Education. Several of you in the audience here belong to this body. I am honored to belong to that Council.

Since the creation of the Council, we have been working on the conference. Because the venue chosen at Antigua was Mexico City, the President of the Council is one of the leaders of physics education in Mexico, Professor Jorge Barojas, of the Universidad Autonoma Metropolitana, at Ixtapalapa, in the Federal district of Mexico.

The problems that we expect to tackle there will not be unfamiliar to any of us. Thus, we shall be dealing with problems of the classroom and the laboratory, how to incorporate technology and new technology into Physics Education, or even how to present new developments such as the problems we have been dealing with in this present conference at Fermilab.

We expect about 250 people, mostly from the Americas, but we hope to host observers from the sponsoring institutions and the rest of the world. Some people from Europe and Asia will thus be there. The attendance, of course, will be divided between people working in the secondary schools and people working in universities. The topic of research in physics education, we hope, will get a good impulse and increasing awareness of its importance. The Conference should serve to promote such research in the Caribbean, in Latin America, and, of course, in the States and the North American continent.

Well, I think that what I have said so far describes mostly what I had to say about the coming conference. I should add who the sponsors are going to be: the AAPT, the Mexican Government, the Organization of the American States, the International Commission on Physics Education, the International Union of Pure and Applied Physics, and UNESCO, though we are not sure about the latter yet. We are also confident that the APS, the National Science Foundation, and the Institute for Theoretical Physics at Trieste will be strong sponsors.

If you have any questions about the Conference, I will be glad to answer them. I brought with me a few leaflets which describe a little more detail what the coming event in Mexico will be, and I will be glad to distribute them among those of you interested. Some of the leaflets are written in English, others in Spanish.

PROFESSOR AUBRECHT: I would like to ask a question. Could you or Professor Kelly tell people here, some of whom are high school teachers and some of whom are university teachers, how this conference might be of interest to either or both of those groups? In other words, why would any of our high school participants want to think about coming to this? Are there going to be special workshops for secondary school teachers? Is the idea of the secondary school teacher's problems going to be integrated?

PROFESSOR LADERA: Yes, they are going to be integrated. The workshops are aimed at both kind of teachers, and we expect groups of people from both fields, high school and universities.

For instance, problems or topics to be treated in the Conference will be physics textbooks, teaching contemporary physics, planning and building low cost laboratory equipment, applications of cognitive processes science to physics teaching, problems of physics teaching, physics for non-science students, and, in addition, relations between the various national physics teaching communities.

So, we expect really enterprising people to go there and work hard, coming from the whole continent. And, of course, we expect a delegation from the States, although surely not a 200-member one. We do hope to have a significant number of teachers from the American secondary schools and universities. And, again, this assertion also applies to Canada and all countries in the southern and central part of our continent, and to the Caribbean region too.

QUESTION: Will the language of the Conference be generally English or Spanish?

PROFESSOR LADERA: Well, we are having some problems there, but we are working hard on them. The style of the conference will be built on the use of several media. There will be plenaries and forums, for which there will be simultaneous translation. For the more technical sessions and the discussion groups, the Mexican organizers, under the leadership of Professor Barojas, are already working hard in order to have two overhead projectors ready for each of such sessions, and ask and help every speaker to produce two sets of transparencies, one in Spanish and the other in English.

One advantage of incorporating university people into the organizing committee of the conference is that most of us have come to English-speaking countries to get our postgraduate degrees. So, I think we will manage somehow to translate for the people who only speak Spanish what is being said if the speaker speaks in English, and vice versa. Thus, we expect to manage and solve those communication troubles quite efficiently. I am pretty confident that our Mexicans colleagues are organized to do just that.

Roundtable Discussion: Innovative Ideas In Teaching Modern Physics

Anthony French, Moderator
Isaac Halpern
Peter Lindenfeld
Joseph Priest
Robert Resnick
Raymond Serway
Frederick Trinklein

Introduction

Anthony P. French
Department of Physics
Massachusetts Institute of Technology
Cambridge, MA 02139

We are going to be very tight for time, as we have been throughout this conference. But probably even more so today. So I'd like to get going, and try to be as tough as I'm capable of being with our various speakers.

My name is Tony French, from M.I.T. And I am going to act as the moderator of this session.

And as you know, the title of this conference is "Quarks, Quasars, and Quandaries." This is the quandaries session of the conference.

Well, for the past 72 hours—I think it probably seems a lot more to most of us—we've been having a feast of presentations about what's new and latest in particle physics and cosmology.

And I can't imagine—and I'm sure I speak for all of you—I can't imagine a finer set of presenters of the information, both in the formal talks to us and in their participation in the informal discussions. It has been absolutely terrific.

It has been a great experience. And in fact, following from that, I can't help being reminded of a story, which I'm told is true, about an Army training manual in World War II. It was a training manual for infantry officers. And there was a question to which there was a unique right answer.

The question was: What should be done whenever possible? Now, some of you may have your own ideas about that. But the correct answer was, you should keep your troops in the shade along the line of march. Probably not what you would have thought of. But we should certainly have this kind of conference as often as possible.

Anyhow, here we've had this terrifically rich feast. Some of us, I expect, are beginning to reach for the Rolaids and the Alka-seltzer and the Maalox, or whatever, at this point. That's natural for a 72-hour feast.

But now, we have to begin to think seriously about what do we do with all this. What *can* we do with all this? What *should* we do with all this?

And again, if I may give a personal anecdote, which I can vouch for: At M.I.T. a number of years ago the inorganic chemistry syllabus was being reorganized. And I heard a talk by a very staid professor of inorganic chemistry.

And he said, "Well, you know, we've been revising this program and we've been bringing in some new stuff. But in the process," he said, "Nickel got squeezed out." And it conjured up for me a picture of nickel being extruded somehow. We, too, have the problem of what to squeeze out.

We are faced with serious problems, with modifications, either major or minor, in how we do our teaching in physics—and of course, that's the main concern from here on for the year or years to come. And it's going to be undoubtedly a debate—which will never have one single conclusion.

But this morning here we have a group of people who are going to address these questions in different ways. And let me just then introduce the members of the panel here.

We have Professor Halpern, "Ying" Halpern, University of Washington, a nuclear physicists. He was a young lieutenant j.g. in the Navy at Los Alamos in 1944 when I went there as a very junior member indeed of the British mission. That's where we first met, so it has been a long-time acquaintance.

Then, we have Peter Lindenfield, from Rutgers, somebody who has been very actively involved in physics education over many, many years, essentially all of his professional career.

We have Robert Resnick, known to all of us, I'm sure. We have Ray Serway, Fred Trinklein, and Joseph Priest, all present, in particular, in their capacity as well-known authors, or co-authors, or both, of widely used text books.

And so, that's our team. And we are going to begin by asking Professor Halpern to give some thoughts of his about what has been happening here.

Some Things I have tried in Modern Physics Classes

Isaac Halpern
Department of Physics
University of Washington
Seattle, WA 98195

I think that there is a sort of complementarity principle in the teaching of physics. It has to do with the conflict between conveying to the student what you either know or believe and letting the students know how you've come to believe it. In the first instance you are trying to teach the student to use an established science. In the second, you want to give the student an impression of how science is made. These are often distinct and somewhat incompatible goals.

The first emphasizes a sort of tightness of mind; mainly the use of logic and deduction to guarantee correct answers. In the second, we are reaching for a looseness of mind. It has components that scientists particularly enjoy: the use of analogies, of apt approximations, plausibility arguments and even good guesses.

In class I often feel the strain of trying to do justice to both viewpoints at once. I find the strain more severe in modern physics than in other branches of physics, because the student does not come to modern physics with a set of familiar, relevant experiences, i.e. with a context which he can use to judge the new information and assimilate it.

For that reason, in my own courses on modern physics, I try to start way back and favor the latter—namely, the "how come" view instead of the one focused on the information itself.

There are some special subjects I take up because they orient the student to problems of understanding and explaining rather than to the 'hard facts.' For example, with cosmology, I generally start by going over with the students some myths from primitive cultures which have to do with natural phenomena. Let me tell you one such myth.

In a Paiute Indian legend that deals with the seasons the sun is carried through the heavens by sun-man. But he is sort of old. And when fall comes around and it starts to get cool, his own son comes to him and says, "Dad, look, you're sort of old, and it's getting cold now. Winter is coming on. Let me do the sun job." And he takes over, carrying the sun through the sky each day until spring. Since sun-man is old and moves slowly, whereas the son is young and strong and moves quickly through the sky, we can understand why the days are longer in summer and shorter in winter.

Now, let me stress at the outset that I do not use these myths to say, "See how primitive and dumb those people were. Contrast that with how clever we are now that we have SCIENCE." Just the opposite. Mythology is science too, rudimentary science. There is a real continuum between early and present- day accounts of natural phenomena. In fact there are more similarities than differences. After all, in this story that I told you, (1) they pick things to talk about that they think are very important, (2) What's more, the stories are based on close observation, (3) And finally, they involve some sort of argument or reasoning. These are the critical ingredients of any science.

What I expect from the students when I tell them such stories is that they discuss them critically. I could not ask them to do so with unfamiliar material. For example, in the discussion of the sun story, the students will invariably pick up the inversion of the length of day and the warmth, pointing out that it is the first which causes the second and not the other way round as the story assumes. It is the sun's presence that makes it warm outside. Think of what happens when a cloud covers the sun. Some students will even point out that the basic observation about the sun's speed was incorrect in the story. The sun does not move more quickly through the sky in the winter; it just has a shorter arc. What I'm trying to do with these stories is to develop in the students an appreciation of science as an activity, an activity where one listens hard, where one tries to identify and criticize bad arguments and learn to justify good ones. I find that without exposing the students to some really crummy arguments, it is difficult to develop their critical judgment and taste in science. If you want them to learn how to engage with real issues, you have to give them a chance to practice at it.

In atomic physics, I also like to start way back with considerations like those of Lucretius. What arguments, I ask them, could have been made a long time ago for the atomic hypothesis? Generally some student will mention that the existence of crystals seems to be such an argument. I then try to get the student to sharpen this argument, by noting that crystals of a certain kind all have the same structure, even though they come in a variety of sizes. This strongly suggests that they have a modular structure and is an important step toward the atomic picture of matter. Then, if you realize that crystals are made out of the same things, i.e. the same chemical elements that other things are made of, that pushes you to recognize that maybe the modular structure is not restricted to crystals, but must be regarded as universal. This is the kind of argument a thoughtful student is capable of carrying through from his own experiences (especially when nudged by the teacher.)

In atomic physics courses, I also try to use some of the early or mid-19th century arguments for the size of the atom, deducing it, as Figure 1 shows, from a comparison of surface tension and the heat of vaporization of a liquid. By comparing the numbers of bonds in a unit cube with those in a single layer you can easily obtain a rough estimate for the size of the atom, as Kelvin and others did. This is an example of an argument based on dimensional scaling. Scaling is a general and powerful technique, especially useful in problems of biological structure.

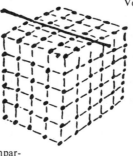

An example of scaling
Volume vs. Surface Energies
Kelvin

Figure 1. Deducing the size of a molecule by comparing surface tension to latent heat of vaporization.

A third subject that I stress in a beginning course in modern physics has to do with familiar symmetries like axial symmetries and bilateral symmetries. One of the things I expect the student to pick up about bilateral symmetry is that because there is a reflection plane, there are two defining directions (those that define the plane) which are responsible for the symmetry. For example for the bicycle, the two directions are that of gravity and the direction that the bicycle is designed to go. What are the critical directions for each of the objects in Figure 2? With bilateral symmetry, one question often raised by students (or by me) is, how come so many living things exhibit bilateral symmetry on the outside, and do not on the inside? (Mammals are an example).

BILATERAL SYMMETRY

Figure 2. Everything with Bilateral Symmetry has two critical design directions (those that define the reflection plane). What are these directions in each of these man-made objects?

One other thing I do in the symmetry business is to try to get the students to play with wallpaper designs, for example to analyze them by locating axes of rotation and deciding whether they are two-fold, three-fold, etc. They should also find the reflection planes (Figure 3). Finally, I also ask them to design some patterns of their own, given certain prescribed symmetries. In the course of these exercises some students manage to discover and even prove some simple theorems, for example, that wallpaper cannot have five-fold rotational symmetry.

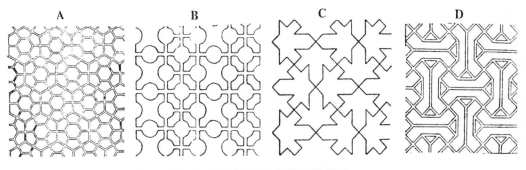

WALLPAPER PATTERNS

Figure 3. Identify reflection planes and points about which there's rotational symmetry. (Specify how many fold.)

Well, so much for the somewhat uncommon subjects that I find it useful to include in modern physics courses. Their shared quality is that they are close to the students own experiences. This makes it possible for the students to manipulate the ideas being discussed. I hope this instills in them a sense and pleasure in pushing ideas around and that it stays with them as we go more deeply into modern physics, taking up less familiar subjects.

Let me finish up with a few things that are not related to specific subject matter. One of the things I've tried to do, consciously, is to have each of the lectures pretty much an independent standing unit with a beginning, and a middle, and an end. If I can, I try to run it like a detective story where by the middle of the hour we have a clear dilemma. If I ever find myself starting a lecture with the words, "As you remember, last time we derived . . " I would know that I am doing it wrong. I understand the pressure to teach physics in this linear one-dimensional way. It comes in part from our perception of the generality and continuity of physics. But I do not think that one necessarily loses the sense of connections and unity by giving compact one-main-subject lectures. It may be better to offer your pearls to the students loose — without insisting on stringing the necklace for them.

The one-dimensionality I speak of is forced on textbook writers, because their chapter five has to join chapters four and six in some way. But I don't think it is the right mode for spoken lectures. And so I try to make sure that the main point of each lecture can be understood from the materials of that particular lecture.

One other thing I do is that whenever I feel that we've consolidated a major point, I will tell the students (after I've summarized the consolidation) that I hope that some of the things that have been questions in their minds have now fallen into place, but that, if things are really going right, this consolidation should almost automatically raise some new questions in their minds and that I would like to hear those questions.

I think it is important to solicit questions. To show how serious I am about it I also do it on exams. (Students are particularly responsive on exams.) I get a lot of surprises from this kind of feedback. Here are some questions that I picked up in a recent atomic physics exam (Figure 4). They do not all have the same degree of cogency to the things we had just covered in class, but I have found even the off-beat questions rewarding for me and for the students.

STUDENT QUESTIONS

What caused the Big Bang? How did all that energy manage to collect in one spot?

If air molecules are as numerous as you say, and are colliding as frequently as you say, why isn't it more noisy?

How come only one electron can exist in each level of an atom? What happens when an electron tries to enter an occupied level?

How can something be truly solid when it is constructed of tiny moving particles?

Do scientists know how long the Universe will last?

What forces keep a proton from blowing itself apart?

Is there anything like a mutant atom?

Is it true that modern physics is beginning to confirm the Biblical notion of a Supreme Creator?

Why have so few women made contributions to physics?

What is the Universe expanding into?

Figure 4. Questions contributed by students taking an atomic physics exam.

In summary, the teaching devices and the special subjects which I have stressed in this talk are meant to help the student realize his obligation to interact with the ideas that come his way, not just to learn them. (Applause)

DR. FRENCH: Thank you for a very thoughtful presentation. And I hope that it is thought-provoking. And we've been discussing how best to conduct this session. If there are any questions relating very directly to what Professor Halpern has said, then we could deal with one or two questions maybe right now.

But of course, he will be here throughout the session. And therefore, I hope we will have time for a more extended discussion later.

If anybody has a very particular question at this point—yes?

QUESTION: Are the students encouraged to raise questions in class as well as on examinations? Do you answer them when they are raised?

DR. HALPERN: Sure, I do. Those that get raised in class, I try to answer at once. Those that get raised on exams, I respond to individually on exam papers. Some of them, as you can recognize from the examples are tough to deal with simply. But that's part of the game.

Radiators and Absorbers

Peter Lindenfeld
Department of Physics and Astronomy
Rutgers — The State University
New Brunswick, NJ 08903

Let me start by asking you to look at the schematic diagram of our enterprise on Figure 1. It shows the teacher radiating knowledge, transmitting skills, aided by a support system, usually very weakly coupled, which includes a miscellaneous collection of professors, computers, books, even a structure meant to look like the building which we are in now with its resident curlyheaded guru. Only a fraction of the radiation gets through the fog, and an even smaller fraction is absorbed by the receivers, the students. Our problem is to improve the matching of emitter and receiver.

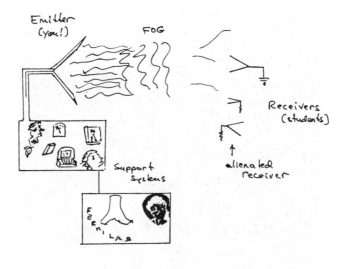

Fig. 1

We all know what we want. We are missionaries, as all teachers are. We know that physics has wonderful attributes of the useful on the one hand, the beautiful on the other hand. It is basic, fundamental, and important for the study of all kinds of other subjects. It can be practical and also abstract and theoretical. Occasionally it is even poetic. (See Fig. 2.)

Fig. 3 shows the cover of *Physics Today* of March 1955. It is a famous graph which shows physics going downhill, at least in terms of the coordinates of the graph, i.e. the percentage of the physics enrollment in American high schools from 1890 to about 1950.

You can have different attitudes towards this result. You can even say, "Maybe that's the way it should be." You will not be surprised that physicists, in general, considered the downward trend to be disastrous and that heroic efforts were initiated to change the situation.

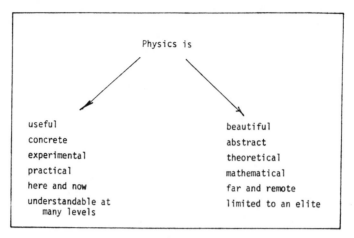

Fig. 2

Fig. 3

It is now more than thirty years later and we can try to see to how successful these efforts were. Fig. 4 shows the same graph carried further, to about 1980. It is rather hard to find the right numbers, but these are comparable, that is, they were gathered in more or less the same way. We see that the downward trend continues.

Part of what has changed is that we started with a system close to that which still exists in a large part of the world where all students have to take physics. Now, however, we live with a kind of demand-feeding system where the students have a considerable amount of choice. Their choice depends to a large extent on what we do.

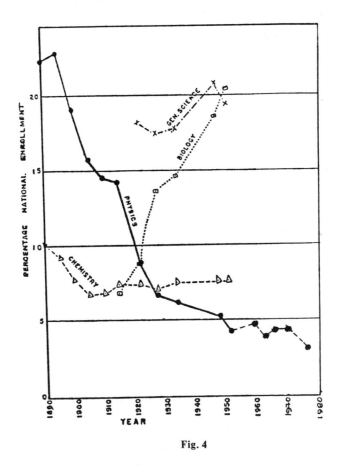

Fig. 4

Fig. 5 shows one of my favorite diagrams. It comes from a presentation of Holton, Rutherford, and Watson in 1967, just before they came out with "Project Physics." Its message is more or less as follows: The old courses are fine for students who come with a strong interest in physics and who also have great ability. For them the system works quite well. But we are failing with the large fraction of the population that does not come within that range.

Holton and company expected to capture a much larger fraction of the spectrum of students. They thought that they would find a new audience among the less mathematical but more humanistically and socially oriented population. Well, the students that they had in mind, if they exist at all, turned out to be elusive. They still don't seem to show up for physics courses.

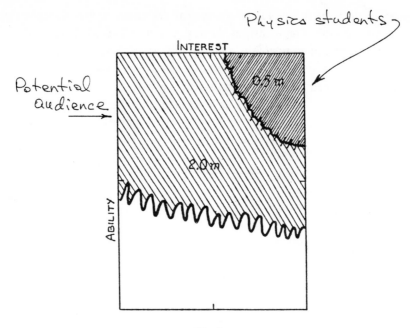

Fig. 5

I think that it is important to continue to try to bring the best of what we can do to the best of the students. This will presumably happen in the Illinois Science Academy which is getting started near here for a specially-chosen elite of students. But I hope we don't leave it at that. I hope we remember that we have a responsibility to a far greater population.

Look at the attributes of physics on Fig. 2. You can go in two different directions: the concrete, the practical, the down-to-earth and experimental on one side, the abstract and mathematical on the other side. My message is: please do not not forget the direction on the left. Of course the two sides are intimately entwined, but I am concerned about the relative emphasis which we place on them.

My hope is that you can make the modern material which is the subject of this conference, and physics in general, accessible to a larger fraction of the students. I suggest that what is needed is more direct experimental experience with physics.

For earlier generations it was easier to grow into physics through experience. Teenagers used to tinker with cars, or make radios. That kind of activity has almost disappeared. Cars have become too complicated. Radios are no longer made with components that one can see and connect.

To some extent the kind of activity which I am thinking of has been taken over by the computer. I have no complaint about that. Computers have generated a variety of activities that a modern young person can get involved with. But the activities cover only a restricted range. They almost never include any consideration of the hardware of the computer. They are primarily for the mathematically inclined. I'm not sure what there is for the teenager who used to tinker with gadgets in the basement, who used to take things apart and perhaps even put them together again. I think that kind of activity is missing and I'd like to see more of it in the physics classroom.

250

There are different kinds of people in this world. Some are attracted by the theoretical and the abstract, and that's terrific. I wouldn't want to do anything to discourage them.

But I don't want to forget those that are on another part of the human spectrum. Those who like to get their hands dirty, whose learning tends to start with the direct experience of their senses rather than through the channels of the brain.

We are now nearing the end of this conference, and it is a little late for me to criticize the program. Nevertheless I don't want to hide my feeling that it is too heavily weighted toward the remote and the abstract. I worry that too many of the concepts are so theoretical that once again we are shutting ourselves off from a large part of our potential audience.

In the marvelous surroundings in which we find ourselves in this building at Fermilab it is easy to say "Physics is Beautiful," and to neglect the subject's more down-to-earth aspects. My plea to you is to remember and to emphasize, regardless of which part of physics you are dealing with, as much as possible of the concrete and the direct, that which can be understood with a minimum of the jargon and special language which we are so fond of.

As an illustration of the direction that we have tried to go in at Rutgers, I would like to describe a project which also deals with the teaching of modern physics. We are developing units which we call "modules," which consist of booklets with kits of material for experiments which form the basis for a week or two of study in a regular physics course. The emphasis is on the experimental and practical and to some extent on the technological aspects. We think that it is a good idea to connect the work to phenomena which the kids have contact with, to the world that they are going to live in.

I am not talking about making it easy. There was a time when there was a great deal of discussion of the word "relevance" in teaching. It got a bad name. In its worst sense it meant not only recognizing the limitation of the student's horizons, but reinforcing those limitations, rather than opening the door to the wider world. If I advocate relevance here it is in a different sense, as a reminder that we are talking about a subject which, while it is perhaps poetic and certainly beautiful, has constant, day-to-day, continuing application to the life of everyone.

We plan to work on five modules this summer. Two are biologically oriented and three are in physics. The one that I am most directly involved with is on semiconductors. Our "team" consists of three teachers, working with Chris Palmström from Bell Communications Research, Walter Brown from AT&T Bell Laboratories, and me. The project director is George Pallrand, who is in the Graduate School of Education at Rutgers. Each of us comes from a different direction. Each has to make adjustments for the others' different experience, background, and approach. Together we think we can produce a better product than any of us could do alone.

Last summer we made rectifying diodes. That turned out to be really easy to do, starting with silicon chips and making contacts either with indium solder or with silver paint. Unfortunately it is not so easy to understand and explain.

This year we want to go a step backward and a few steps forward. We are preparing a box with a holder to which the sample chips can be attached and where they can also be heated. That will allow us a look at some of the most fundamental properties of metals and semiconductors. We will be able to show how conduction varies with temperature, and relate the differences between different materials to the differences in their structure.

The next step is to use one of the new, strong, little permanent magnets to produce a voltage, the Hall voltage, at right angles to the current. The arrangement will allow us to study the effect of a magnetic field on moving charges and to measure their sign and concentration. That should also be easy to do and not too hard to understand.

Finally we want to do something that may turn out to be harder, and make a field effect transistor. It is the most common transistors today, and it works like a triode. We are talking about really modern physics, and at the same time about something that will remind those of my generation of devices that they grew up with and were familiar with long ago.

I can't promise that all of this will work. But I think you see what we are trying to do. We are trying to bring a piece of the real world to the classroom. When kids use their computer they will then know that it is made of familiar parts that they understand. We hope that the direct experimental experience will make the material more meaningful, and more relevant, if I can use that word again.

There is one aspect of our program that I want to emphasize very strongly. In the past decades a number of educational innovations have been developed by professors and research scientists, and then "handed down." Some of what was produced that way is very good, but some of it is badly matched to the needs of the teacher, to the context of the classroom, to the background of the students.

We are working with the teachers at each step, planning, building, writing, and experimenting together. We hope to include them in everything that we do, and we hope that they will include us in a continuing, vital mutual support system between the schools, the colleges, and the research laboratories.

The teachers are the ones that will have to make the program work. They are are ones who will have to show that there is some overlap between the emitting function of the teacher and the receiving function of the student. They are the ones who have to dispel the fog, to get the signal through. That's why it is so important that they help to match the wavelengths, and help to try to raise the absorption on the student's side far beyond the present limits.

QUESTION: I was wondering if you had published something about the project that you described or how we could get more information on the specifics, such as how to make diodes.

DR. LINDENFELD: I am glad you asked me: I brought some reprints of our *Physics Today* article, which appeared in November 1985. If you need more, please write to me.

To make diodes take a chip of silicon, and make contacts with indium or with silver paint. Each will be rectifying. The problem is that if they are the same the two rectifying contacts will cancel each other and there will be no rectification. The trick is to make one much smaller than the other, a big blob on one side and a little blob on the other side. The big one should have so much area that its resistance in either direction will be small enough to be negligible compared to that of the other contact.

I am willing to describe this in more detail. And eventually we hope to have kits and booklets.

QUESTION: Professor Lindenfeld, you showed a graph that showed the physics enrollment going down, down, down.

At one point you said, "here we made a change and let students choose whether they wanted physics or not," suggesting that before we gave them that choice, things were better.

Is part of the answer that physics should be a required subject?

DR. LINDENFELD: I suggested that one of the reasons for the downward trend was that the curriculum became looser. I didn't say whether that was good or bad.

The physics enrollment would be larger if everybody were forced to take physics. But I don't think that's in the cards, and I'm not sure if it were that I would be in favor of that.

There are many factors involved in the declining enrollment. Students have a choice. We too have to make some tough choices on the amount and kind of physics that we teach.

In the reprint which I mentioned earlier there is a section called "Freedom Unused," whose message is that high school teachers often feel themselves far too constrained by a perceived pressure to include lots of subjects. It seemed to us that teachers should feel free to cover some topics in depth and to leave some out. Textbook writers want to put in everything possible, because they are afraid that someone will say, "Why is there no fluid dynamics?" or "What about the third law of thermodynamics?" Many teachers feel that they have to "cover" every page. The test makers follow with questions on every subject. This is a trend that should be opposed and, if necessary, ignored.

DR. FRENCH: I think I will call a halt right now to further questions. I would like to say, however, that already in the planning is indeed a conference, a counterpart of this one, which will be concerned primarily with the areas of modern physics not dealt with here.

And now, I'd like to ask Bob Resnick to talk. I didn't mention, but I'm sure you know that he's the new vice-president of AAPT. And in that capacity, he has all the concerns with physics education in general that are represented by his claims as an author.

Overcoming Constraints and Problems

Robert Resnick
Department of Physics
Rensselaer Polytechnic Institute
Troy, NY 12181

I'm grateful to be able to follow the previous two speakers. What I had in mind, and sketched out only last night, was to try to address very concretely how to overcome the constraints and the problems that I heard people mention at this conference arising from trying to get all this modern physics into our course. I wanted to be very concrete and specific about that. But I felt very constrained by the limited time and realized that I couldn't make some necessary background comments. Fortunately for me, however, the other speakers already have made them.

What I'm going to do is tell you how, at RPI—at Rensselaer Polytechnic Institute—we face some of these same problems. And I'm very cognizant of the fact that we're not typical of the "real world" out there. We had a four-semester course, a two-year course, in introductory physics, and the last semester, 25 percent of the course, was devoted to what you're calling modern physics. The real world hit us when we were forced to cut back to three semesters. I know that still sounds luxurious, but that's the kind of cut, isn't it, which we all faced around the country? So there we were, faced with one of the constraints I was going to talk about, *the time constraint*. The length of the course limits how much, if any, modern physics you can get into the course.

Well, we had a belief at RPI, illustrated by a comment that I first heard maybe 35 years ago from Walt Michels, that we really shouldn't try to "cover" physics, we should try to "uncover" it. That meant that we did not want an encyclopaedic coverage that was shallow and just got you the vocabulary, which you then might easily forget anyhow, leading to dilettantism and sophistry. But we would like to achieve some real depth of understanding. We didn't want to squeeze four semesters of work into three. Our first temptation was just to forget that last quarter of the course, no modern physics.

Instead, with this terrible time constraint, we performed a different surgery. We decided to eliminate part of classical physics. It was less important, perhaps, to treat some classical physics than it was to get the modern physics in. We made the decision to eliminate thermodynamics and fluid motion. Thereby, we gained about half a semester of time. And in the last half of the third semester of the course, we now have time for atomic, nuclear, and solid-state physics. I'll come back to why we made the choice to add only those areas of modern physics.

Now, why did we drop thermodynamics? Well, frankly, the kids were least interested in that of any part of the course. Moreover, we're a technological school. All the science students take a year of chemistry as freshmen, where they hear a lot of the vocabulary of thermodynamics; all the engineers get a year's course in thermo as sophomores, and so forth. We were quite sure that down the line the students would at least hear some of those concepts, much as we hated to give them up in our first course.

Please understand, I don't propose that you all follow this particular model. Because, if you do, in a few years the NSF will sponsor a conference on bringing thermodynamics back into introductory physics courses. You can make your *own* choice as to what areas to drop from classical physics and what areas to add from modern physics in order to do justice to everything you teach. So that's one idea.

What's another constraint I hear? Well, we don't really have the freedom to do what we want in the course. We teach mostly engineers and pre-meds. We have to pay attention to what the chairmen of those major departments say. We want to enroll their students in our course. Amongst the reasons for that is that we want to keep our jobs. We're influenced also by the fact that the students, too, have ideas on what's relevant, what they like. And if you want to win them over to physics, if you want to keep them in the course, if you want the word to go out that it's a good course to take, you bend with the wind on what interests them.

I just completed conducting a survey of the physics department chairs in the country for the APS Education Committee, and the overwhelming consensus was that we should move towards applied physics and compete for those students that engineering and computer science are getting. And if I read anything out of that, that would say, for heaven's sakes forget about general relativity and elementary particle physics, but go for nuclear, solid state and atomic physics. Now, these physics chairmen might be wrong, but that's the mood out there. And so there is a constraint you have, the *enrollment constraint*. If you want to face this, what do you do about it?

Well, first of all, in response to that, we chose to include some modern physics—and again, as I said, to do it with some depth, at least the depth with which we covered the classical topics. We realized that we first needed a certain grounding in the ideas of quantum physics. So we gave that first and then we applied these ideas to atoms, solids, and nuclei. And that's it. We stopped there.

What about the rest of modern physics? Well, we invented a technique at our school, which could be useful elsewhere also. We realized that, while students are studying the meat and potatoes of physics, they might be looking ahead to the dessert part of the course at the end or to desserts we don't provide. But in China, I found, the dessert was right in the middle of the course, i.e. the meals, it wasn't at the end. And there was no reason why we had to wait until the end either.

So we created what we call one-credit courses: one lecture a week for 14 weeks a term, pass-fail, no final exam, just turn in some report. And you come and you listen to someone lecture—for 14 times—on some dessert-like subject. And you get credit for it. It isn't too much to add to your curriculum, either. What were these topics? Lasers and holography, tour of the solar system, cosmology, general relativity, bio-materials, medical physics; things of that sort. The faculty was enthusiastic about it. These courses were at the level of *Scientific American*, or perhaps a little lower, and reading references were given for students who wanted more.

Well, the kids flocked to those courses. I was amazed. When we analyzed the senior graduating class two years ago, it turned out that 75 percent of the engineering students took at least one of these one-credit dessert courses. Many took several. And so, there we were, getting the word across that physics is exciting, on the forefront, and here it is, and we're not going to test you on it or give you a grade on this stuff, just show you how applicable it is and where it's heading. I don't know if you can use that technique, but it worked beautifully for us.

Then, there's another constraint. It might be called the *foundation constraint*. I define modern physics as that physics which will require an understanding of relativity and quantum mechanics, or quantum ideas, in order to be understood. Now, that being the case, you immediately run into the problem that both of those theories presuppose a knowledge of concepts or ideas evolved in classical physics, whether wave concepts, or conservation ideas, or field concepts, or ideas such as symmetry and unification, and things of that sort. You need a good background to get securely into that, content-wise. In addition, these areas are remote from our immediate sense perceptions. They're not part of the ordinary macroscopic world. It's the high-speed world, the small-scale world. And therefore, it intrinsically becomes more abstract to the student. So you have to provide the time for the foundations, and the extensions are honestly a difficult thing to do. We have to face up to that.

Well, here, I guess what we do at our school can be described by three words: *overview*, *sprinkle*, and *broaden*.

Overview. We begin each segment of the course, the mechanics, let's say, or optics, E&M, wave motion, whatever you do, with an overview of the field: What we're going to do, where we're going to wind up at the end, what's its relevance to other areas of physics, why we need to know it, the unifying ideas. Then we go into it. And then, at the end, we tell them what it was, summarize. I was impressed with some of the high school work here, overview ideas, that I heard from an excellent teacher in California, Arthur Farmer, where he does this at a much deeper level. But that idea works. Let them know where you are, where it fits in, and what it applies to out there in the great wide world. So now, here's why you've got to do these concrete things in our course and learn how to solve these problems we give you, and so forth.

Then *sprinkle*. What do I mean by that? We can sprinkle a modern physics unit into the classical physics, and we've been doing that all the time. There are all sorts of examples of conservation principles that can come from nuclear physics and elementary particle physics, for example. You surely know that. When you discuss resonance ideas, you can bring in modern concepts. Wave ideas, field concepts, unifying ideas. There's no time for me to go through all this, but as you go through the course, you can allude to all of these things and bring in these modern areas that you aren't going to treat coherently at that point.

We also have guest lecturers. There's a lot more to the world than textbooks. You know that very well. The lecturers have a great deal more freedom. And so, along with the course, we allow some time for a guest lecturer to come in and maybe give a talk on lasers and holography in the optics part of the course, maybe on the physical imaging methods used in medicine after we talk about magnetic resonance, things like that. When we finish gravitation, in comes one of our astronomers to talk about general relativity. It won't be on a quiz. We won't test you. But right at that point, it fits into where we are in the course. And we show films. We're lucky that we have a few good ones. Fred Leitner made one on superconductivity. That comes right in while we're discussing Faraday's law, and so forth.

Well, you've got the idea for sprinkle. And I think that that's a constructive thing to do. In recitation sections, I understand we're distinguished from most schools in that we assign thought questions for discussion; not just homework problems, thought questions. And those questions often point towards microscopic physics, or conceptual things like symmetry and conservation laws, and other things of that sort. It critically depends on who your instructor is, how well that can go across. But there again, you can get all these ideas in. We even go so far as trying to test on that, by having the equivalent of thought questions on the quiz.

And then, *broaden*. We give references to students. We know that not too many are going to do this reading. But they get excited about a book about Einstein, or the latest thing on elementary particles, and things of this sort. As a matter of fact, at this conference I disappeared for a while, to read a book, *The Second Creation*, by Crease and Mann. It is referred to positively by Georgi and other people. It's a marvelous, anecdotal book about physicists and the way they think. And as you read it, you learn physics. It's all in there. We can do more of that sort of thing, promoting that kind of book. And there are out-of-course activities which we do. I don't know if you have analogies to that. We have a physics club, and an astronomy club, and they'll have invited speakers, or they'll have exhibits, or perhaps they'll take people up to the observatory to look at Halley's comet and things of that sort. All this is outside the course, but it sure is modern and up to date. Also, some of our sophomores, while in the introductory courses, were involved in undergraduate research participation programs. And the word gets back that, "Hey, in physics you can get involved in that stuff." And so all of these things broaden—I don't want to overdo it, but you get the picture.

This then is overview, the sprinkling of ideas, and the broadening of things. You can get in general relativity, gravitation, elementary particles, and things like that. You can get the smell and the taste and the feel of it without necessarily giving them quantitative problem solving or a formal development.

What's the last constraint that I detect here? People said, "We don't have good materials for modern physics. And I don't understand it that well." Well, that's pretty honest, you know. *Knowledge constraint, materials constraint.* And that's where this conference fits in, I think. The educational materials available for modern physics at the introductorty level really are limited, certainly as far as the subjects that we discussed here are concerned. And that's one of the goals of this conference, to produce materials that we can use.

But I must say, there's more available out there already than many people think. One of the troubles is that it has a short half-life. And that's not our fault. Because what you write about elementary particles today, for example, is different than what was said two years ago. This is a rapidly changing subject. And I can't fault any textbook author for not having the latest word. Look at the date when the book was published; it's not continously variable after all. In any case, there are many paperbacks out there. There is no reason to constrict yourself to *one* book. You can get paperbacks. You can get all sorts of materials like that which can be kept up. There are xerox machines, too, and current articles can be reproduced. There are even some films available, such as ones on symmetry. And I realize, when I go back I'm going to have to put my money where my mouth is, and show people that there really are some materials already available. Each generation thinks the previous one didn't do anything, but at least that motivates them to do something on their own.

But, even if these materials are inadequate, the crucial element is still how much the teacher, the lecturer, understands this stuff, and what enthusiasm he or she has for it. And that's another goal of this conference. To get us to learn more about this, to get us excited and enthusiastic about this, to keep us learning this stuff. And in a very constructive way, this conference is trying to meet this last constraint that I see head-on, to produce the materials, to get us to learn, to develop some enthusiasm for it. And then, somehow, by all these techniques I was talking about, we'll get something into the course.

Finally, I want to say, remember that, although classical physics may seem old hat to you, it's not necessarily old hat to your students. Don't downgrade the classical physics that you teach in your frustration that you can't cover all of modern physics. It would be self-defeating, it really would be. Instead, I think you could express the excitement that's there in classical physics, too. I get the kids all excited in gravitation about space travel, and about communication theory when we get into electromagnetic waves—and there are so many phenomena in optics that they get excited about.

We need to recharge ourselves all the time. I realize that. Only if we teach enthusiastically will we be good. But I'm reminding you that we can teach the classical stuff with enthusiasm too. And the kids are interested in that as well. Moreover, if you do many of the things I have alluded to here, you can show the unity of classical physics with the modern as well.

So the chief thing is that, while you continue to learn—and we all do, all of us, all the time—do what you can do best, what you're most comfortable with, what you can do with some enthusiasm, and your physics course will be the best you can teach. It will be the best for your students. Listen to the so-called experts, but then go home and make up your own mind about what's best to do. And it will work best.

I'm going to finish with a story. I give talks about Einstein, and I emphasize that I'm not an expert about him. Although I may have written a relativity book, I didn't know the man, I didn't do research in relativity and so forth. But I add, that Einstein was skeptical of experts anyhow. And you ought to be.

And that reminds me of the story that Bill Fowler told about experts. It seems he cut his thumb very badly, and he had to have it taped and bandaged. And he despaired of that nuisance for several days. So he went to the doctor, and said, "Can't I get rid of this damned bandage?" And the doctor took it off, and looked at the healing thumb, and said, "I guess you don't need the bandage. But I want you to dunk your thumb in cold water three or four times a day." "*Cold water*," said Fowler, My mother told me to dunk it in *warm water*." "Well, your mother's wrong," said the doctor. "*My* mother told me to dunk it in cold water."

So much for the experts. We get our Ph.D. and then, with that authority, we teach others what our mother taught us. (APPLAUSE)

Dr. French: I think, in view of the shortness of time, I'd like to carry right on with the presentation by the other panelists.

Bob Resnick has rightly said that all kinds of resources other than books exist by which we can teach. But I think we all know that the textbook continues to be perhaps the most dominant single influence in what happens in our teaching of the subject.

And it is therefore particularly valuable, I think, to have three very successful and good textbook writers here with us today to give their own thoughts about—and their reactions to—what has been happening here these last three days, and what that makes them think about in correction with their role as authors.

Overview

Raymond A. Serway
Department of Physics
James Madison University
Harrisonburg, VA 22807

I was asked to join this panel and give a brief overview of the proceedings, where I think things are going, and how these ideas may affect the content of future textbooks in introductory physics.

This has been a very exciting conference. Many excellent ideas for bringing current topics in physics into the classroom have been presented and shared among you. I myself had the pleasure of being a member of one of the working groups. I am sure that you, as participants in this conference, will take many of these ideas back to your institutions and share them with your students and colleagues. In my opinion, if you fail to do so you will really be short-changing yourself and the long-term benefits of this meeting. You should also make other neighboring institutions aware of these proceedings. And by all means, get involved with regional meetings. In Virginia, for example, James Madison University hosted the Virginia Academy of Sciences this year. Most of the faculty in my department contributed papers to this meeting. Some papers represented reports of ongoing research projects, while others involved matters of educational interest.

First, I would like to comment on how future textbooks in introductory physics might incorporate some of the more exciting recent developments in the various areas of research. I wish that this conference had been held about two years ago, because the second editions of my calculus-based texts were recently published and could not benefit from this conference. However, I do plan to take many of these ideas into consideration in the future. It is impossible to say how this will be implemented at this time.

One excellent idea just proposed by Bob Resnick is to take these materials generated at this conference and sprinkle them throughout the course, wherever appropriate. This is an excellent motivational tool and one that will generate more interest in the course and stimulate classroom discussions. Another motivational feature which I have used in the second edition of my book is the addition of a number of essays written by guest authors. Most of these essays cover topics of current interest to scientists and engineers and are intended as supplemental readings for the students. Many texts also include optional special topic sections in order to expose the student to various practical and interesting applications of physical principles. In my opinion, however, the main objectives of any introductory physics textbook should be to provide the student with a clear and logical presentation of the basic concepts of physics, and to strengthen the understanding of such concepts through a broad range of interesting applications to the real world.

The manner in which textbooks will change in the future will depend to some extent upon the feedback we get from users of the books, as well as reviewers' comments, and so on. However the most important factor which will affect the content of textbooks will continue to be the constraints placed upon us by the curriculum. As educators, we are expected to cover a certain minimum amount of material. At the same time, we must satisfy the various engineering departments, and cover basic material of interest to other science majors. In the end, one usually tries to cover a broad range of topics in as much detail as time permits.

I also want to make a few remarks that represent some of my own philosophies as they pertain to instruction in introductory physics. First, it is very important to understand and be sensitive to the limitations of the students. For example, if you intend to "sprinkle" many topics in modern physics throughout your courses, make sure you do not overwhelm the students with more detail than they can comprehend. The same can be said of most topics in a course. Too much detail may cause confusion and cloud the main points of the subject matter. As someone once said, too much rigor leads to *rigor mortis*. Second, it is very important to present the students with a clear and logical discussion of the basic principles before you deal with many applications. Third, you should make use of the student's past experiences with the real world when trying to teach various concepts and principles. Finally, you should introduce the mathematics slowly so that the mathematical level at any time during the course is consistent with the development and interests of the students. I have found it useful to sometimes give a brief review of selected topics in mathematics, such as series expansions, approximation techniques, etc. I believe these are key factors which often separate the outstanding teacher (and course) from the average ones.

Finally, I would like to comment on the serious problem we face regarding the shortage of competent, committed science teachers. Professor Weisskopf has already pointed out the sad statistics on the opening day of this conference. You know you have a serious problem when you hear that there are only about ten teachers certified to teach physics in a given state. (I won't mention the state because I am not sure that the data is accurate.) Part of the problem of this shortage is the ever-present salary situation which needs to be more competitive with industrial salaries. We must also find means for increasing public interest in science, and convince the public of the need for improved science education at all levels. We as teachers may be able to improve the situation by encouraging our students to enter the profession of teaching and pointing out the various advantages of a career in teaching. I strongly urge you to do so.

Incorporating Modern Physics

Joseph Priest
Department of Physics
Miami University
Oxford, OH 45056

I have been trying to think of ways that we could incorporate the things that we've learned here at the conference, both into textbooks and into curricula at school. I've asked as many people as I could about ideas about this. I must admit, I didn't get a lot of help from them.

I just wanted to offer to you a couple of specific suggestions of ways that I think that some of these things can be introduced into a text and into curricula, and the ways that I'm going to explore when I go home.

My own experience in teaching university physics at the freshman level off and on for about 25 years is that one of the most exciting topics to students is special relativity. But when you sit down to be an author, you find that there's a big question about whether you should even include special relativity in your text, and if you do, where you should include it.

There is no question that students who will never say anything in the course up until the point where you get to special relativity, suddenly become very, very excited about that. The subject raises a whole host of questions.

I really see this material that was presented to us on general relativity as possibly doing the same sort of thing. Many of the derivations were elegantly simple and don't require a lot more than what's involved in beginning university physics. There are really beautiful experimental tests of those ideas. They employ very modern technology, which is appealing to people. And most of all, I think it really beautifully illustrates the scientific method in the development of theories and the testing of theories.

So I think that is one area that can be gotten into, both in textbooks and curricula.

The second area is this. I don't think that in beginning mechanics we make as much use of the idea of the bound state and bound systems as we should. We talk about the moon being bound to the earth, and the earth being bound to the sun, and what-have-you. But I think that you can extend that mechanical idea a long way.

We certainly do when we talk about the hydrogen atom. We calculate the binding energy and talk about that. When we get to the deuteron, we do the same things in terms of the nuclear force.

I think that, with a little bit of effort, you can also do this with some two-body quark systems. There are others as well. So I think there is a big opportunity to take advantage of some fairly simple principles of mechanics to at least get the ideas across and to instill some of the excitement about the particle physics that we've seen at this conference.

I'm sure that there are many other things that could be included, but these are two ideas that come to my mind, and two ideas which I'm going to explore when I go home.

Quandaries in Science

Frederick Trinklein
Long Island Lutheran High School
Brookville, NY 11545

Someone has said that it is more worthwhile not to understand something worthwhile than it is not to understand something that is *not* worthwhile. In keeping with that adage, I want to tell you that this conference has been eminently worthwhile for me. The question now remains as to what I should, in the words of the previous speakers, "sprinkle and vignette" into a high school textbook that is the old guy on the block. (*Modern Physics* was first published in 1922 and has been updated about every four years since.)

I was impressed by several things during the past few days. First of all, I find that I have a decision to make as an author. Since the lead time for a textbook is about two years, I have to try to figure out which of the things we've heard will be around for a while and which of them will not. So I asked several of the speakers in private whether their presentations were set in concrete. I won't mention any names, but the answers were surprisingly similar. "Yes," each one said, "my material will definitely be around, but I have serious doubts about what the other guys said." And yet, as one of the speakers said, there is an increasing interdependence between cosmology and particle physics. But that interdependence doesn't seem to extend to the credibility level.

So I am faced with a quandary. But, come to think of it, that is part of the title of this conference.

A principle that I have been using in my teaching and that has not been mentioned in this conference, however, is the Principle of Le Chatelier. In essence, Le Chatelier said that whenever you do something to nature, nature will resist what you are doing. This truth is as much philosophical as it is scientific, but I find that a great many laws of physics can be made more understandable through it. In fact, when you stop to think about it, even teacher-student relationships are corollaries of Le Chatelier's Principle.

This brings me to the main point of what I'd like to say in these few minutes. In high school, we face the problem of how to put across not only the ability of science but, more importantly, the *credibility* of science. In our enthusiasm for the field, I'm afraid we sometimes oversell our discipline.

Recently I attended a physics seminar at Brookhaven National Laboratory (if I can mention that institution here at Fermilab). The speaker was Dr. Max Dresden and his topic was the alleged fifth force. Dr. Dresden spoke for a long time, and when he was finished someone wanted to know whether the information in his lecture should be taught in introductory physics courses. Dr. Dresden gave a profound answer: "In everything I have said today, I want you to remember that you should *teach* it but not *believe* it." Science teachers have a tendency to forget that truth.

In teaching science, we should follow the adage that is often used in the travel business "Getting There is Half the Fun." In fact, since in science we never really get completely "there," the process of searching for the laws of the universe is practically *all* the fun.

(There are exceptions to the travel adage, however. Earlier this month, I led a group of amateur astronomers to Peru to see Halley's Comet. As it turned out, getting there was no fun at all. Within a week, we were involved in a train derailment, an earhtquake, military curfews, and a bomb scare. When somebody asks me how Halley's Comet looked from Peru, my answer is, "What comet?")

But back to the limitations of science and science teaching. Some years ago, I conducted a series of personal interviews of Nobel laureates in science and other leading scientists in eight countries to find out where these people felt the limits of the scientific method were. The results of that effort, which were published in a book entitled *The God of Science*, showed that the world's foremost scientists do not easily fall into a set pattern. But they did agree on the fact that there is a limit to the questions that can be investigated through the methods of science and that students should be told about that limit.

I think this is what Dr. Schramm was referring to when he said that there is a point where he stands in awe of the fine-tuning ability of the engineer of the universe. This awe seems to universally pervade the pioneers on the frontiers of science.

One of the people I interviewed for my book was Dr. Walter Brattain, one of the three scientists to receive the Nobel Prize in Physics for developing the transistor at Bell Laboratories. Dr. Brattain told me that I should always tell my students at the beginning of a science course that there are two kinds of questions in the world, "how" questions and "why" questions. "How" questions can be addressed in scientific laboratories, but "why" questions can, in the final analysis, be answered only by saying, "God only knows!" I like to sprinkle "why" questions into my tests now and then, just to see whether my students got the point.

(Dr. Brattain and I once started writing a physics text together that would embody the approach I have just discussed. But the book didn't "fit into the projected program" of the publishers we contacted, as pink slips have a way of putting it. To an author, of course, a book for which you cannot get a publisher is "ahead of its time.")

Students know that there is more to life than science. And they look at us with a jaundiced eye when we pretend that science is more than it is. I suggest that we can raise the credibility level of our courses if we keep the limitations of science in mind in our teaching.

I read in the paper on the way to this conference that 20% of the American population is functionally illiterate. If that appalling rate holds true for general knowledge and skills, I wonder how high the rate would be for functional illiteracy in science. There are, of course, two ways in which to change that situation. For example, if you want to reduce the crime rate, you can either hire more law-enforcing personnel or—repeal the laws. Similarly, one way to reduce scientific illiteracy is to lower our standards and then bring the textbooks and our teaching down to a level where they will satisfy those lower expectations.

This is, of course, the wrong solution to the problem. This conference has reinforced my conviction that we must work hard, in writing textbooks and in teaching, to raise our students to the level of the subject and not lower the teaching to the level of the uninformed student. But I can tell you that one of my biggest problems as an author is to convince publishers of this truth. "Why not make the book easier so that more schools will buy it?" is apt to be the argument. And there certainly are textbooks on the market that are following this line of reasoning.

But I think the trend is reversing. And I want to thank the people at AAPT and Fermilab for recharging my batteries toward that end during these past few days. And even though I don't understand all of the worthwhile things that were said, I take comfort in Professor Lindenfeld's advice that it is better to be exciting and a little inaccurate than to be accurate and dull. After all, our students will remember much longer whether we were exciting in our teaching than whether we were absolutely accurate. Anyway, what is absolutely accurate today will probably change slightly tomorrow. It is, of course, a challenging assignment to maintain the proper level in a textbook while, at the same time, heeding the warning of Occam's razor to keep the presentation as straightforward as possible.

I want to close with another piece of advice I got during my interviews of leading scientists. Dr. Hubert Alyea, the renowned professor of chemistry at Princeton University, told me that, on the first day of class, I should tell my students that the universe is made of "matter, energy, and blurps." When they ask me what blurps are, I should tell them that they haven't been discovered yet. Then, long after I'm gone and blurps are finally detected, I will be remembered as a brilliant teacher.

So my goal as a science educator is to make my subject exciting as well as reasonably accurate. Thank you very much.

DR. FRENCH: At this point, we have probably about ten minutes free for initial questions and discussions from the floor, or of course among the panelists.

So I'd like to ask for any questions directed at particular panelists. And if they aren't directed at particular people—if they are, please say to whom.

Yes?

QUESTION: Do the engineers complain that you've dropped thermodynamics from your beginning course?

DR. RESNICK: They don't say that to us. In fact, they even cut the chemistry course, for example, which shows that thermodynamics is a subject that they least would like us to spend time on in the introductory courses. Now, maybe that's not true in general around the country. In our own department, we're concerned about that, however. Our astronomy majors, for example, surely have to know thermo, and so we use one of our other devices: we offer a one-credit course in thermodynamics.

The engineers don't talk to us that way about electromagnetism, or optics, or mechanics, or wave concepts; nor about atomic, nuclear physics. We get a lot of support from nuclear engineering and materials engineering. Relativity, they will fight all the time. And yet, we find agreement with what the last speaker said, that it's quite an exciting subject. And in fact, as you probably know, that's what led to my writing a relativity text. And we took almost a whole month of the course to treat that. Then, when that was forced out, in it went as a one-credit course. Two hundred and fifty students enrolled to hear 14 lectures in relativity. We didn't lose too much thereby.

But is that addressing the question? Yes? Okay.

QUESTION: Professor Lindenfeld, you showed a graph at one point that showed physics enrollment going down, down, down.

The kids that get into the best colleges are not necessarily the ones that memorize some formula of some obscure type. And I hope that we get away from the feeling that we are making them do this kind of thing (Inaudible).

DR. RESNICK: I just wanted to add that the concepts I mentioned of overview, and sprinkle, and broaden, are what you could apply in your books, without covering it in the sense of test items, but getting the ideas and concepts and the excitement in.

DR. FRENCH: I'll have to bring this session to a close. I don't know how many of you have had time to look at the little pamphlet, collection of caricatures and so forth. But it's available from this conference.

But I couldn't help noticing one, of the theoretician in front of three filled blackboards. And he's saying, "I'm getting close to the answer, but I can't for the life of me remember what the problem was."

I'm afraid that we're not yet at that happy state of being close to the answer. But there are some ingredients of both problems and answers very much present, and of course will continue to be. And, of course, there's going to be lots and lots of further discussion.

Summary Reports

Report of the Subgroup on Lectures

Paul J. Nienaber
Department of Physics
Eastern Illinois University
Charleston, IL 61920-3099

There were five of us involved in this summary presentation on lecture preparation, one of the most difficult areas to address. We broke down our comments into two principal areas: one, a list of general philosophical comments; the other, specific recommendations for use of the resource material we have gathered at this conference. On the general approach, we felt that there is a need for material in three principal areas we identified: 1) Self-contained worked examples, 2) fully-developed modules, and 3) background materials.

The first idea is the need for short, self-contained worked examples and problems, highlighting ideas of contemporary physics, the hot topics of today, which can be put into the curriculum at various points to illustrate material and to give the kids an idea that all of this physics that we are talking about wasn't done in the 1600s.

Possible examples of these might be side-bar comments on text ideas. Some of these are in the more contemporary textbooks, the ones that have been published in the last several years. For an example of a side-bar comment on text concepts, consider the situation when you talk about Newton's universal law of gravitation. You might explain why the m that goes into Gmm'/r^2 is the same m that goes into F equals ma. How do we know that those two ms are the same? The idea is that of the inertial as opposed to the gravitational mass, and that of course ties into the Eötvös experiment discussed by Prof. Will.

Another idea which might be even more useful, especially for people who want to insert things as examples into their lectures, is to solve problems using situations from contemporary physics. When we talk about the expanding universe, one of the pieces of evidence we have for that is the red shift of light from distant galaxies. That light tells us that the galaxies are receding by use of the Doppler shift.

That is a place where we could really use some materials. I'm not sure that those are going to get into the textbooks right away, and I'm not sure that that would be a great idea in any case. This kind of fund of contemporary problems could be a continually evolving set.

A second idea is that we could develop fully some modules, something equivalent to a text chapter. Two areas in which the subgroup felt that this might be particularly exciting to kids would be in the areas of general relativity and particle physics. I think we have seen from the presentations given here that there is a lot of material that can be done at that level. Such material is available both in things that people have written here and in contemporary articles and books. If we develop articles, pamphlets, and booklets, we might look to the state-level associations or the national level to help us get some of these promulgated.

A third proposal, the most crucial one of all, is the idea of resource materials. We need some background materials. It is out there but must be dug out; both the background for us as teachers and supplemental materials for us to hand out to the students.

One special thing of use is annotated bibliographies. Someone should take a pass at what is there, and throw out the junk. That might give us an idea of what portion of all the available material is most comprehensible, useful, or most appropriate. We need somebody out there to provide quality control. I know I am not expert enough. If I pick up something in Scientific American or the New York Times and read about cosmology, I am going to run down the hall and ask our local general relativist, "Jim, is this stuff any good?" And he may say, "No, that's all garbage, you don't want to believe that." But those of you who are out in the high schools may be the only person within a radius of a hundred miles who knows the difference between cosmology and cosmetology. (Laughter)

You need somebody around to help give you background material and advice. I think that is again something for which we look to the national organizations for support.

At this point, I would really like to praise the people at Fermilab to the skies. They have done an outstanding job.

Not only have they given us people who are really knowledgable and are active in the fields, but people who can unpack the material for us, can lay it out and say, "Well, this is really just like this and you really already know it, but it is just a new kind of wrinkle on it." That is a great kind of explanatory power. I think they have gained a lot of experience from doing the Saturday high school program, and I look to this laboratory as a disseminator and a provider of expertise.

Those are some of the philosophical issues. Those first two, the ideas of the fully worked problems and the textbook chapter module, are especially important to the high school teacher. The high school teachers don't have somebody next door who can act as a support network, someone to whom you can say, "Gee, I really don't understand this, can you explain it to me?" You may be the only person in the school, and your next door chemist still thinks that electrons are tiny grapefruit spinning on their axes. That is the other problem with quality control, that you don't want to teach the kids something wrong.

I am getting really tired of unteaching chemists who say, "Well, the electron is really like a tiny grapefruit spinning on its axis," which is how you get electron spin resonance, and say, "Well, no, I don't really know how a point particle could have a spatial spin anyway."

The report recommendations are correlated to the materials that we have produced at this conference.

We have broken our considerations into four principal areas: symmetry; general relativity; cosmology; and particle physics, GUTs, the standard model, or whatever you want to call it. We have sifted through stacks and stacks of xeroxed papers and tried to pick out and highlight for you. I don't envy you your task. You are going to be given each a tree to take home. Make sure it doesn't gather dust on your shelf or be used as your plant stand, but that is about the size of it.

We were trying to point to some areas where you might profitably look for some quality materials that are coming out of that draft presentation. We must all recognize that those first order presentations are going to be rough and are going to have rough edges and may indeed have some wrong physics in them. That is life in the Big Ten, as we always say.

Symmetry problems: In the symmetry area, in your packet there is a summary of the ideas from the symmetry subgroups. There were five groups, each of five groups presented us five blue papers. One really useful thing that happened, I guess it was Thursday night, after the lecture that a number of us got together in the evening sessions, informal discussions, and kind of boiled and sifted and discarded and questioned and probed and summarized some of the principles. We located some of the best ideas we thought were coming out, some of the most useful ideas, and there is a sheet that is entitled "The Summary Report," and I think that might draw your attention first of all.

One idea that is in there, that is an important distinction, is the business of the use of symmetry as a tool to simplify and to gain insight into some physical situations, as opposed to using it in the very abstract and subtle and powerful sense of Noether's Theorem, in talking about the connection between continuous symmetry and conservation laws. Noether's theorem is certainly important and deep and subtle, but somehow I think I would have an easier time of telling a kid that, "Well, see, if I rotate the problem, I still have the same problem, right, so I should get the same answer." I think that is a little bit easier than explaining some subtle connection.

General relativity: There is a draft lecture report from the general relativity subgroup that is probably worth your attention. That is rather well written; further, deponent sayeth not.

And also included with your registration materials is the "blue book," in which we recommend two complete presentations.[1] One is a complete lecture by Tucker Hyatt called "The Birth of the Universe," which also contains (I was reading it last night, it is quite good, despite what I was telling Tucker last night.) a marvelous bibliography of reference materials; it is not your basic rinky-dink bibliography. There is some lower level stuff, but it points in the right directions. If you want to tell a kid, go on and look at something like, Eisberg and Resnick, there are some reference chapters in there. I look a little askance at people who expect high school teachers to have a complete set of Landau and Lifschitz on their shelves, but I think some of those might be useful to you and that is recommended to you.

There is a set of transparencies of a lecture on the big bang theory included in that package, which we also recommend to your attention.

There is a short sprinkling paper on how to sprinkle some ideas from cosmology into places like gravitational red shift, et cetera, into the curriculum that is also in there. You might do well to look at that.

One thing that we noted is that it is probably easier to pick up some of these ideas in general relativity, in symmetry especially. Symmetry is a hammer the kids should have in their tool kits, and you should make sure you explain to them what you mean by that.

I always like the story of the student who raised his hand. The professor was always saying, "for all practical purposes, for all practical purposes" during the course of her lectures, and one student raised his hand and said, "Excuse me, Professor, would you explain to us what you mean by 'for all practical purposes'?" And the professor smiled and said, "Certainly. You have a male student standing on one side of the room and a female student standing on the other side of the room, and every ten seconds they halve the distance between them. Now, you all know from calculus that they will never meet, but soon they will be close enough together for all practical purposes."

So defining the terms is useful in telling kids what you mean by symmetry arguments. This might be a good thing to do early on in the section in dimensional analysis, rather than invoking it as sort of an ansatz from on high in the section on Gauss' theorem, where everything is done by symmetry. They think that is one of God's names.

In the section on GUTs and the standard model, we included both the stuff of the grand unified theories and the standard model of electrons and quarks together. There is quite a bit there.

Paul J. Nienaber

None of it particularly stood out as being terrible, but none of it stood out as being particularly excellent. This is probably one of the topics where you can find lots of stuff in the contemporary literature. Everybody is trying to explain to you why $SU(3) \otimes SU(3) \otimes U(1)$ is really $SO(10)$.

Also a final comment: It is probably harder to find ideas from GUTs and $SU(3)$ and supersymmetry to plug into the standard curriculum, except possibly sections on Pauli exclusion, and it is probably hardest of all to put those kinds of topics into the introductory curriculum. I think it is easier to talk about the Eötvös experiment, — maybe not with all its subtleties, or the Doppler shift, than to do something like bringing some of these ideas and have people calculate the binding energy of the Y or something like that.

1. G. Aubrecht, ed. *Papers Prepared for the Conference on the Teaching of Modern Physics,* AAPT, College Park, 1986.

Report on Demonstrations

David A. Cornell
Department of Physics,
Principia College,
Elsah, Ilinois 62028

We could stand here and give you some general discussions and observations, but we thought probably those would be manifold from the other groups, so we will try to do some physics with you that we would do in our classroom. I guarantee you, what I will share with you I would take home and do on Monday, if possible, and I think the rest of our group will feel the same. There may be some time during questions if you would like to ask us about our more phi-lospohical points of view.

Thanks to Earl Zwicker, IIT, I discovered that in my back pocket there is a small part of a physics storeroom, and he loaned me this (takes out two small sheets of polaroid film). Can you see those two little bits of film on there? (points to film on overhead projector) Those are two sheets which Earl carries around in his wallet. Now, if we place them on top of one another, they show a well known property of polaroid sheets, transmitting one way, and if I rotate one on top, you see the alternating brightening and dimming. Now, return and reset, as our examplar Chris Hill pointed out. You would expect to see some other symmetry shown here. How about if I turn it or rotate it about a horizontal axis.

What do you expect to happen if I turn it about a horizontal axis parallel to a side? It should become dark or light? No change. Okay. Here I go. Rotate about a horizontal axis. No light is transmitted. Try it again. Reset. Turn one about a horizontal axis. I honestly am doing it about a horizontal axis.

All right, reset.

Turn it about a diagonal axis. Here we go. Move the apparatus. Am I doing magic to make the light disappear? Do you believe it? It is reproducible, so there must be something there.

To demonstrate this symmetry, you might take what you and I would be culturally used to thinking about. We have something here which seems to demonstrate what you could call a property or a hidden symmetry in the material, and we could represent the hidden symmetry by this arrow (Figure 1). And so we try rotating it, we rotate it by 90 degrees about the axis, which is perpendicular to our display, and you saw that the property changed, didn't you? However, when we rotate about the horizontal axis parallel to a side, this should predict no change in the property. This diagram (Figure 2) would predict no change in the property, rotating about axis B. So somehow this particular symmetry picture doesn't apply to the situation we are describing, does it?

All right, let's take another picture which may help you solve the puzzle. I frankly will admit, I didn't solve this puzzle until I was told it. The geometric symmetry of the sheet may have suggested something to you that wasn't there. Is there any reason why the hidden symmetry has to be aligned along the geometrical sides? (Figure 3a.) Now, if I try rotating clockwise by $\pi/2$, you do get a change, (Figure 3b), but also if I rotate about a horizontal axis now by π, (Figure 3c) what happens? If I rotate about the diagonal axis by π, nothing happens (Figure 3d). Isn't that lovely? All right, so we have a way to demonstrate hidden symmetry.

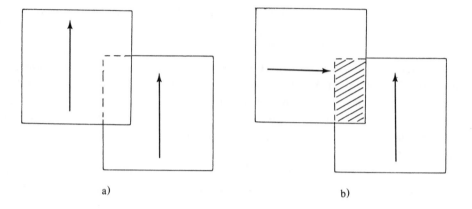

a) b)

Fig. 1. The arrow points in the assumed direction of the polaroid's hidden symmetry. a) If a piece of polaroid is arranged with arrows parallel, light is transmitted. b) If one polaroid is rotated by $\pi/2$, no light is transmitted.

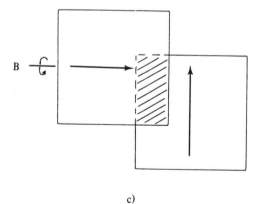

c)

Fig. 2. Predicted result of rotating the apparatus of Fig. 1b about axis B by π.

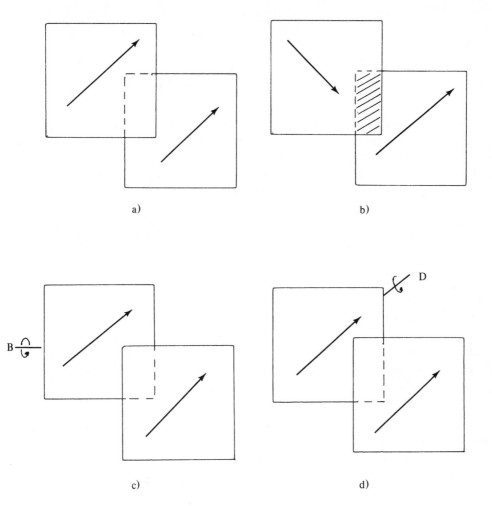

Fig. 3. a) The hidden symmetry may lie along the diagonal. b) A rotation of $\pi/2$ in the plane of the polaroids results in lack of transmitted light. c) A rotation about axis B by π changes the orientation to allow light to pass. d) Rotation by π about a diagonal axis D leaves light passing through both polaroids.

Margaret Hill
Vianney High School,
St. Louis, Missouri, 63122-7299

Basically, it is really hard to bring cosmology into the classroom unless you are creative. Since science teachers always have things in their pockets, I brought several galaxies along with me. So we pick one galaxy. You are here in this galaxy, and this is very simple and quick, if you are worried about the time on getting these things into your curriculum. This is not going to take long. I have a rubber band with several safety pins through it.

Pick a galaxy (one of the safety pins) for yourself, and then watch the relative rates of recession of those galaxies as you stretch the band (Figure 4). The galaxy is moving around a little bit, but you can take a look at that and the students will be able to see the relative rates. I don't know about calculating the speeds of recession, but I would think that you could do that also. Take a look at the motion, mark how far the furthest galaxy goes, and measure the amount of time that it is moving versus those that are closer in. That is one thing that you can do quickly.

I think one thing we have to watch is not to be, in the first place, too sophisticated. I know a lot of my high school kids are afraid to voice opinions or they are afraid to do experiments because they are afraid that people will make fun of them, or that they are not sophisticated enough. I think a lot of the things that we ask them to do are very sophisticated, but there has got to be that childlike sense of wonder involved to involve them in the activity.

The second thing to remember is that any model we make is not going to be sufficient. A model is simply a model, and I think one of the things we can do for kids at the high school level is to help them see that models have limits. If they can begin to take a look at those models and see where those limits are and how models fall short, that is going to help in their analysis of any way we choose to look at the universe.

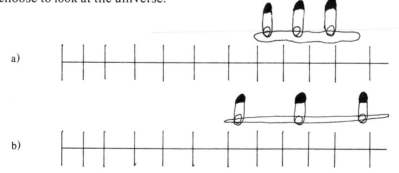

Fig. 4. Rubber band with safety pins through it. a) unstretched. b) stretched; separation increase is proportional to distance of separation.

Charles Lang
Omaha Westside High School,
Omaha, Nebraska 68144

Very quickly, I will discuss general relativity. First of all, you cannot find a globe in this entire facility. I had to have Gordon go out and buy this beach ball for me.

The toughest thing I think, if you are going to try to introduce general relativity is more a question of the students thinking in Euclidian geometry, you are asking some non-Euclidian geometry out of them. I would suggest it works really well in a high school because the social studies department usually has several globes, so you really turn this demonstration into sort of an experiment. But you simply take a string and, you know, go along the geodesic here or along the longitudinal line with your string, then start with another string, and then ask them to simply extend this parallel section (Figure 5).

Of course, what is going to happen is that all longitudinal lines are going to meet, but this kind of thing is not highly obvious. Yesterday we saw two enlargements on this, one of these being the example of the triangle in three-dimensional space having interior angles totaling more than 180 degrees.

I haven't tried this yet, but I believe I can just add the adaptation of putting masking tape right on the globe, taking it off of the globe and laying it down, and then having the students measure the angles. They should notice that the angles are more than 180 degrees.

And then this last one Professor Will talked about yesterday, is the ruler and moving it along. First of all, you do it, of course, the Euclidian geometry way, in two dimensions, and then moving it along the triangle and noticing a difference in orientation of 90 degrees. So that is a quickie on general relativity.

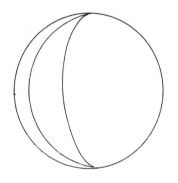

Fig. 5. It is impossible for strings on a sphere to remain parallel.

Eugenio Ley-Koo
Departamento de Fisica,
Universidad Nacional Autonoma de Mexico,
Mexico, D.F.

This is also described in the blue book.[1] There are several games that we can play, depending on your interest. There are also other papers on his kind of thing. There is one by Jack Hobson about the composition of quadrants and I will play the simplest version of the game here.

Here are the cards. (Figure 6) You have the up and down cards and the anti-up and anti-down cards. This is to play the game of an ordinary matter, which is made up of the neutron and the proton.

Of course, you can have the neutron. Remember, the properties of the charge and the baryon number, and you can add up those numbers, have the students add them up and convince themselves that they have zero charge and that one is the baryon number.

The general rule for this game is that you need three quarks to make up a baryon. You can also substitute for a down-quark in the neutron an up-quark, and that will change the particle into a proton, and you can also make sure you have a charge of one and that one is the baryon number.

For those who want to go on and make it more sophisticated, they can say this contains the SU(2) transformations. The SU(3) transformations are changing the u-quark to a d-quark and a d-quark into a u-quark. Of course, you could take more quarks. Here you have one with 3 ds and here you have one with 3 us, and of course you can also have other combinations. Some of these baryons have spin one-half, but other ones have spin three-halves. These have the same quark composition, but the spins are different.

In order to play the meson composition game, you need not only the quark cards but also the antiquark cards. The mesons are characterized by having a zero baryon number, so you need a quark and and antiquark. You understand mesons are made up of a quark and an antiquark, so with d- and u-quarks and antiquarks, you can make these combinations and you will realize that there are two possibilities of getting a particle with zero charge, and the situation for the ρ° is more complicated, as Professor Georgi pointed out, but this definitely enters into its composition.

Now, you can also bring in color, that is, flavor on color cards. Of course, there are also the other cards, and you can play the strange game, the charm game, the beauty game and the truth game.

In order to introduce or to see the need of color, you must put in the Pauli exclusion principle. For the nucleon resonances you need all the spins to be parallel. We must take one-half plus one-half plus one-half, in order to get the three-halves, but then immediately you see here the contradiction with the exclusion principle. The quarks are fermions, so they cannot be of the same quantum state. Here you have three quarks but that you can save the situation if you give

278

them an additional property, and that was how color was introduced as an idea in 1964. Of course, it took some time to accept it.

Here you have the combinations in order to have the color. In the paper I describe different games, and I think you can have your students do them as an outside activity, as a motivation to see the history and the physics behind it.

1. G. Aubrecht, ed. *Papers Prepared for the Conference on the Teaching of Modern Physics,* AAPT, College Park, 1986.

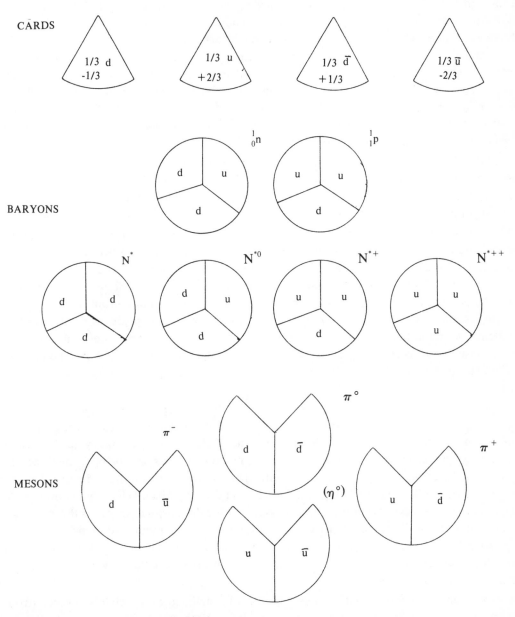

Fig. 6. Cards for the game described in the text.

Don Schaefer
Bettendorf High School,
Bettendorf, Iowa, 52722

In one form or another each Fermilab group that discussed general relativity would probably come up with something like this; but, we are suggesting that the following idea might be used as a *demonstration* so an entire class can participate. The idea is that a frame can be made which fits the stage of an overhead projector, and Saran Wrap (or some kind of clear plastic) be tightly fastened across the top of the open frame. (An embroidery hoop can be used in place of a home-made frame. These hoops can be a foot or more in diameter, and distorted or torn plastic can readily be replaced.) If a small ramp is set on one side of the frame, and a small steel ball run down the ramp, the ball will run straight across the Saran Wrap (plastic). What is observed on the screen is of course straight-line-motion. If a much heavier steel ball is placed in the center of the plastic sheet, it will distort the plastic somewhat.

(We have found that Saran Wrap will work, but once used it tends to be permanently distorted...so, a heavier grade of plastic or heavy duty Saran Wrap might be a better choice for repeated demonstrations. Such heavier grades of plastic are readily available at hardware stores, and are used as substitute storm windows.)

When the small steel ball is again run down the ramp, it will appear to curve inward toward the large center ball as it rolls past. What the students see on the screen is a two-dimensional interpretation of a three-dimensional event. (We would suggest that some kind of sight-barrier be placed in front of the overhead so students can't see what is actually on the overhead, and will see *only* the projected image.) If this demonstration is used to extrapolate from work with universal gravitation (including observed accelerations and subsequent postulation of forces to cause them), we think that the mind-set of the students at this point might well cause them to explain what they see as an attraction between the large, stationary center ball and the moving ball. Their suggestion might be that the center ball is really attracting the smaller ball.....possibly it is magnetic.

If students are then allowed to see what is actually happening, the obsevation is then explained as the ball simply following the curvature of the 3 dimensional plastic "space," and not as some kind of attraction between the large and small balls. The accompanying discussion might include the suggestion that if the observer were a "two-dimensional being," the projected view is all that he could see. But three-dimensional beings might explain the same event as the moving ball simply following the "curved space" around the larger ball. What if you observe the motions of small particles moving past huge masses in three dimensional space? Is there a possibility that there is a 4-dimensional space-time continuum? What might 4-dimensional beings see? How might *they* interpret what *we* explain as an acceleration that results from a gravitational force? Might they interpret the motion as the particle simply folowing the curved 4-dimensional space around the massive object?

A final recommendation is that a suggestion might be included with each topic or idea (from the Fermilab conference) as to just where in the regular high school physics course each *might* be included. In case a teacher wishes to use each topic as an extension of a presently used section (instead of treating the topics as a separate unit), such suggestions would be especially useful.

As an example: we would suggest that the "4-dimensional space" analogy might be used right after development of the Newtonian gravitational theory. Other teachers might include this in an extension relating to symmetry. Still others might choose to use this as a relativistic extension of "frame of reference."

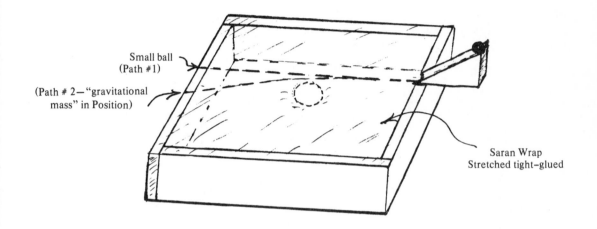

Small ball
(Path #1)

(Path # 2—"gravitational mass" in Position)

Saran Wrap
Stretched tight–glued

Class observations: (1) Projected (2D) View only—suggest cause;
Class observations: (2) Observe "actual" (3-D) view—suggest cause;
Discussion: Extrapolate from 3-D view [small mass, passing (close) by sun]. To "4-D" explanation;
Extension: Add grid lines (marking pen) (Large mass changes "space" itself?)

Fig. 7. Model of gravitational attraction in two dimensions. (2D).

Report of the Group on Software and Audiovisual Aids

Dean Zollman
Department of Physics
Kansas State University,
Manhattan, Kansas, 66505

When we think of audio-visual and software, of course, we think of lots of different things and people who do this like to work with gadgets. I try to describe here (Figure 8) the various gadgets that we included: film, computers, the videodisc—it is hard to tell that that is a vidodisc, I know—videotape, overhead projectors, slides, and, of course, books and all sorts of other things that are audio-visual in some way. We hope that all of those are somehow going to be connected to the student, who I put there in the middle, and that the student will somehow learn a great deal.

As shown in the figure I have noticed recently on our campus that there is symmetry breaking now on students. They all have something hanging off one side. I suppose at RPI that this is probably a calculator, but on our campus it's a Walkman. I added the symmetry-breaking Walkman for a reason. One of the things that we don't think about too much when we think about audiovisual materials are some of the simpler ways to do it. Because approximately one-third of our students seem to be "plugged into" audio tapes, we ought to be doing something with audio tape. Maybe then, instead of listening to whomever they are listening to, they would listen to us occasionally.

Basically, the various groups did a couple of things that were quite similar. One was to look at existing media. (Look at means look at in our minds for the most part.) We did have one person who did a little research in the Fermilab library, and tried to come up with some ideas of things that are already out there and that might be useful. It turns out there is a rather large number of them.

I just put down a few representatives ones from the list of course films and videotapes.

One that we didn't think of (at least we don't think anybody thought of until late last night) was an old film by Judith Bergman of Brooklyn Polytechnic called "Symmetry." For a long time it was more or less distributed by IBM. I don't know where it is available now, but it is a very artistic look at symmetry, and certainly is a nice thing to get your hands on, if you can. And then, of course, parts of "The Ascent of Man," parts of "NOVA," and parts of "The Mechanical Universe" can be quite useful.

"The Edge of Time" is a film that I hadn't heard of before, so I guess that is why it got on the list. Lots of those things are out there, and they are out there in ways that you can get hold of, usually fairly easily. Unfortunately, frequently they are long programs and you probably don't want to use them. Fortunately videotape allows you to edit them a little bit and get through them.

Another suggestion is some film strips, and AVNA has a couple of cosmology and general relativity film strips that somebody remembered. Also, another medium that we probably don't think about too much is bubble chamber pictures. They are all over the place in this building. There are probably a few million nobody wants any more. The old Project Physics supplement on elementary particles contained some that are really quite good.

So, all of those things are out there. Perhaps what is really needed at this point is somebody to spend a year collecting all of them, looking at all of them, and cataloging them.

Now, my own prejudice is we should look at them and decide what we can put on videodisc. Videodisc is a much more convenient way to deal with material and a much easier way to edit. Since I was elected spokesman for my group, you have to get my personal prejudice as well.

Then the one thing that there is a little bit of, but not a lot of, is computer software. At the moment, Cross Software, I believe, has a couple of relativity programs, and then there are a few other odds and ends of things around, but not a great deal. (Some physics teachers have developed programs, but they are not available commercially.) So some of the groups spent some time thinking about various types of software you might want to put together, and I am just going to draw a couple of quick examples, one of which I actually enhanced a little bit last night as I was writing up the transparency, so it doesn't quite appear in the packet you will get in this format.

Other members of the group have not heard this, so if you don't like it, it is my fault. If you do like it it is my fault, too. In any case, the idea of putting bubble chambers on a computer or videodisc program and having the students actually analyze that picture would be a nice one.

One thing wrong with these bubble chamber pictures is there is too much stuff in them. I am sure that is a problem for particle physicists as well, but it is certainly a problem for the students. So it would be nice just to take out all but the essentials. Then have the students go through the kinematic analysis the best they can. Last night it occurred to me that we could animate them. We can't do that in real time, but on a computer videodisc we can slow down the time, and we can unfold, in time, one of these things. In this way particles come in, and then something happens. In this case I was even thinking about using different colors to help focus the students on the events at which we want them to look. Then it unfolds and some more things happen. The students then could analyze at least part of this and go through it. Perhaps we could have a few programs that help them analyze it along the way.

We think there are some possibilities there. We don't think anybody has done it, but there might be some ways to actually put this in, say, some kinematics or some E&M course.

One of the problems with something like this is the need to do the programming. I have students who program for me all the time. Every time I present an idea to them, they say that it will be done in two weeks. Then in two weeks the good ones come back and say "I haven't got it done yet;" the bad ones come back in a month and tell me "I haven't got it done yet."

So another thing I think one ought to look at is how to use some of the existing software in this sort of thing. I have one thing in the booklet that you already have on using data bases. We discussed that in one of our groups and decided that it has some problems. Perhaps I will work on it a little bit more.

Another one I came up with yesterday or the day before was try to use a spreadsheet to do a little bit with symmetry breaking. In fact, I think this idea might fit nicely right after you have your students do the card game that was just discussed.

You give the students the quark data which you lay this out on Lotus 1;2;3 or Visicalc or whatever. Then, you say "Okay, the quarks all have the same mass. Now try to build your particles." The idea is that you do something like a standard baryon chart here and you point out by the asterisk where the particles are, and they are supposed to build them out of the quarks.

If they build them out of this nice symmetric system, they have a whole set of baryons, all of which have the same masses, and doesn't quite fit. So, you tell them fiddle with the masses, which we now know is called breaking the symmetry instead of fiddling with the masses. You let them play with those masses and see how well they can do. It gives them an opportunity to explore with this idea how you might be able to break symmetry a little bit here and come up with a spectrum of baryons that look something close to the right one. Whether or not it will really work, I don't know. If you play around, it probably won't come out exactly right, but in any case, the idea will be there.

So those are just a couple of quick ideas that we had out of this various collection and, as I said, there is a lot more out there.

Fig. 8. Audiovisual and software aids.

Report on Test Questions and Problems

Betty Preece
Melbourne High School,
Melbourne, Florida 32901

The Test Questions Group talked about a sprinkling module; we talked about a more in-depth module, and then we tried to see how we might fit evaluation questions into the things that would be useful to you. We came up with three types of questions.

The first type listed resources: where you could go to find questions that might be suitable for what you want to do; where you could get some ideas about questions and have an answer; where you could look at a way you might lay a problem out to a student and say, "Gee, this does say both what I want to say and what I want the student to get out of it."

We looked at a second kind of a question which we called "definitive." For these questions there were specific answers that you submitted. You do a calculation and when you get finished there is an answer. We received quite a number of these.

The third type of question that we found was one that we called "food for thought." Many of these were strictly open-ended questions: "What would happen if— ?", "What do you think this means?", "How could we discuss this topic?"

So we are asking you to help us put together a test bank of these 3 types of questions. I have talked with Gordon Aubrecht and, if you wish, you may submit questions like this to AAPT through the modem or through the mail.

We have decided that we will try to meet in San Francisco and sift through these questions to come up with some reasonably good ones that we would like to share with you.

We are asking you to do a couple of things that could help us immensely. If you send in a question, it would be useful if you could give us some idea of which one of these three topics you feel it fits best. If you could type it or at least write legibly something that could reproduced, especially diagrams, this would mean that we can pick those out that would seem to be best for you. Please give us an answer or what you think is an answer. Even if somebody can prove it is wrong, at least we have a starting place.

We have discussed with Jim Nelson, who has done the AAPT-NSTA physics test, whether or not we could submit some of these questions to him. He has indicated that right now there are no questions on that test about the kinds of modern physics that we have discussed here. He, or whoever is going to head that committee, will be open to suggestions from the test bank.

We talked to Art Eisenkraft about the International Physics Olympiad test that is given to select the U.S. participants. These questions would also be available to that group, if they so desired.

It is our hope that, by your input, we might be able to submit some of these questions to the SAT and similar groups and suggest that they might be included as some the test questions there. When these kinds of questions appear on standardized tests, teachers tend to be aware that they are being asked and they then make an effort to include some of these topics in their curricula. We felt that, while this is sort of a backhanded way to do it, it was a way we could point out that some introduction of this kind of material would be desirable in the curriculum, particularly at the high school level.

We did not directly address the college level because we felt that those questions from the test bank could be used by college-level introductory courses just as well as the high schools.

I think this pretty well ties together what we have. It is our hope that we might continue to refine the test bank over a couple of years to the point where we might produce a small booklet of test bank questions.

Question: Is it true that some of this material is already available in test banks that text book authors supply?

Mrs. Preece: Yes.

Question: I am thinking, the one that springs to mind immediately, of the test bank for Hewitt which is available in software that we would use. I think several of the other authors are coming out with multiple choice test banks that include some of these topics, so if you're stuck that might be a place to start.

Mrs. Preece: We had hoped under the resource section up here that you would give us some information like this. We checked in the big fat book that we got, and there are several tests in there from which you could immediately draw questions covering some of these topics.

Report of the Lab Subgoups

Eric W. Danielson
Department of Physics
Hartford College for Women
Hartford, CT 06105

I think our enthusiasm for the topics in this conference and our zeal to install them into the physics curriculum reminds me of the observation that if you give a child a hammer then the whole world looks like a nail.

We would like to suggest a few nails at which, perhaps, you can aim. There is a lot of parallel evolution I think that has gone on here in the last few days in the small groups and so coming fifth in this list of six, basically most of the topics that we have dealt with have been described in either condensed form in terms of demonstrations, or verbally in terms of literature material. What I am going to do is really give you a shopping list of lab suggestions that we have dealt with. Let me go through this quickly and then we can discuss or go from there.

I first want to point out that we all received this "blue book," in which there are ten labs that are presented in some detail. They cover the following topics. There is a lab on Hubble's Law and the determination of distance to a quasar; there are Doppler shift measurements; there is determination of space curvature by counting isotropically distributed objects; there are a couple of microcomputer simulations; the Milliken experiment; and radioactive decay, construction and use of simple cloud chambers. There are two or three labs that deal with quarks. In fact, the card games that you have just seen demonstrated a moment or two ago could certainly be used in a lab situation.

A variant model of the Milliken experiment illustrating the three-quark structure of baryons called the milli-can experiment since it involves putting objects in small cans, identification and classification of various symmetries, and interpretation of bubble chamber tracks. So there are those already.

Then there are various ones that were suggested and submitted during the times of our meeting in the small groups. Some of these you will recognize as pretty much direct steals from lecture presentations. Others are a little more disguised, so let's just quickly go down through the list that we have here.

There are four different labs, whose materials I guess you will be running across eventually here, in which the student is identifying and classifying both natural and manmade different types of symmetries and events, and also manipulating objects to investigate the symmetrical configurations and cataloging these operations required to achieve each of these symmetries.

Solutions of various kinds of force table problems by symmetry arguments only, using no calculations, and then verifying the predictions using the force table, is a matter of different approach rather then of breaking new ground. I think it an interesting idea to apply.

Look for rotational symmetries of shadows of an unknown object. Let me talk about this in a bit of detail. The thought is to take a flask, a frosted flask with an opening at the top and shine a nice bright parallel beam through it, through the opening, using a laser or any other bright source, and then at the base of this flask you have a shaft that goes through which has skewered on it some sort of object that may be symmetrical or may not. The student, not knowing what is inside, can observe only the shadow pattern change and thereby determine the repetition rate, the frequency of the symmetry that evolves, and eventually perhaps make some guess about what the object itself might be inside.

Something that was suggested earlier is use of prints of bubble chamber tracks, Fermilab prints or prints from other laboratories. There is a marvelous parallel in the Palomar Sky Survey prints which you can purchase for lab uses. These are 48-inch telescope photos of various pieces of the sky in both blue and red. They are wonderful lab materials to work with, and if we had something parallel to that for particle interactions with helpful hints, just chosen to illustrate various types of ineractions, I think that could be a very exciting source of lab materials and lab experiences for students.

The suggestion of doing the Compton scattering experiment, as long as you have a multi-channel analyzer, and Fred Domann has a homemade version, a model essentially which he included in the writeups that you will be getting, and he is eager to share and talk about it.

A one-dimensional expanding universe model, was something that two or three groups came up with, using various kinds of materials, very similar to the rubber band and pin arrangement that you saw here. The suggestions for doing this are using surgical tubing and also using a Slinky. If you stretch out a Slinky, for example, and then just locate the positions of galaxies, for example, on various points and students do the very same sort of thing that Peggy was suggesting, measuring from their own postion the distance to these various galaxies at different times, then the isotropy of this one-dimensional space can be observed because all students are observing the same kind of thing from their reference points. In fact, by measuring the dislocation of the galaxies at different times, you can come up with speeds of recession, and plotting a speed versus distance diagram, you can actually end up with a line which will give you the "Hubble constant" for the Slinky and then you can use this for predictions. For instance, one can determine the age since the birth of the slinky by working it in a backwards direction. There are also interesting discussions that you can develop from this as to what it is out there that is pulling the slinky.

Labs can deal with spherical trig relations to determine curvature via making angle measurements on triangles of various sizes, or on spheres of different sizes. Then students see how they can converge in the limit down to a 180° measurement, but are evidence, at larger sizes, of some sort of curvature effect.

Use of an Atwood machine to demonstrate the equivalence principle is a thought that came forward and there is a little writeup and diagram that you will be getting on that.

One more example is of illustrating the gravitational force interpreted as a distortion of space time by using some sort of stretch material, and suggestions for using licro or expandex which are available in various fabric stores. They come with grids already on them, and they are nice stretchable materials to use, so there are some interesting possibilities there.

If you want to illustrate the opposite kind of effect, some sort of central repulsion, of course, you can draw the object up in the middle instead of stretching it down, and then you get the opposite effect. I'm not sure you want to do that if you are trying to talk about general relativity, but it is bound to come up anyway.

The pendulum period measurements using bobs of different types of matter á la the suggestion of Chris Hill yesterday is another possible lab.

I have a couple of comments at the end of this list. Most of what I have just described are model types of labs, rather than true hands-on labs measuring real things. I think the reasons for that is that the latter are much more difficult to conceive and also to develop. I hope that there is a gestation time that we all go through in which these other ones will come forth, but I think one of the frustrations of our group was to find that we have relatively little in the way of real physics that we are doing here with the kinds of labs that are suggested. I certainly welcome the suggestions of the very first group, the group of authors and professors, I guess they are called, in developing the lab materials that might take us in that direction.

I think our group agreed that it would be enormously useful to have some sort of literature search and cataloging take place, in order that we develop some sort of coherent collection of what is already done. There is going to be a lot of reinvention of the wheel, and if we can identify and catalog and collect the materials that have already been put out there, many which are no longer used perhaps because they have just fallen out of favor. There are tremendous amounts of material there. Then we can focus especially on filling gaps and maybe correcting other problems with the existing ones. But that kind of cataloging and literature search would be an enormously important next step, we feel.

Report of the Exercise Subgroups

Scott Cameron
San Marino High School
San Marino, CA

We are just two miles south of Cal Tech, and I don't think there is much relationship physically or mentally between what goes on there and what goes on in our high school. I got Dr. Richard Feynman of Cal Tech, probably one of the most famous puzzle-solvers of our century, to come to my high school. In talking with him for a few minutes before he addressed my physics students, I asked him, "What are you doing besides your research? Do you have a lot of departmental responsibilites?" He made the statement to me that he had decided some time ago to be consciously actively irresponsible; that is, he was not going to do anything besides physics, and he wasn't going to be involved in anything like this conference. His job was to solve puzzles that he liked about nature. So the only reason that he would come to a high school was to talk about nature with kids, and to solve puzzles that might be in their minds.

Many of you know that he accepted the President's request that he go and investigate what happened with the Space Shuttle Challenger. That was a real departure for him because that was a committee, but he served on it. In an article in the Los Angeles Times Magazine, he is quoted as saying "When you have to get up there and you get involved, you just have to do it and it is exciting once you are stuck in it. It is like asking somebody who is almost having an automobile accident whether it is exciting. You're damn tootin' it's fun to steer between the cars, isn't it?"

I have several things to say about what we have discussed which I divide into four items:

The first one deals with the distinction between exercises and problems. There is a difference and we probably ought to keep this in mind. *Exercises* are those things in which the student probably can read and recognize the question right away. The exercise is grunt work to get to the answer. In *problems,* the student probably spends more time figuring out what the question is. Once you figure that out, you can do it. However, there may be a trick to it, such as symmetry. You search for an easy way to do the problem; otherwise you could have to integrate and do a bunch of fancy math, and you don't want to do that if you can get the answer another way. Exercises reinforce basic facts and information, while problems challenge the known to reveal the unknown.

The second item has to do with the types of exercises. Professor Lindenfeld alluded to the difference between *loose* and *tight* problems, which have to do with the culture of our discipline. Among the loose categories, we have global or divergent problems. These might reflect more a pedagogical style that is wide open; examples of those include talking about problems of symmetry in architecture and music. There are the lecture extensions which could be either way, in which you just go one step past the lecture, and the one that jumped out at our subgroup is the symmetry with the triangle. Everybody has been talking about that, but that was just a lecture extension. Building up the hadrons from quark constituents could be just an extension of what we saw during the lecture. You could do things with poetry, as we see in *The Physics Teacher,* which I think would fall under this category.

The tight categories I see in two different ways: The Fermi-type questions; and the more rigorous math questions. An example of the Fermi-type questions could be what we saw to open the conference (which I thought just fantastic), Professor Weisskopf showing us how you do these back-of-envelope calculations to get answers. You compare them with your experience to see if they are close. We have one in our big manual[1] on calculating the masses of the quarks or the hadrons based upon the quarks, and I thought that was very valuable. It is something that I, as a high school physics teacher, can use in class, and my kids can actually calculate something about modern physics. That is why I like that example. You can see the two types of questions I am talking about: the ones that you can do with quick algebraic calculations; and the more rigorous ones in which you might have to do an integral as the only way you are going to get the answer.

The third item we talked about is the excitement of doing problem-solving. I don't think people get in this business of physics if they don't like to solve problems. If you are in physics, you like to solve problems one way or the other, and there had to be hours you spent on one problem. That is part of what draws you to the subject matter.

Oftentimes, I think that when we teach, we don't show our students the excitement of that problem-solving. To give you an example, last night we were supposed to be doing modern physics problems. We talked about breaking of symmetry, in an old problem in which there is a disc and it is a classic hypergeometric problem I saw a long time ago. We couldn't remember how to do it, and so we argued. We spent half an hour talking about the problem in which you have a disc with a hole in it. You are given the size of the hole and the fact that it is R/2 from the center. Where is the center of mass?

You can do this problem by symmetry, or you could do hairy mass calculation integrals. We had four or five people walk by and take a look at this and they said, "Oh, yeah, I know how to do that," and "No, they didn't know how to do it," they couldn't remember. So they started getting into this. The key point is that the sharing of how to solve this problem was there. I think if we would do that with our students it can be very special.

I want to mention two problems that we came up with, which probably won't be in the handouts, that illustrate the kinds of things we were thinking about. You could take a mylar strip or aluminum foil folded into a cylinder, forming a reflective surface, and you could take a piece of paper out away from it, and ask them what kind of figure they would draw on the piece of paper such that when you looked at the image in the cylinder reflector it would come out to be a triangle, say, or a square. It is an interesting symmetry question, one explored in anamorphic art of the Eighteenth Century.

Another question which has more to do with what we saw given in the lectures. We actually went upstairs in a little group, barged right in, and interrupted Professor Quigg. We walked right in, we thought this was the most important question in the world. We had seen this graph three different ways illustrating the points at which the three forces, strong, electromagnetic, and weak, come together at very small distances. This could be either a distance graph or an energy graph. The U(1), SU(2) and SU(3) all seem to come together at one point. We asked how we could have students calculate this. It is one thing to read the *Scientific American* and see this graph, but it is another thing to give the students some algebra so they can understand how the coupling strengths merge.

He just opened his book and gave us the expressions, they fit on less than one line; you can give them to your students. It is in his book on gauge theories,[2] in the back, page 283 or 294, and you can actually have your students calculate them. I think that is important. High school students can do the calculation.

This leads me to emphasize that it takes time to develop good problems and exercises. You all know, as Betty Preece said, that to develop good tests and good evaluation instruments takes a long time. You have to test them and it takes time to determine whether they are adequate. We did a couple of examples and these were the kinds of things we were talking about.

What am I going to do with this when I go home? I can tell you that next year, the first week, I am going to talk about where I think physics is today, based upon the conference and what I read. I am excited about it! I want to do some things on the general theory of relativity, on modern particle physics, and so on. I want to use slides and films, anything I can get my hands on, to show them where physics is today.

The last item in closing is that Hewlett-Packard has a "PICS call," and I know other computer companies have such a function, too. I don't know exactly what PICS stands for, but I could say person in charge, and I think we have six areas here that we talked about in the conference.

If we could have one person in charge of each one of these areas to filter the information that we get, we could make PICS calls to this person in charge or write letters to that one person. The PIC could sift through this information and of say, "We're strong in these areas, we're weak here." The PIC could communicate back to those groups that have that area and they could give us some more input. We could have somebody call when there is a problem. It would enable us to make a PICS call on modern physics.

1. G. Aubrecht, ed., *Papers for the Conference on the Teaching of Modern Physics,* AAPT, College Park, 1986.
2. C. Quigg, *Gauge Theories of the Strong, Weak, and Electromagnetic Interactions,* Benjamin/Cummings, Menlo Park, Ca, 1983.

Closing Session

The Job Ahead

Gordon J. Aubrecht
American Association of Physics Teachers
University of Maryland
College Park, MD 20742
(on leave from Ohio State University)

BEYOND THE END — PREPARATION FOR THE FOLLOW-UP MEETING

Here you are. We are almost at the end of this conference. What are we going to do from now on?

The school year will have ended by June. While it's difficult to imagine doing much between now and the end of this school year, if you do do anything between now and the end of the school year, please share it with me.

Between now and the end of this school year, there is time to devote more thought to modern physics. Share your thoughts and achievements through me with everyone else. How is that going to happen? It will only happen if you communicate with me, or with one of those people who have volunteered to be people in the center.

In June of 1986, the summer meeting of the American Association of Physics Teachers takes place at Ohio State in Columbus. We are going to try to get a room for people to meet together to talk about introducing more modern physics into the curriculum in more detail, to use the additional book that you're going to get today, and discuss whatever else has come out in the meantime. The idea is to share those ideas, and to get more people involved. After all, we're all charged up now. We ought to go out there and interact with other people. It will be a strong interaction, and we ought to be able to get more people involved.

During the summer, we have some time for work and thought on this problem. In the fall and winter of 1986, we need to plan for trial runs based on these materials we have prepared here.

What do I mean by trial runs? Someone trying something would be helpful. Those of you who are going to try something should let me know what you're going to try, and then let me know how it came out. We should then share the information among us. It is nine months, a very long time (the gestation period of some well-known mammals) between this conference and our follow-up in January.

I think that we ought to keep in touch. We have to keep together, or assuredly, we'll just all fall apart. In January of 1987, we have the follow-up at the joint AAPT/APS winter meeting in San Francisco. The format for the follow-up will be partly based on the recommendations that you will make. You will remember that we handed out suggestion and evaluation forms. I really want to know what kinds of suggestions you have. We will consider them seriously, and do the best job we can of doing what you want.

After January of 1987, we will no longer have any monetary support from the National Science Foundation for our work. January, even after all of this work, is, after all, only a beginning.

By then, we will have had a chance to look at some of these things, to have tried some of these techniques and ideas, to have filtered some of these ideas, and to have had time to think about some of them. Part of the purpose of our January meeting is to decide what sorts of pedagogical ideas are suitable for achieving the goal of bringing particle physics and cosmology into the classroom. After that the real development job begins. All of you who have ever developed any materials know that first you have to come up with some ideas that seem like they can withstand the test of time, and then you have to carry them through.

RESEARCH ON TEACHING MODERN PHYSICS

Some of you have felt acutely uncomfortable at times about the lack of explicit guidance at this meeting. We have not specified how we are to produce materials on any of the topics considered by the subgroups: lectures, demonstrations, audiovisual materials, problems, examinations, and laboratories. Instead of "convergent" thinking and behavior, we have made all of you participate in "divergent" thinking and behavior. This is not our usual experience. It seems an impossible, onerous task. But think about how one does modern physics of any sort, whether it is particle physics, or cosmology, or whatever:

First, there's a puzzle (Figure 1). After all, people don't get interested in things unless they're puzzled about it. There has to be a conundrum of some sort. Without a conundrum, you can't do much of anything.

Once you have a puzzle, what do you do? Well, the puzzle might be something like, is a particular meson resonance split? Whether the A_2-meson was actually one or two resonances was a burning issue in particle physics back in the early 1970s. It was a worthy puzzle.

How do you find out whether the A_2 is split or not? Well, you generate ideas to try to solve the puzzle. You probably don't have any basis for intuition about such splitting. You have some problem, it's about the physical world. Now how do you solve it? What do you do? It is these sorts of problems which must be addressed in any physics research.

Because of that, everything is self-directed. Since you have no set ideas of the solution to the puzzle, you shoot into the dark. That is, you don't really have guidelines for how to solve such problems. No one has ever done this particular problem before. How do you solve it? It is very difficult to handle such a divergent question in the absence of intuition.

All of the ideas which have arisen through the shots in the dark and the self-direction have to be tested. Most of the ideas that are generated aren't right.

You have heard about charmonium, which was discovered back in the mid-1970s. When charmonium was discovered experimentally in the psi meson system there were hundreds of ideas attempting to explain what it was (of course, this is the way physics goes). Only one of those ideas, more or less one, withstood the rigorous tests of physics. Some ideas have merit; others don't. But unless we have those competing ideas to test, we will not be certain of how to explain what we observe.

Clifford Will yesterday mentioned the importance of the Brans-Dicke theory in the development of experimental general relativity. With Einstein's publication of the theory of general relativity, we had only one theory available. Nobody was very excited about it, because, they would ask, what could you test? With the advent of the Brans-Dicke theory, suddenly there were two competing theories. There was a puzzle. There were things that were predicted to be different according to which theory was chosen. We could begin to think about how to test those predictions.

So, some guesses have merit, and, of course, those which have merit provoke other possibilities. Once you have started testing the ideas, you have to work on them more, to really work hard, to hunker down and try to get someting done. Finally, after all of the testing and disagreement and dissention, there is a consensus on the solutions—or the solution, depending on whether there is one or there are many—and there are consequently newer corollary puzzles that come out of that. With the appearance of these new puzzles, we must begin the cycle again, with a new puzzle replacing the one which was just resolved.

What is the relevance of this paradigm? We are attending a conference on the teaching of modern physics, not one on particle physics. But I assert that everything that I said about doing modern physics can be said about developing methods to teach modern physics. That is why we're here. We have a puzzle. The puzzle is to describe modern physics, which is happening, in which people are engaged right now, and to try to find a way to communicate the ideas and the excitement of this physics research to other people.

How do we communicate modern physics to the public and to students? It's a big puzzle. We don't know how to accomplish it. If we knew, none of us would need to have been here. We would then have been able to set out something saying, "This is how you do it."

But we don't know how to do it, so we're trying to generate ideas to solve the puzzle. That's where we are right now. We are at that second stage, that of generation of ideas to solve the puzzle, of making those shots in the dark.

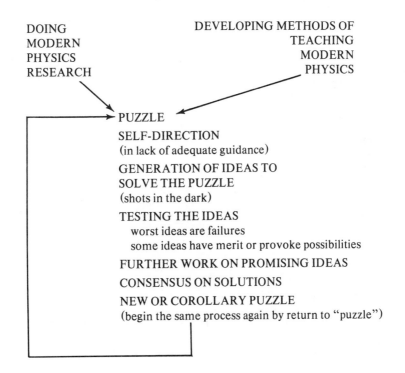

Figure 1. How doing physics research proceeds and how doing research on teaching modern physics proceeds.

Where are we going to go from here? Well, that depends on all of us. This is self-direction. We don't really have guidance. Nobody has ever managed to communicate the ideas of particle physics and cosmology in the way we would like to. We have to be self-directed.

Naturally, there are examples and paradigms from other areas, but still this is essentially self-directed work. We have to test the ideas generated. We have to accept that most of these ideas are going to be failures. But some of the ideas we have developed or will develop as a result of this conference are going to be successful.

We have to do further work on those promising ideas, and at some stage we will have to reach a consensus on the solutions. So, if you were wondering where we were going, and why we didn't already know where we were going: We didn't know where we were going because, in research on the teaching of modern physics we really are doing things parallel to the way physics research is done. We did't know how to get there, because we've never gone there before. We are blazing that trail.

OTHER TACTICS FOR INTRODUCING NEW PHYSICS IDEAS

I would like to share with you a couple of ideas that have not yet been mentioned here. These ideas are not mine alone, and many have been discussed in the literature. Some of the ideas have been discussed in meetings at the Users Center here.

In talking to students, in talking to people of that age, I have found that they are very similar to me when I was that age: they liked science fiction, they like to stretch their minds in ways that aren't as tough as actual hard physics problems. It may actually be a way to persuade them to want to do actual hard physics problems.

There are several books which can serve as supplemental readings in a physics class, or just as books in which the science is fairly honest. The first book I will recommend is not science fiction. It is a book about physics and science fiction. Amit Goswami and Maggie Goswami, friends of mine, wrote *The Cosmic Dancers*. I should admit that I reviewed the manuscript for them and tried to contribute to its success. I think it is a useful excercise, a useful book. A lot of the physics ideas that we talked about during this conference are discussed in *The Cosmic Dancers* from a physics point of view, but in connection with science fiction.

There are also science fiction novels which are examples of various ways that topics from actual modern physics has worked its way into science fiction. All of these people whose names are in Table 1 write what's called *hard* science fiction; that is, science fiction directly connected in some way or another to physics or other natural sciences.

In *The Gods Themselves,* Isaac Asimov, one of the most famous practitioners of hard science fiction, talked about the four forces of nature and what would happen if there were two bubble universes in which the strengths of the forces were slightly different. This is a thoughtful, provoking idea, based on real physics. Recent ideas about the evolution of our universe make the idea even more plausible today than when *The Gods Themselves* was written in 1972. In the "new inflationary universe" models, the universe undergoes exponential expansion for a period of time during which the radius of our universe grows by many orders of magnitude. An analogy for this is the growth of bubbles of steam within water as it boils. Other, unconnected, sectors of the universe (other "bubbles") also grow.

In *Timescape,* Gregory Benford, who is chairman of the physics department at the University of California at Irvine, wrote about tachyons and the effect that followed the discovery of a way to communicate with the future by exchange of tachyons (advanced waves). Benford is a low temperature physicist, and a lot of the book has to do with low temperature physics. I couldn't tell where the boundary was between what was physics and what was science fiction. I asked a colleague who is a low temperature physicist and a science fiction fan, and *he* didn't know where the physics ended and the science fiction began.

In addition, there are a couple of books that have to do with the Schrödinger equation, the basic equation of quantum mechanics which describes the time evolution of the wavefunction $P(\vec{x})$. We believe that one describes the universe by such wavefunctions and that the absolute square of the wave functions $|\psi(\vec{x})|^2$ describes the probability $P(\vec{x})$ that the particle described by $\psi(\vec{x})$ will be at $\vec{x}$. In this theory, there are operators which describe physical measurements. When the respective physical quantity is measured by applying the operator to produce an expectation value, the wavefunction picks out the physical quantity; it "collapses." The act of measurement causes the collapse. Before measurement, the answer to a physics question is probabalistic; after measurement, the physical quantity has a specific measured value. Measurement "picks out" only one of the possiblities. Which possibility is picked out is random. Many answers could have been found, but only one is realized. The probability distribution could describe the results of many identical measurements. Each measurement corresponds to collapsing the wavefunction in one particular way. There is thus a multiplicity of possible collapsed wave functions in the universe. Everett[1] has hypothesized that each collapse produces its own universe. There is therefore an infinite multiplicity of universes. The number grows with each collapse of the wavefunction.

Two books were recently published which are based on Everett's idea: Frederick Pohl's *Coming of the Quantum Cats* and James Hogan's *The Proteus Operation*.

Paul Preuss has written *Broken Symmetries*. The book discusses the consequences of the existence of a new, "inner," quark which is unstable—its symmetry is broken. The book is a provocative look at the interface between scientific and military research, in addition to being good hard science fiction.

Some of you may have met Richard Carrigan, a Fermilab physicist who took some of you around on the Fermilab tour. He and his wife wrote a book back in 1972 about muon catalysis. About three or four years ago, a parallel idea, monopole catalysis, was all the rage in Physical Review Letters.

Others who are doing similar sorts of hard science fiction that have some connection to physics include Poul Anderson, David Brin, Hal Clement, Gordon Dickson, William Gibson, Robert Heinlein, Larry Niven, and Jerry Pournelle. I mention those names so that you may recognize them if you wish to look at some of those things. If you decide it is worthwhile, you may wish to get some of your students involved. This is still another way of educating. We are teachers, and the best way to teach is to get people where their interest lies. Science fiction is where some of our students' interests lie.

SHARING IDEAS

Another way that we can get to other people has to do with dealing with the ideas that have been generated here. There is an AAPT bulletin board in College Park. Those of you with micro-

TABLE 1

MY IDIOSYNCRATIC, INCOMPLETE VIEW OF "HARD"
SCIENCE FICTION AUTHORS

USUALLY WRITE HARD SF:

Poul Anderson	Jeffrey Landis
Isaac Asimov	Larry Niven
Gregory Benford	Jerry Pournelle
Ben Bova	Paul Preuss
David Brin	Stanley Schmidt
Richard and Nancy Carrigan	Charles Sheffield
Joseph H. Delaney	G. Harry Stine (as Lee Correy)
Robert Forward	H. C. Stubbs (as Hal Clement)
John Gribben	Jules Verne
Robert Heinlein (also as Anson MacDonald)	Vernor Vinge
James Hogan	H. G. Wells
Fred Hoyle	James White
Michael Kube-McDowell	Timothy Zahn

OFTEN WRITE HARD SF:

Greg Bear	Joe Haldeman
James Blish	Jeff Hecht
John Brunner	Colin Kapp
Orson Scott Card	Mack Reynolds
Gordon Dickson	Eric Frank Russell
William Gibson	Norman Spinrad
Tom Godwin	Michael Swanwick
Will Jenkins (as Murray Leinster)	John Varley
	Walter John Williams

Analog Science Fact/Science Fiction Magazine is the home of hard core SF.

computers and modems may wish to use the bulletin board to keep in touch with one another. In addition to those, I want to remind you that there are some things that already exist as a network for sharing information among physics teachers and physicists in general.

The Physics Teacher (TPT) is a respected journal among physics teachers. Many physics educators read TPT every month for its physics explanations and for its "how to" articles. Notes and articles based on this conference would fit perfectly TPT's aims. I and eleven others are on its editorial board. These people can serve as bridges between you and the editor, or you may choose to submit articles or notes directly to TPT.

The Science Teacher is published by NSTA and also publishes scientific articles. Many high school physics teachers, including most of those whose work is distributed among biology, chemistry, math (to a lesser extent), and physics, read The Science Teacher.

The *AAPT Announcer* does not ordinarily publish articles, but it will be used to keep participants in touch with one another. People may write letters to the editor or submit other material for publication. This journal is more informal than the others mentioned

The *American Journal of Physics* publishes more technical articles, and has many articles on teaching quantum mechanics. Conference participants wishing to write more technical descriptions of their work may find AJP the most appropriate organ for the dissemanation of the information.

The *Journal of Physics Education,* which Professor Jossem showed to you during the International Discussions, is published in Great Britain. It is similar in aim to TPT. The *European Journal of Physics* is a Europhysics journal similar to AJP. *Physics Today* has many articles at a level slightly more technical than *Scientific American* or *American Scientist.* All the journals listed have their editorial office addresses near the table of contents.

You may write some of the articles which will be appearing in these journals in the future. After all, we are interested in sharing this information. Those of you who have great interest in spreading information on teaching modern physics may write about it, and get it out to other people through the medium of the physics journals. If you do publish, you could mention the generosity of the National Science Foundation if you were supported by the NSF, the support of the Department of Energy, and the hospitality of Fermilab. This conference could not have taken place without their efforts.

Now, I'd like to step out of my role as some sort of organizer of this Conference, and speak to you person-to-person with no official capacity whatsoever. This is a personal and unofficial request from me.

You know that the state of funding for science education is not always very good. One way of doing something about that is acknowledging things that have been useful and helpful to you and to me as we teach, and how they were helpful: what you've learned, how it helped you personally, what might happen as a result of all of this for your teaching, how it could not have happened without the support of NSF, AAPT, the Department of Energy, Fermilab, all of the tremendous people who have been working so hard on your behalf, and on all of you who have been working so hard in an attempt to improve physics teaching.

I would like to ask all of you to think about doing this. Because something like this, with the details—not, "I had such a wonderful time," because they won't understand what you mean by that; but rather, "I had such a wonderful time creating these materials that I'm going to use in my classroom and that hundreds of other teachers are going to use in their classrooms." That is something that could really help the funding of science in this country.

DISCUSSION SESSION

I will now answer any questions that you may still have about what we expect, in the follow-up between now and January. Then I hope that some of the people who have been so helpful will be here. No?

Drasko Jovanovic: They are still busy. If you want your copies, they are in the Comitium. But they will accept all the acclamations in absentia.

Dr. Aubrecht: You may see up here in the front of the stage we have some flowers for some people I would like to mention, as soon as I deal with your questions. Those of you who wish to come with me after this is closed will be welcome to go up there and see them respond to this. Are there any questions that I can answer for you?

Lila Adair: Like you said, we obviously need more time to put together stuff we might want to share. The problem with this is that we've listed our school addresses, and we won't be back until September. Is it possible you could get a list of home addresses and mail it?

Dr. Aubrecht: It is possible that I can get a list of home addresses. What Lila has said is that the addresses are all school addresses, and it makes it difficult to communicate during the summer. I will, to the extent possible, use home addresses. Not everyone has given me a home address. So if you know you have not given me a home address, please send it to me. I will generate such a list and get it out to you as soon as possible.

QUESTION: I just want to clarify a point. The Ray Echerle film called "mathematical peep show"—it has excerpts of different fields of mathematics. It does have a section called symmetry. It would be wise if you would add that to the audiovisuals list, too.

QUESTION: Are we going to be put into groups, or subgroups or can we work on anything we feel like?

Dr Aubrecht: You can work on anything you feel like. I would appreciate it if you would let me know what you're working on, so I in turn can inform everyone else. It might be good for some people to work on the little triangles and the transformations, but if it turns out that everyone is working on those little triangles it probably wouldn't be very useful. One of the reasons that I want to keep communication going is so that not everyone condenses into the same state.

QUESTION: There's a great book by Harald Fritzch, *The Stuff of Matter*. It's full of ideas, and it's a useful resource. About a year ago, I tried to find a copy on the market, and found several people who had copies, and libraries such as at Washington University, but I was frustrated trying to find it.

QUESTION: It's already available in paperback. I have it in paper.

Dr. Aubrecht: It is a very good book. There's also a book called *The Cosmic Code*, by Heinz Pagels, which is an interesting resource book.

QUESTION: Saturday Morning Lectures, from Fermilab, that are available—they have a free place to get it if you go to this museum. You have to physically present yourself at the museum? Or can you mail that?

Dr. Aubrecht: Ted Ansbacher is connected with the museum. Could you answer that question, please?

Mr. Ansbacher: We do have those tapes. We will duplicate them. If you will send a blank tape in, and cover the postage, we will mail them.

QUESTION: Are those available in all formats? Or just beta and three-quarter inch?

Mr. Ansbacher: No, it is available only in half-inch VHS.

QUESTION: Only half-inch VHS. Are they available in beta from Fermilab?

Mr. Ansbacher: I don't know who else might have them available.

Dr. Aubrecht: Thank you. Any other questions?

QUESTION: Yes, the spiral-bound book that goes with the Saturday morning series says they're available in three-quarter inch. It depends on where you get it. There are four or five places where you can get them. I still have some of those little forms upstairs.

Dr. Aubrecht: Are there any other comments or questions?

(No response)

Dr. Aubrecht: All right. I'd like to thank Lisa, Pat, Marilyn, and Eva—Marilyn especially. Marilyn Smith is Drasko Jovanovic's secretary. They have done a tremendous job of typing and the results will be in your hands before you go.

In addition, many of you have seen Marge Bardeen, who is associated with Friends of Fermilab, helping out. She flew here from Germany to help with this conference. She did a tremendous job. And we have something here for her, too. Judy Zielinski, I'm sure almost all of you have seen. She is the striking blonde who sat at the registration desk and was so patient with all my problems, and all your problems, and who has helped me so much in organizing this. She deserves an incredible round of applause. If you will just join me—I know she can't hear it—

(Applause)

Dr. Aubrecht: Stanka Jovanovic has been very supportive in getting us help with the Latin Americans. We could not have had all the Latin Americans who were able to get here come without her help. She helped in the organization. She helped with the choice of the group leaders. She has just been tremendous. And, again, Stanka, I thank you, although you're not here. I wish you could hear this.

(Applause)

Dr. Aubrecht: The group leaders themselves are apparently also up in the Comitium working on getting those things out to you. They have worked long and hard. In fact, after this session is over, they're going to have a meeting with me, so they're going to keep working. And they have done just a tremendous job—you know it; I know it. I really appreciate all their work. I wish they were here to hear the round of applause for them.

(Applause)

Dr. Aubrecht: All of the other gifts were given in absentia. However, I've mentioned Drasko's secretary and Drasko's wife. But I haven't mentioned Drasko Jovanovic. Drasko has done labor above and beyond the call of any duty. He's just been tremendous. He's been helpful. He's worked hard. We couldn't have had this without Drasko. And Drasko, as a token of our appreciation—at least you're here to get something—as a token of our appreciation, I have here something called novic slivovic. This slivovic is an original Serbian plum brandy. Now, as you may notice the name Jovanovic is Yugoslavian. I hope that I have my information right—that you like slivovic, and that this is a good one. I hope you like it, Drasko. Thank you very much.

(Applause)

Dr. Aubrecht: In addition, some of our speakers were here until just a few minutes ago. They have done a tremendous job. The chairmen of the various sessions, some of whom are sitting here, have done a tremendous job. All of you have done a trememdous job of putting up with me. I congratulate you. My wife doesn't know how you did it. Thank you very much for coming. It's been great. And we'll see one another again. Vaya con dios.

(Applause)

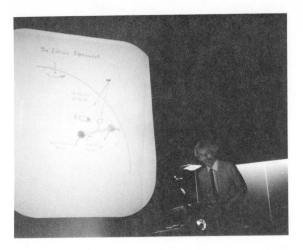

Participants

William T.Achor was born in Birmingham, Alabama in 1929. He holds a B.S. in physics from Auburn University and an M.S. and Ph.D. in physics from Vanderbilt University. He is married and has two daughters. He was an instructor in physics at Western Reserve University from 1957–58; a member of the technical staff at R.C.A Laboratories from 1959–62; and an assistant professor of Physics at Earlman College from 1963–65. He has been a professor of physics and department chairman at Western Maryland College since 1965. His earliest research interests were in low-energy nuclear physics. He also has worked in physical electronics, and most recently in environmental acoustics. His primary teaching interest has been in the development of laboratory instruction and of courses for non-science majors. He is presently involved in a program to prepare high-school science teachers who already are certified in one area to achieve certification to teach physics.

Lila M. Adair is the science department chairman and physics teacher at Central Gwinnett High School in Lawrenceville, Georgia. She is the 1985 Georgia Science Teacher of the Year and recipient of the Presidential Award For Excellence in Science Teaching. Lila holds a M.S. in physics and is completing an educational specialist degree in leadership. She is a native of Georgia and has taught school for 19 years. Lila has served as state director of the Georgia Junior Academy of Science and is actively involved in science youth activities in Georgia, and is serving this year as the Junior Academy director of the Science Olympiad. She sponsors the science fair and award-winning science team at Central. She and her students have written several curriculum guides for use in high school physics classes: environmental effects of nuclear power plants, high technology, and lasers and optics. Lila's current interests involve the development of innovative and creative techniques for the teaching of high school physics.

Theodore H. Ansbacher received a degree in metallurgy from M.I.T. and, after a brief period of work in industry, returned to graduate school, receiving a Ph.D. in physics from the University of Vermont. During a teaching career at Worcester (Mass.) Polytechnic Institute, New College(Sarasota, Fla.), and Lafayette College (Easton, Pa.), his interests evolved towards teaching non-science majors, resulting eventually in his moving from academia to the Museum of Science and Industry(Chicago). There, as director of education, he has been involved with increasing public understanding of science through informal education, bringing some of the museum's approach to learning to bear in the schools,and training science teachers.

Ron Apra lives in San Jose, California, and teaches physics, math, and computer science at Pioneer High School. Four years ago, he was offered the opportunity to teach physics, and quickly discovered it to be the most exciting subject he had taught in 17 years. The physics room had been used in an off-and-on mode for the past several years, and contained a lot of old equipment dating back to 1959 and earlier. Some of the equipment was in beautifully finished hardwood boxes, and produced rather interesting results when plugged in. His goal this year is to continue to acquire more funding and materials to modernize both the lecture and lab curriculum. His hobbies are camping with his family and long-distance swimming.

Gordon Aubrecht is a high energy theorist who lately has been doing phenomenology with the Fermilab group E-531. He attended Rutgers University, from which he managed to get a B.A., and Princeton University, which gave him a Ph.D. in spite of his consuming interest in S-Matrix theory. Gordon has spent this year on sabbatical leave from Ohio State at AAPT at the University of Maryland. He is married to Michelle A. Aubrecht, and has three daughters, Laurie (25), Dacia (16), and Katarina (20 months), and a granddaughter, Alex (14 months).

Lawrence J. Badar is a native of the Cleveland area and still a Browns' fan. He has been at Rocky River High School for 24 years and teaches physics and ninth-grade honors science. Earlier, he taught at St. Bonaventure U. and as Teaching Fellow at St. Louis U. and Case Western Reserve U. He is a Graduate of John Carroll U., with an M.S. from St. Louis U. He has done additional graduate work at Case Western Reserve, Ohio State, Cleveland State, and Cornell. He served two years in the U.S. Army Chemical Corps at Edgewood, Maryland. He was the first president of the AAPT Ohio Section and is presently Section representative. He is the 1985 Ohio Presidental Award Winner and an AAPT Physics Teaching Resource Agent. He and his wife Dolores have six children.

Jeannette Letjer de Bascones was born in Santa Teresa del Tuy, Estado miranda, Venezuela on February 12, 1934. She graduated in 1953 with a B.S. in physics and then taught math and physics at the secondary-school level. In 1959 she enrolled in a masters' degree program in the Physics department at Cornell University, but never finished because of marriage and three children. Back home in 1960 she continued working as a physics teacher until 1979. In the meantime she took several courses in physics and education in Venezuela and abroad. In 1979 she went once again to Cornell as a visiting teacher in the Department of Education. The author of two books and several articles, right now she is doing educational research about instructional strategies for overcoming students' conceptual difficulties in learning physics. She is also in the process of developing laboratory prototypes for teaching experimental physics with a grant from CONICIT. (National Council of Scientific and Technological research).

Earl N. Benitz was born November 23, 1940, in Petersburg, Alaska. He went to high school in Petersburg and graduated in 1958. He earned a B.S. in physics from the University of Alaska in 1962. After some time in the U.S. Reserves, he was married and went back to the University of Alaska for a secondary teaching certificate. He has taught 19 years in Alaska—7 years in Cordova, Alaska, and 12 years in Petersburg. He has two sons, and his hobbies are hunting and fishing.

John M. Boblick, science resource teacher at Poolesville Jr/Sr High School, Poolesville, Maryland, has taught physics and chemistry in the Montgomery County (MD) Public Schools for the last 24 years. A physical science major at Indiana University of Pennsylvania, where he earned his Bachelor's and Master's degrees, Mr. Boblick taught in his native Pennsylvania before moving to the Washington, DC, suburbs. Mr. Boblick also studied physics, chemistry, and education in the graduate schools of the University of Maryland, Western Maryland College, and Miami (Ohio) University. Mr. Boblick has contributed to numerous curriculum development efforts in his school system, including the writing of the county's physics and earth science courses of study. He has had extensive experience in the use of computers for instruction in physics and chemistry, reaching from the use of mainframes in the 1960s to the microcomputers of today. Mr. Boblick has received the NSTA STAR award and the Leo Schubert Memorial Award for outstanding teaching of Chemistry. He was cited by the National Science Foundation in 1985 for his contributions to science education.

Jennifer Bond, from Andover, Massachusetts, is a 1982 graduate of Wellesley College with a B.A. in both physics and astronomy. She is currently working toward a 1988 completion of a Masters of Natural Science at Worcester Polytechnic Institute. In 1982 she began teaching physics and astronomy at the Pomfret School in Pomfret, Connecticut, a small private boarding school of about 300 students. She is now nearing the completion of her first year at Phillips Academy, Andover, where she has been teaching physics, astronomy, and cosmology as well as running the astronomy club and guest lecturing on Halley's comet.

Jeffrey J. Braun was born August 7, 1945, in Elgin, Illinois and grew up on a dairy farm. He received his B.A. in physics from Knox College (Galesburg, IL) in 1967 and his Ph.D from Oklahoma State University in 1974. For the next four years, he taught a variety of physical science courses at a small community college in southern Kansas, and then moved to Springfield, IL, where he taught physics at Lincoln Land Community College and Sangamon State University until 1984. He is now completing his second year in the physics department at the University of Evansville. He developed a modern physics laboratory program while at LLCC, and is currently attempting to do the same at the U of E. He is also directing a year-long series of workshops in science education for primary-grade teachers in his area. His family includes his wife, Diane, and three children.

Larry G. Brown was born and reared on a Kansas wheat farm. He graduated from Ottawa University (Kansas), with a B.S. in physics (1963). He received his M.S. and Ph.D. in physics (1968 and 1973) from the Illinois Institute of Technology. He taught part-time at Roosevelt University and Moraine Valley Junior College. From 1968 to the present, he has been a faculty member at Morgan Park Academy. He is the science department chairman at Morgan Park, and while there he has established the high school computer education curriculum, hardware, and maintenance; created and installed Development Office and Library computer system; and created the 9th grade chemistry-physics and 12th-grade physics courses. He is also the president and owner of CRIS, Inc. (Computer consultants, sales and services of IBM and Altos Microcomputers for industry and education). He is married and has two children.

Wendell Brown has taught physics and computer programming (Pascal structured BASIC) for six years at Kentridge High School, in Kent, Washington, near Seattle. Prior to that, he taught a variety of math and science classes at Mountain Empire Junior–Senior High School, east of San Diego. It was at that small rural school that he received his first experience in teaching physics. His M.S. in applications of physics, from the University of Washington, was completed just last year. At San Diego State University, he received a B.A. in mathematics (1974) and one in psychology (1972), with a minor in biology. His teaching of physics has been mostly limited to the traditional Newtonian approach emphasized in the PSSC text. He is looking forward to learning ways to bring modern physics into the high-school classroom.

Carlos I. Calle received his Ph.D. in physics at Ohio University in 1984. He is chairman of the physics department at Sweet Briar College in Sweet Briar, Virginia, where he is also an assistant professor of physics and has been since 1981. His Ph.D. dissertation was entitled: " A Microscopic Treatment of Nucleon Scattering from Complex Nuclei." He has authored many papers and a book entitled "La Gran Aventura del Universo" (The Great Adventure of the Universe).

Scott Cameron received his undergraduate degree in physics from the University of California at Santa Barbara in 1969. He holds an M.S. in physics from California State University at Los Angeles. He has taught physics in San Marino since 1972, except for a year's leave (1983–84) to do computer work. During several summers he has done research in the Advanced Systems Group of Xerox Electro-Optical systems in Pasadena. The work involved fiber optic multiplexing, special films, laser penetration of smoke, and simulators using LEDs. Currently, he is science department chairman and teaches AP calculus as well as two levels of physics. For fun he coaches basketball and volleyball teams.

313

Dave A. Campbell, Ph.D. Physics, has 15 years of diversified teaching experience and has taught at Saddleback College; University of California-Irvine; University of California Irvine, College of Medicine; several science enrichment programs for K–12 teachers, including the UCI Summer Science Institute; and at the Massachusetts Institute of Technology. He is also a member of the core development team at the California Institute of Technology on *The Mechanical Universe* T.V. series, funded by the Annenberg/CPB Project. Dr. Campbell has extensive experience in the field of computers and has had previous National Science Foundation Grants. He is a specialist in Interdisciplinary Studies, developing courses titled "Individualism: Man's Search for Meaning," "Gods, Clocks and Visions," and "Living with Technology." He has taught physics, mathematics, astronomy, science fiction, interdisciplinary studies, earth science, and medical physics.

Luis Canderle was born in Villa Mercedes (San Luis)-Argentina on October 25, 1954. There, he studied at the elementary school and high school obtaining a Bachelor's degree. In 1973 he went on to obtain the degrees of FISICO (1976) and LICENCIADO EN FISICA (1978) from the San Luis University. He studied dielectric theory with Dr. Mechetti at San Luis University. In 1979 he moved to La Pampa University (Argentina) as adjunct professor. He has received different fellowships in Argentina to attend graduate courses on physics, and has obtained (1982, 1983, 1986) international fellowships to attend the Summer Institute of Physics at Valdivia (Chile). He has also obtained the "Manuel Noriega Morales" fellowship from OEA and another from the Swedish Institute for Quantum Chemistry at Uppsala (Sweden) but could not accept. Now he is a teacher of Quantum Mechanics at La Pampa University and has started working on electro-optics properties of matter. He is also very interested in the problems of teaching Physics at the high school level and introductory physics courses at the university level.

Francisco Claro was born in Santiago, Chile. He did his undergraduate work at the Catholic University of Chile and obtained his Ph.D. in Physics at the University of Oregon in 1972. His permanent appointment is at the Catholic University in Santiago, where he is a professor of physics and has taught extensively. For a number of years he has directed the graduate program in physics of that university. He has had visiting appointments at the Universities of Illinois, Purdue, Indiana, and Oregon and since 1980 has been an associate member of the International Centre for Theoretical Physics in Trieste, Italy. He has served terms as Vice President and President of the Chilean Physical Society. His scientific publications are mostly in the field of condensed matter physics but he has also published articles on the creative act in physics and the use of multimedia in teaching.

Kathleen Conn is presently completing her thesis on methyl group dynamics, investigated via pulsed NMR on organic solid molecules, for a joint degree in physics/biology from Bryn Mawr College. Her undergraduate work was in physics, and her Master's from UCLA in radiology/medical physics. She has taught at the college level in biology, physics and environmental science for about ten years. In September She will begin teaching physics and environmental science at Agnes Irwin School in Rosemont, Pennsylvania.

William K. Conway was born in Marshfield, Wisconsin, on November 4, 1932. He received his M.S. and Ph.D. degrees in science education (physics minors) from the University of Wisconsin–Madison. Additional study was done under NSF sponsorship at the University of Minnesota, Illinois Institute of Technology, and the University of Wisconsin. His research was in electronic pumping (physics) and physics enrollments (science education). He taught instrumental music and mathematics at the Stratford Wisconsin High School, served two years in the U.S. Army, taught science and mathematics at Merrill Wisconsin Junior High School, and now teaches physics at Lake Forest High School in Lake Forest, Illinois, and also at the College of Lake County, Grayslake, IL. Bill was a reader of advanced placement examinations; consultant in A.P. physics for the College Board; and has received numerous NSF fellowships, an Allis-Chalmers Fellowship, AAPT national recognition, ISAAPT physics teacher of the year, Illinois master teacher, ISTA distinguished teacher award, presidential scholars excellence award, Fermilab fellowship, and a 1985 AAPT Physics Teaching Resource Agent award. Bill lives with Gloria, his wife of thirty years, in Lake Forest, Illinois. They have three children.

David A. Cornell earned a Ph.D. degree in physics in June 1964, at the University of California–Berkeley. There he worked under Walter D. Knight on nuclear magnetic resonance investigations of liquid metals. Since graduation, he has been teaching physics at Principia College, Elsah, Illinois. He is a member of AAAS and AAPT, having served the latter as chair of its Illinois Section in the cycle 1983–86. He performed research during summers and sabbaticals at McDonnell-Douglas Research Laboratory, University of Warwick (England), Washington University (St. Louis), Iowa State University, and Monsanto Company. He attended Chautauqua short courses on holography, the special theory of relativity, and elementary particles.

Frank E. Crawley, III is an associate professor in the Science Education Center at the University of Texas at Austin and has been on the faculty for 11 years. He received a B.S. degree in physics from the Virginia Military Institute, an M.S.T. in physics education from the University of Wisconsin-Superior, and an Ed.D. in science education from the University of Georgia-Athens. He has six years of experience teaching secondary school physics, advanced-placement physics, and physical science in the states of Virginia and Georgia. He has taught physics and professional education courses at UT-Austin to prospective secondary science teachers. He has presented papers at numerous state, regional, national, and international conferences; has published 25 articles in major science education journals; and has served as a consultant to state and national curriculum developers, test developers, and textbook publishers. Recently he received a summer training grant from the Coordinating Board of the Texas College and University System to offer a "Summer Institute in Science" program, consisting of five courses for elementary and secondary science teachers in Texas.

Eric W. Danielson received his A.B. in 1962 from Harvard College in Astronomy. He received an M.Sc. from McGill University in 1969 in Meteorology. Between degrees he spent five years as a weather forecaster in the U.S. Air Force. From 1968–1983 he taught astronomy, meteorology, physics, and computer science to students of all ages at the Talcott Mountain Science Center in Avon, Connecticut. Since 1983 he has been assistant professor of science at Hartford College for Women, where he teaches the same subjects. He has authored a variety of publications, most recently a junior high earth science textbook. He lives with his wife and four children on a mini-farm in Canton, Connecticut, in a solar house largely of their own design and construction.

Doug Davis received a B.S. in physics and mathematics from Wichita State University in 1966. It was there that he realized he wanted to go into university teaching which required a Ph.D. He received his Ph.D. in theoretical nuclear physics from UCLA in 1970. With a strong interest in teaching, he accepted a position as Eastern Illinois University that fall. At the request of his draft board he joined the Navy. He returned to Eastern Illinois University when he was released from the Navy in 1976. He has taught a variety of courses there–Physical Science for El Ed Majors; "Adventures in Physics," a liberal arts survey course, both algebra-based and calculus-based introductory physics; classical mechanics; optics; and quantum mechanics. Looking for effective teaching techniques has led to his interest in computer-assisted instruction (CAI). His attempts at CAI have included microcomputers, a local mainframe, and PLATO. His book on Classical Mechanics has just been published. He is presently working on another book, Applied Physics, which is designed for a technical physics course.

Paul T. Debevec is a professor of physics at the University of Illinois at Urbana–Champaign. His undergraduate work in physics and philosophy was done at MIT. His graduate work in physics was done at Princeton. He has been a research associate at Argonne and a faculty member of Indiana University. His research has been in experimental nuclear physics. He has been a member of the Nuclear Science Advisory Committee, Subcommittee on Electromagnetic Interactions, and the Long Range Plan for Nuclear Science Workshop. He has also been a member of the Nuclear Physics Panel of the Physics Survey conducted by the National Academy of Science.

Karen C. Dietrich is a member of the Congregation of the Sisters of St. Joseph of Chestnut Hill, Philadelphia, Pennsylvania. She teaches the biological and physical sciences at St. Rose High School in Belmar, New Jersey. Keenly interested in interdisciplinary education, and especially integration of biology, chemistry, and physics, she has co-authored an article "From Atom to Eve: An Interdisciplinary Unit on Origins." It will be published in the September issue of *Momentum* (a national educational journal). Receiving a B.S. in Biology from Chestnut Hill College, Philadelphia, she studied anatomy at The Hahnemann Medical College in Pennsylvania and completed her M.S. in biology at Villanova University in Villanova, PA.

Fred Domann was born in 1937 on a small farm in southeastern Wisconsin. After graduating from Kewashum High School in 1955 he enlisted in the Air Force and was discharged in 1959. He began his college education in 1961 as an electrical engineering student, and graduated in 1965 with a major in math and physics. After teaching for one year he entered graduate school and earned his M.S. at UW–Milwaukee in 1968, and his Ph.D. at the University of Vermont in 1974. In 1976 he joined the Physics Department at University of Wisconsin–Platteville.

Arthur Eisenkraft is the physics teacher and science coordinator at Fox Lane High School in Bedford, New York. He received his B.S. and M.A. degrees from SUNY at Stony Brook and his Ph.D. from New York University. He is the author of *Physical Optics*, a laboratory manual for laser optics experiments; co-author of *Moments of Discovery*, an AIP-sponsored project about the discovery of fission and the discovery of the optical pulsar; and was awarded a U.S. Patent for a vision testing system. He has served on the editorial board of The *Physics Teacher* magazine and the *Journal of Computers in Mathematics and Science Teaching*. Presently, he is the chairman of the nationwide Duracell Science Scholarship Competition, as well as the U.S. Academic Director for the International Physics Olympiad. His teaching career began as a Peace Corps volunteer in Nepal.

Andria Erzberger was born and raised in Iowa, where she participated in everything in a tiny high school. Receiving a National Merit Scholarship, she majored in math and physics at the University of Northern Iowa. When she received a graduate fellowship for Cornell or Stanford, she chose Cornell. She received her M.S. from Cornell, got married, and went to California, where she has been ever since. She worked one year for a NASA contractor and accepted a job teaching math and science to Palo Alto high school students. She has been teaching physics at Palo Alto High for eight years and is a mentor teacher this year. Her hobbies are reading, traveling, sewing, gardening, and telling physics jokes.

Arthur V. Farmer was educated at Rensselaer Polytechnic Institute where he also taught while majoring in hypersonic aerodynamics in graduate school. Six years as a research scientist in industry convinced him that he wanted to return to academics. At his wife's urging, he tried teaching in high school and has remained in teaching for 20 years. In 1982 his program was selected as an exemplary program by NSTA. In 1983, he was awarded the Presidental Award for Excellence in Science Teaching. He is currently teaching at Gunn High School in Palo Alto, California, and writing high school physics materials under a grant from Fairchild Semiconductor Corporation.

Allan P. Feldman is currently teaching physics at Germantown Friends School, which is a Quaker independent school in Philadelphia. This is his 14th year of teaching and he has had experience in both private and public schools, and in college. He has an M.A. from Columbia U. and a B.A. from NYU. He has been an active member of the Southeastern Pennsylvania section of AAPT since it was formed and has served as one of its officers. Although he is not a member of the Religious Society of Friends, he is sympathetic to many of their teachings, especially those that relate to world peace and respect of others, and has worked to further those aims. He is married and lives in an old house in Philadelphia. During the summer he reads, writes, bike rides, and renovates houses.

Neil Fleishon was born in Philadelphia in 1952. He notes that "My career in the sciences was virtually assured from the start by the fact that both of my parents were high school English teachers." As soon as he could, he said goodbye to Philadelphia and pursued his world line through Boston (MIT) and Berkeley (UC). Then followed several years (UW–Seattle) when he was "poised on the event horizon surrounding the black hole of research in quark phenomenology and strong interactions before pulling away and returning to the main sequence." Today he is at Cal Poly in San Luis Obispo, teaching undergraduate physics, always learning new things, and enjoying the central coast of California.

Kenneth L. Frazier was born February 4, 1938, in Pennsylvania. He received his B.S. and M.Ed. degrees from Indiana University of Pennsylvania and an M.S. from Marquette University. He attended NSF Institutes at Hope College (chemistry), the University of Arkansas (nuclear physics) and the Pennsylvania State University (history of science), plus did additional graduate work at other colleges and universities. He has authored five articles for publication. He has been very active in local, state, and national science education organizations and was elected twice to the NSTA Board of Directors.

He chaired a regional science convention in Cleveland, Ohio, in 1980. He has presented papers and workshops and chaired many sessions at various conferences. He has taught at North Olmsted High School in Ohio since 1961, and presently teaches physics and is also science department coordinator. His hobbies include reading, stamp collecting, fine arts programs, and outdoor activities.

Anthony P. French has been a Professor of Physics at MIT since 1964. He was educated at Cambridge University, and was originally a nuclear physicist. His chief interest for many years has been in physics education. He was President of AAPT from 1985–86.

Joan C. Fu is an assistant professor of physics at Los Angeles Harbor College in Wilmington, California. She completed her undergraduate studies at Central Michigan University in Mt. Pleasant, Michigan, and her graduate training at the University of Massachusetts in Amherst. Upon graduation from the University of Mass., she did research and development work at the University of Southern California, and then at Hughes Aircraft Company's Electro-Optical and Data Systems Group, before joining the Los Angeles Community College District on a full-time teaching basis in the fall of 1984.

Anthony J. Giancola served as an aviation electronics technician in the U.S. Navy from 1955 to 1958. He graduated from the University of Bridgeport with a B.S. in electrical engineering in 1963. He worked in industry for six and one-half years. He received his Teaching Certificate and M.S. degree in 1972 and received a Certificate of Advanced Study in 1974. He has eight years of experience as a high school physics teacher and three years as a PIMMS Fellow at Wesleyan University. He was an AAPT Physics Teacher Resource Agent in 1985. He is married with two children in college.

Isaac Halpern received his B.S. degree from the City College of New York in 1943, and his Ph.D. from MIT in 1948. He has been Instructor at Princeton University (1944); Staff Member at Los Alamos (1945–46); and Staff Member in the Laboratory for Nuclear Science at MIT (1948–53). Professor Halpern has held visiting positions at Niels Bohr Institute in Copenhagen, at CERN in Geneva, and CEN in Saclay. He is currently located at the Nuclear Physics Laboratory at the University of Washington in Seattle. His primary research interest is nuclear physics, with emphasis on fission, statistical decay, giant resonances, photon and pion induced reactions.

Michael J. Harding received his undergraduate education at the University of California–Santa Barbara, leading to a B.S. in geology in 1966. Following graduation he entered a secondary credential program at San Jose State University, completing those requirements in 1966 with a teaching major in physical science and a minor in mathematics. In the fall of 1966 he began teaching at Los Altos High School where he was to remain for 15 years. His teaching assignments have included physics, earth science, astronomy, geology, chemistry, physical science, futuristics, and algebra. His primary assignment, however, has been in physics and earth science. Since 1972 he has been teaching Project Physics, a course with which he agrees philosophically, but which he doesn't believe achieves its intended goal. For the past seven years his duties have included the job of science department coordinator, which he accepted because of his interest in curriculum development. He recently served on an industry/education math–science task force, looking for ways to improve secondary education in California. In 1984 he completed the requirements for an M.S. in geology at S.J.S.U. with a concentration in geophysics. His current projects include searching for a new physics text, working on increasing physics enrollment, and wrestling with an ever-decreasing budget.

David Lee Van Harlingen was born in Dayton, Ohio, on September 8, 1948. He has a B.S. in physics and mathematics from the Ohio State University, an M. S. (in experimental high energy physics) from Rutgers, and a doctorate in science education with specialization in physics education research, also from Rutgers University. He was assistant to the editor (1972–76) for the book series *Adventures in Experimental Physics* (World Science Education). He has ten years of experience as a graduate teaching and research assistant in physics and education. He has extensive experience working with underachieving students. Since 1981 he has been assistant professor of physics and engineering at Somerset County College in New Jersey. Currently, he is chairperson of the College Senate. He is the author of the study guide for *PHYSICS* by Weidner . He was President of the Hemophilia Association of New Jersey (1978–79). His memberships include: AAPT, Sigma Xi (Scientific Research Society), Kappa Delta Pi Education Honorary, Cousteau Society, Sierra Club, and the Union of Concerned Scientists.

C.T. Harper was born and raised in New Zealand, where he graduated from Otago University with a M.S. in geology and chemistry. He attended Oxford University, ostensibly to study mineral petrogenesis, but his interests were soon diverted from petrology to radiometric age determination. He was awarded a doctorate at Oxford for his work on $^{40}K/^{40}Ar$ dating. In 1965 he accepted a two-year appointment as research associate in geophysics with the Department of Physics at the University of Toronto. For the following eight years he taught at Florida State University, working in association with an active group in nuclear science, developing facilities for age determination using a variety of naturally occurring, long-lived radio-isotopes. This work culminated in the publication of a monograph on "Geochronology." At Florida State he became more and more interested in the problems associated with perception and comprehension in the physical sciences. After a summer with the Institute of Nuclear Sciences in Wellington, New Zealand, he accepted a teaching positon at a large independent secondary school there. He returned to the United States in 1983, and is currently teaching at Phillips Exeter Academy in New Hampshire, where he is responsible for the AP physics program and courses in astronomy and chemistry. His appointment at Phillips Exeter has coincided with a school-wide curriculum review involving an extension of the science requirement and he is now deeply involved in developing a new interdisciplinary approach effective for teaching modern physics at the high school level.

Ward Haselhorst receieved his B.S. in physics from the University of Illinois in 1964. In 1970 he received his M.Ed. from the University of Illinois and in 1972 an M.S. in physics at Purdue (NSF Summer Institute). He has taught physical science, physics, and advanced physics at Proviso East High School in Maywood, Illinois, since 1965. He attended a Summer Institute for physics teachers at Fermilab in 1985.

Charles E. Hawkins was born on November 29, 1941, in Pontiac, Michigan. He graduated from Pontiac Nothern High School in January 1960. He received an A.B. in physics from Greenville College in June 1964, and then did his graduate work at Dartmouth College, and receiving a Ph.D. in 1971. His research area was plasma physics, and his thesis involved the interaction of an electron beam with a plasma. During the summers of 1977–1984, he was involved in spacecraft propulsion research at NASA's Lewis Research Center in Cleveland, Ohio. This work involved electric propulsion (the "ion engine") and a proposed microwave scheme for heating a propellant gas. He taught physics at Spring Arbor College from 1969 through 1980, and he has been teaching physics at Northern Kentucky University since 1980. He has attended NSF Chautauqua short courses on the origins of life and on microcomputers as lab tools. He has taught modern physics on a regular basis.

Tucker Hiatt is a 31-year-old physics and astronomy teacher. Since 1978, he has practiced his art at San Francisco University High School, a private, non-denominational school of 360 students. Tucker received his two physics degrees from The University of California: an A.B. from the Berkeley campus in 1976, and an M.S. in experimental solid state physics from the Santa Cruz campus in 1980. University High School lies just nine blocks from The Exploratorium, San Francisco's renowned museum of science and perception. Tucker has incorporated many of The Exploratorium's hands-on exhibits into his coursework in physics, astronomy, and A.P. physics. He spent the summers of '84 and '85 as an instructor and exhibit-builder at The Exploratorium.

Margaret "Peggy" Hill received her B.S. in physics from The College of William and Mary in 1976 and began teaching science at Hampton Roads Academy in Newport News, Virginia, where she taught physics, physical science, and coached the school's fencing team. In 1981 she moved to St. Louis, where she spent three years teaching physical science in a public junior high school. She is currently teaching physics, college physics (through St. Louis University), and geometry at Vianney High School. She also works with the Gifted Resource Council, a non-profit alternative educational program for gifted children. She helped the Council develop curriculum and taught a summer program entitled "Space Academy" (now in its third summer), a humanities oriented science enrichment program based loosely on NASA's Space Camp. In 1985 she began work with the St. Louis Science Center, presenting teacher workshops and in-service training for elementary and junior high school teachers in giving effective lecture-demonstrations in science. An avid amatuer astronomer, she is awaiting the arrival of the optics necessary to build a 13.1" reflecting telescope, her second telescope building project.

Patrick Hogan has been teaching physics at Colorado Academy for the past eight years. He came to this position after several years' research involving laser-induced fluorescence of gaseous free radicals and multiphoton ionization of gaseous atoms. Although he has enjoyed the transition to working wih high school students, he is frustrated with the lack of attention which is given to modern physics in most textbooks at this level. Four years ago he retaliated by establishing a two-trimester course called "Physics Beyond Newton," which is designed for seniors who have already had a year of classical physics. Quantum mechanics, relativity, and particle physics are the main topics of this course, and it gets good reviews, as well as apparently having served the purpose of convincing several of his brighter students to major in physics in college. In his spare time he "runs the science department, does college counseling, coaches mountaineering, plays a so-so orchestral string bass, and tries to understand his pre-teen daughter."

Linda Huetinck started her college education at the University of Kansas and then moved to California where she received a B.S. and M.A. from California State University at L.A. In 1965, after two years as an aerospace engineer, Linda began her present physics teaching position at Glendora High School. She was awarded the Exemplary Physics Program for California, and in 1985 an Ohaus national award for creative curriculum design. For the past three years she has been one of 20 educators on the California Assessment Program in charge of writing a statewide eighth-grade science exam. Linda has contributed to her field through presentations at local olympics, computer applications, and recently as a presenter at an Advanced Placement workshop. As science department chairperson, her current project is implementing a $20,000 grant for an interfacing computer-science laboratory. In the past year she has written teacher/student materials for two Television programs—*Creation of the Universe* and *Planet Earth*. She currently serves on the steering committees of the L.A. County Science Fair, California State Science Fair, and Magic Mountain Physics Student Day. She is married to an engineer and they have two teenage sons. Her hobbies include travel, sewing, reading, and skiing.

Stephen P. Jacobsen graduated from Montclair State College in New Jersey in 1963. He taught in West Orange, and worked as an educator/supervisor at The Franklin Institute in Philadelphia. He then returned to graduate school at Northeastern University and obtained a Ph.D. in Experimental Solid-State Physics in 1973. He has been teaching physics and other sciences at Lewisburg High School in Pennsylvania since 1972, with a three-year break for a Master's program in Educational Psychology. He has also taught part-time at Bucknell University in the physics and electrical engineering departments. Presently he is working on the adaptation of a prevocational physics course for his high school. His other interests include backpacking, bicycling, and childrearing.

John W. Jewett, Jr., was born in Yokosuka, Japan on September 18, 1947. His education includes a B.S. from Drexel University in 1969; an M.S. and Ph.D. from Ohio State University in 1974. From 1974 - 1984 he worked at Stockton State College in Pomona, New Jersey. Since 1984 he has been Associate Professor of Physics at California State Polytechnic University in Pomona, CA. His Thesis was "EPR and Luminescence of Rare Earth Ions in Single Crystal La_2O_2S." His specialty areas include: physics education, solid state physics, musical acoustics, and energy calculations in buildings.

323

JoAnn Johnson is a resident of Warrenville, Illinois. She was born and raised in the Chicago area. She graduated from the College of St. Teresa in 1971 with a B.A. in physics and a minor in education, and began teaching physical science and general science at Willowbrook High School in 1971. After one year she moved to Glenbard North High School and taught physics, physical science and advanced physics for nine years. By 1975 she received her M.S. in education physics from Northern Illinois University. She retired from teaching for three years to raise her family, but continued to work as a technical recruiter for the chemical industries in the Midwest area. She returned to teaching in 1983 at Wheaton North High School where she wrote a new curriculum for Physics and Chemistry to be taught at the ninth and tenth grade levels, and to teach physics and physical science. This year the physics team at Wheaton North is in the process of rewriting the entire physics curriculum. She is married and has two children.

W.H. Johnson

Herman Keith was born July 30, 1940 in Tucumcari, New Mexico. He was educated in Hillsboro, Texas. Upon graduating from Hillsboro High, he was named Thomas J. Watson IBM National Merit Scholar in physics and Rice Liberal Arts Scholar at Rice University. After a peripatetic academic career he was graduated from the University of Houston with a major in chemistry and a minor in physics. Two years of experience teaching high school physics led him to return to school for a M.S. in physics from East Texas State University. After a northern sojourn (study at Stony Brook, N.Y., and teaching on Long Island) he returned to Texas to work as an industrial engineer and chemist. Returning to teaching, he has taught physics at Booker T. Washington High School for the Engineering Professions since 1980. Married to Margaret Schumann Keith, he is the proud father of two sons, Seumas Marshall, and Ian Houston. He was elected to membership in the American Physical Society, American Chemical Society, and Sigma Pi Sigma. Among his hobbies and interests are classical music, history, electronics, mathematics, and books.

William Clark Kelly was born in Braddock Pennsylvania in March of 1922. He received his B.S. in 1943, M.S. in 1946, and Ph.D. in 1951 at the University of Pittsburg where he was an associate professor until 1958. He was director of the department of education and manpower at the American Institute of Physics between 1958 and 1965; director of fellowships NRC in Washington from 1965-67; director of the Office of Science Personnel from 1967-74; executive director of the Commission of Human Resources from 1974-82; director of the Office of Science and Engineering Personnel from 1982-83; and the secretary of the International Commission of Physics Education of the International Union of Pure and Applied Physics from 1966-72, chairman from 1972-75.

Lois Kieffaber has a M.S. in nuclear engineering from Columbia University and a Ph.D. in physics from the University of New Mexico. Her research interests are atmospheric physics (observing the infrared airglow) and zodiacal light. She has made measurements of the F-Corona at four total solar eclipses, has flown airglow experiments aboard NASA's Convair 990 Airborne Laboratory during Space Shuttle Simulation flights, and has been Co-Investigator on five NASA, NSF, and AFGL grants. She has interfaced airglow photometers to a computer for data acquisition, developed programs for analysis of airglow and eclipse data, and modeled ionization states and residence times for coronal constituents. Her teaching experience includes two years at a teacher's training college in Malaysia (with the U.S. Peace Corps), ten years at a large research university, and five years at undergraduate liberal arts institutions. She currently heads up the physics program at Whitworth College, where she is administering an NSF CSIP grant which has placed computers in the undergraduate laboratory for data aquisition and analysis. She is the author of 19 published papers and a *Study Guide* for general physics students. Her current interests include the preparation of high school physics teachers and encouraging women to pursue careers in science.

Sung K. Kim is a native of Korea and was educated at Davidson College and Duke University. He began his teaching career at Macalester College in 1965. He has since held visiting appointments at the University of California-Irvine and at Bethel College in St. Paul, and has taught high school students at the Twin City Institute for Talented Youth. In 1980–81 he was a visiting scholar at the Astronomy and Astrophysics Center of the University of Chicago.

Jaime L. Klapp was born in Mexico City on July 10, 1952. He studied for a First Degree in physics in the University of Mexico (1971–76), Part III of the applied mathematical tripos in Cambridge University, England (1976–77) and a Ph.D. in theoretical astrophysics in Oxford University, England (1977–82). He worked at the University of Florida (1983–84) as a postdoctoral associate. He is presently working in the Unversidad Autonoma Metropolitana, Mexico. His research involves problems in theoretical astrophysics, general relativity, and cosmology.

Leonard E. Klein graduated from Oakland University and then completed his graduate degrees at Wayne State University and Purdue University. Presently he is science chairman for Birmingham (Michigan) Schools. He reports that he tries to make the courses he teaches exciting vehicles for the study of nature by making them inquiry in approach, experiment-based, and fun. When asked why he teaches high school physics, he responds "Simple, I love both the subject and the students!" He has made presentations at national and local meetings concerning Birmingham Science Programs, computer applications in science, and physics experiments and demonstrations. He is married and has two children. He enjoys skiing, catamaran racing, and photography.

John Kolena reports that "Although I've been interested in astronomy ever since I can remember, I majored in physics as an undergraduate at Case Institute of Technology (my parents said they wouldn't pay my tuition if I was going to learn how to do horoscopes)." He spent his graduate-student years at Indiana University, where, in addition to earning a Ph.D. in Astrophysics (his thesis dealt with molecular radio emission from circumstellar shells), he managed political campaigns, co-founded the world's largest computing-dating service, and worked for the Kinsey Institute for Sex Research. While at IU, he also discovered that he "liked teaching and was reasonably good at it." He has spent the last eight years teaching various astronomy and physics courses at Duke University of North Carolina–Chapel Hill, and the North Carolina School of Science and Mathematics.

Gilbert Kuipers (B.S. in chemistry from North Georgia College, Ph.D. in physical chemistry from Georgia Tech) is an assistant professor of science at Valley City State College in North Dakota. He teaches nonmajors chemistry and physical chemistry in addition to various levels of introductory physics.

Charles Lang is a high school teacher in Omaha, Nebraska. He received his Ph.D. from Kansas State University. When asked what he has learned from life, he responds "What I have learned from life is that I don't know very much." His hobbies are thoroughbred horses and classic cars, and he notes "Looking at my job and hobbies you can see I've never grown up!" His future plans are "to see as much of the world as I can."

Luis Lauro Cantu was born on November 9, 1943, in Agualeguas, Nuevo Leon, Mexico. After graduating from preparatory school from Universidad de Nuevo Leon in June 1962, he entered the Instituto Tecnologico y de Estudios Superiores de Monterrey and there received a B.S. in physics and mathematics in 1967. He received his M.S. in physics from the University of Wisconsin in 1972, and his Ph.D. in science from Purdue University in 1977. He is currently a physics professor, teaching at the Instituto Tecnologico y de Estudios Superiores de Monterrey in Mexico, where he has taught for the past 19 years.

Eugenio Ley-Koo was born in Mexico City on March 20, 1939. He received his undergraduate degree in physics at the University of Puebla (Mexico) in 1959. His M.S. (1962) and Ph.D. (1964), degrees were earned at Indiana University, Bloomington. He has done research in nuclear, atomic, and mathematical physics at Institito de Fisica and Professor of Physics at Facultad de Ciencies, Universidad Nacional Autonoma de Mexico since 1966. He has taught high-school, undergraduate, and graduate physics courses, and nuclear techniques to medical doctors, biologists, chemists and engineers, in more than 20 universities and research institutions in Mexico, as well as at the University of Chile 1969, Central American Physics Course, Guatemala 1972, Insituto de Fisica Teorica, Sao Paulo, Brasil 1979 and China University of Science and Technology, Hefei, China 1981.

G. Samuel Lightner is presently professor of physics at Westminster College in New Wilmington, Pennsylvania, where he has been for 16 years. He was born in Virginia and received his B.S. in physics from Randolph-Macon College in Ashland, Virginia. His professional interests include atomic physics, computers in physics, meteorology, science and sports, and teaching elementary physics. He is actively involved in AAPT. He is married, and his hobbies include photography, sports, and travel.

Peter Lindenfeld received his Ph.D. in 1954 from Columbia University in New York. He has taught at Rutgers University since 1953, as professor since 1966. He has authored publications in low temperature physics, superconductivity, and physics-teaching activities. He was the chairman of the executive committee and co-editor of the *Proceedings* of the AAPT/APS-sponsored Conference on Teacher Institutes and Workshops, held in April 1984 in Washington, DC. He was a member of the APS Committee on Education from 1982–85. He is the chairman of the Executive Committee of the Center for Mathematics, Science, and Computer Education at Rutgers University, and is a member of the Joint APS-AAPT Committee on College--High School Interaction. He is an APS Fellow and an honorary life member of the New Jersey section of AAPT.

Mary Beth Livingston teaches physics and honors physics at Summit Country Day Upper School in Cincinnati, Ohio. She received her B.S. in physics in 1973 from Thomas More College in Covington, Kentucky, and her M.S. in biophysics in 1975 from Ohio State University in Columbus, Ohio. She is a member of NSTA and the Southern Ohio Section of AAPT. Mary Beth has attended workshops given by AAPT and was selected to participate in an NSF-sponsored graduate program for physics teachers at Xavier University in Cincinnati, Ohio, during the summer of 1985. She is also active in a group of Cincinnati-area physics teachers who meet several times a year to share ideas and discuss topics of common interest. She is married and has two children.

Carl Manning was born in Chicago, Illinois. He did his undergraduate work at the University of Illinois and graduate work at the University of Washington. His area of interest was particle physics. He has been a member of the faculty of St. Martin's College for the past 17 years, where he has taught a variety of physics and mathematics courses. He currently chairs the Divsion of Science and Mathematics. He has taught Auto-Tutorial Physics courses on the Keller Plan. And recently, he team-taught a course on the "Last Fifty Years of American History." He is active as a member of the Pacific Northwest Association for College Physics—a regional organization dedicated to the promotion and improvement of teaching college physics. His outside interests include golf, travel, and wine making.

Harry Manos received his B.A. and M.A. from California State University at Los Angeles. He has served four terms as vice president of the Southern California Section of AAPT. He is science department chairman and teacher of physics and calculus at George Miller Schurr High School in Montebello, California. Mr. Manos is a regular contributor to *The Physics Teacher,* and he was a member of the panel that wrote the student handouts and teacher's guide which were distributed nationally for the Public Broadcasting System special *Creation of the Universe* that aired in November 1985.

Michael J. Matkovich was born and raised in Chicago, attended local schools, and received his B.S. degree in physics from De Paul University in 1961. He then earned his M.S. in physics from Boston College and after working in industry and education for 13 years earned a M.S. in mathematics from Northern Illinois University in Chicago. He has attended numerous seminars and has completed the Professional Training Program at Oak Ridge National Laboratory. His professional experience includes doing research with Dr. R.J. Van de Graaf at High Voltage Engineering Corporation in Burlington, Massachusetts, and also co-authoring a research paper on infrared radiation and high temperature materials while working for the U.S. Army Material Research Agency in Watertown, MA. He also worked for Welch Scientific Co. (now Sargent-Welch Scientific Co.) doing engineering and development on scientific apparatus for teaching physics, and as a sales engineer for Gaertner Scientific Co. In 1970, he was one of the first faculty hired when Oakton Community College was founded. Since then, he has taught every level of physics and mathematics courses offered by the school. In addition, he has been an active member of the Chicago Section of the AAPT, serving as President in 1980 and as Secretary for the last five years. He has also done consulting work for Simpson Electric Co., Little Fuse Inc., Crane Packing Co., numerous publishing companies, and has taught part-time at College of Lake County and Great Lakes Naval Training Center.

Anthea Maton was educated at a high school in the south of England. Then, after some time in France, she trained as a radiotherapy radiographer, receiving her diploma in 1965. During her training, she was lucky enough to have as her professors various physicists, all specializing in radiation physics. Many had been in laboratories working on early accelerators and cyclotrons, and through them she was taken to Harwell where she spent time in the isotope laboratories. Experience with isotopes was also gained while working in the Royal Free Hospital's new Radioactive Scanning Department, a new venture in 1965. After qualifying, she married, had two children, and continued to work as a radiotherapy radiographer using X and gamma ray beams for radiation therapy. In 1970 she went to the Hammersmith Hospital, and in 1972 started working in the cyclotron unit where she was responsible for planning treatments using fast neutrons. She started teaching for this unit in 1973, and in 1974 began formal training as a teacher with the College of Radiographers. She received her higher diploma in 1976 and her teacher's diploma in 1977. Later that year she became Principal of the School of Radiotherapy at the Royal Free Hospital and was elected radiotherapy teaching representative to the National Health Services Northwest London Committee on Radiography. She left there in 1980 to take up a post as physics teacher in a London high school after the birth of her third child, and has been teaching physics since then. Her interests outside physics and teaching are art and music. She currently lives in Bethesda, Maryland, with her husband and youngest child.

Stephen Mautner

Ron McLachlan received his B.A. degree in philosophy from Immaculate Conception Seminary at Conception, Missouri, in 1970. After spending a year in Conception Abbey Monastery, Ron joined the faculty of Mount de Sales High School in Macon, Georgia, teaching physics and English. After a few years of summer school at Mercer University in Macon, Georgia, and the University of Georgia in Athens, Georgia, he began teaching chemistry as well as physics at Mount de Sales. In 1978 he received his M.Ed. in science education from the University of Georgia. Since then he has attended summer classes at the College of New Rochelle in New Rochelle, New York, and the University of Missouri at Kansas City in Kansas City, Missouri.

329

Jim Minstrell was born in April 1939, in Reno, Nevada. He has a B.A. in math education (UW), an M.S. in science education (U Penn), and a Ph.D. in science education (UW). He presently teaches physics and integrated physics/mathematics. He does educational consulting and research in cognitive skills and implications for curriculum and instruction. His instructional approach focuses on inquiry, development of thinking skills, conceptual understanding, and problem solving. He has worked on *Project Physics* and numerous science and mathematics teacher-training institutes and workshops. He was awarded NIE and NSF Cognitive Skills Research Grants 1979–1986; Sigma Xi (Research Honorary) Teacher of the Year (1981); Search for Excellence in Science Education: Physical Science (1982); and the Presidential Award for Excellence in Science Teaching (1985).

Daniel P. Mioduszewski was born in Buffalo, New York, in 1941. He received his B.S. in 1963 from Canisius College, his M.S. in 1965 and a Ph.D. in 1968 from Pennsylvania State University, where he worked at the Ionosphere Research Lab. He has taught at the Lawrence Institute of Technology as an associate professor of physics from 1968 to the present.

Connie DeLong Moore was born in Atlanta, Georgia. She later moved to Mississippi and graduated from Pearl High School in 1967. Connie graduated from the University of Southern Mississippi in 1971. While at USM, she received the Outstanding Physics Student Award. She has taught physics and mathematics at her alma mater, Pearl, and at Reed Junior High in Springfield, Missouri. In 1974, Connie returned to USM to pursue graduate work in physics. With the birth of her first son, Guy, now 10, she accepted a teaching position at Oak Grove High School in Hattiesburg, Mississippi, where she has been for the past ten years. Joshua, age seven, completes the family. Connie attended a NSF-sponsored Institute in Astronomy in 1973. Other honors include STAR Teacher 1980, 1982, 1984; Outstanding Young Woman of America, 1984; Outstanding Physics Teacher of Mississippi, 1984; and Outstanding Student Council Advisor, 1980. Her most recent accomplishment was her selection as Mississippi finalist in the NASA Teacher in Space Project. Connie is a member of Sigma Pi Sigma, AAPT, MAP, NSTA, MSTA, MSTM, NASAA; is on the executive committee of MSASC; and is charter member of TISF.

Thomas A.G. Moore is the junior member of a four-person Physics Department at Luther College in Decorah, Iowa. He received a B.A. in physics from Carleton College in 1976, an M. Phil. from Yale University in 1979, and a Ph.D. from Yale in 1981 for a dissertation in relativistic astrophysics concerning the generation of gravitational waves by collapsing stars. He taught at Carleton for four years before taking up his current position at Luther. He is also a 1976 Danforth Fellow. His current interests in physics include research in general relativity theory and relativistic astrophysics, as well as developing interest in quantum field theory. His hobbies include Irish music, contra-dancing, the game of Go, hiking, Biblical studies, and movies. He is 31 and married, but has no children.

Donald E. Morningstar is a native of York, Pennsylvania. Upon completion of high school he entered the U.S. Navy where he served his enlistment as an aerial photographer with the U.S. Navy hurricane hunting squadron. He then enrolled at Shippensburg University, receiving the B.S.Ed. in physics in 1958. While teaching high-school physics, he continued his education by attending Purdue University during the summers, receiving the M.S. in physics in 1964. He completed two additional years of graduate studies in physical science at Pennsylvania State University. In 1962, after four years of high school teaching, he accepted a position at Shippensburg University as instructor of physics. He was promoted to assistant professor and appointed to chair the physics department in 1969 where he served until 1973. In that year he was promoted to associate professor and appointed to the position of acting dean of the School of Science and Mathematics. In 1974 he returned full-time to the classroom, where his major responsibilities are the teaching of calculus-level introductory physics and advising the secondary education physics majors. For a number of years, he has served as a consultant to the Pennsylvania Department of Education. Most recently, he accepted an appointment as chairperson of the Physics Competency Task Force. He holds memberships in AAPT, the AAPT Central Pennsylvania Section, Sigma Pi Sigma, the Association of Pennsylvania State College & University Faculties, and the American Association of University Professors.

Robert D. Murphy received his B.S. in astrophysics from Michigan State University in 1980. He is a graduate student in the University of Minnesota M. Ed. program. He was a participant in the NSF-sponsored "Honored Physics Teachers in Minnesota" in June to August 1985, at the University of Minnesota. He has been a science teacher for five years at Spring Lake Park, and has also taught at the Science Museum of Minnesota. Robert is 27 years old and married to Marie Luck. His interests include astronomy, science fiction, history, hiking, and other outdoor activities.

Clifton Murray is a physics instructor at Stark Technical College in Canton, Ohio. After earning B.A. and M.A. degrees in philosophy, two cunning University of Akron professors—perceiving that physics was his first and true love—talked him into pursuing an M.S. physics degree. He found this endeavor "challenging, to put it mildly," but loved it nonetheless, and finds teaching physics highly fulfilling. His other interests include logic and reasoning, flying (student pilot), music (rhythm guitar), running, and swimming.

Jim Nelson received his B.S. in physics from Lebanon Valley College in Pennsylvania. Since 1961, he has been teaching physics at Harriton High School in Rosemont, Pennsylvania. He spent a one-year sabbatical as the visiting fellow at AAPT during 1984–85, where, among other things, he helped direct the first Physics Teaching Resource Agent Institute. In 1985 he received the Presidential Award for Science Teaching from Pennsylvania. He is currently assistant editor of *The Physics Teacher*.

Paul J. Nienaber was born on January 25, 1955 in Covington, Kentucky. He attended parochial schools in Covington, and graudated cum laude from Thomas More College in Fort Mitchell, Kentucky, in 1975, having earned a B.A. in physics and mathematics. He then entered the University of Illinois at Urbana–Champaign, and received an M.S. in physics from UIUC in 1976. He is currently attempting to complete his Ph.D. thesis in experimental high-energy physics at UIUC. His thesis work involved single π^0 production by neutrinos; the experiment was performed at the Alternating-Gradient Synchrotron (AGS) at Brookhaven National Laboratory. He has been a member of the physics faculty at Eastern Illinois University since January 1983. His interests at Eastern have been in computer-assisted instruction, and have also centered around curriculum development in the area of the physics of sound and music. He is a member of AAPT.

David R. Ober is a native of Indiana, having recieved a B.A. degree from Manchester College and a M.S.–Ph.D. degrees from Purdue University. He joined the Ball State University Department of Physics and Astronomy in 1968, and since that time he has conducted research in low-energy nuclear physics. Since 1975 he has directed summer enrichment (Student Science Training) programs for high-ability high-school students. For the past six summers, workshops have been presented for high-school physics teachers. Teaching responsibilities at Ball State have been at both the graduate and undergraduate levels; they currently include the beginning general physics course for physics and pre-engineering students, the nuclear techniques course, and the laser-optics courses. He has also previously served as the University Radiation Safety Officer.

Sister Mary Ethel Parrott teaches physical sciences and computer programming at Notre Dame Academy, a girls' high school in Covington, Kentucky. She is a Science Club moderator and has been very active in encouraging original student research. She received her B.A. in physics from Thomas More College, and her M.S. in physics from the University of Kentucky. Her thesis research in biophysics was done at the Albert B. Chandler Medical Center. In recent years she has been very involved in presenting workshop sessions for both elementary and high school teachers. She is an avid fan of most sports and enjoys reading, especially international espionage.

Betty P. Preece is a member of the physics faculty at Melbourne High School, Melbourne, Florida. She holds a B.S. in Electrical Engineering from the University of Kentucky and an M.S. Science Education from Florida Institute of Technology. She was admitted to the doctoral program in science education as a physics major at Florida Institute of Technology. She is an adjunct faculty member of the Florida Institute of Technology. She was previously the project engineer, surveillance systems, at the Air Force Eastern Space & Missile Center. Her memberships include: Senior & Charter Member, Society of Women Engineers; Women's Engineering Society of Great Britain; Missile, Space & Range Pioneers; AAPT; NSTA; Association of Women in Science; American Institute of Aeronautics & Astronautics; Florida Academy of Sciences. Her presentations and publications have covered topics such as career guidance, education, and science for professional engineering and educational organizations. Her honors include: Eta Kappa Nu, Delta Kappa Gamma, and the Woodrow Wilson National Fellowship Foundation 1985 Summer Physics Institute.

Frederick S. Priebe, chairman of the science department at Palmyra Area Senior High School, Palmyra, Pennsylvania, has taught physics and other science for the past 20 years. A native of the southeastern Pennsylvania area, Priebe holds degrees in physics education from that state's universities at West Chester and Millersville. He is a member of AAPT, NSTA, and the Pennsylvania Science Supervisors Association. In response to student fears following the 1979 TMI accident, Priebe founded and continues to advise his school's unique Radiation Monitoring Group. Concerning his choice of a physics teaching career, he notes, "I always had a curiosity as to how everything in the universe worked. It is fair to say that that was an extension of my fascination with the nature of the Creator of it all. There was also a desire, and some ability, to articulate that which I'd learned for the benefit of others. Physics teaching allowed me to be subsidized for following those natural inclinations, a fairly ideal situation."

Priebe and his wife, Patricia, along with daughters Pamela and Barbara, reside in Palmyra.

Joseph Priest is Professor of Physics at Miami University in Oxford, Ohio. His research interests include low-energy nuclear physics and physics education. He is author of three editions of *Energy: Principles, Problems, Alternatives,* published by Addison-Wesley Publishing Company, and is coauthor of *University Physics,* published by Academic Press, Inc.

Jonathan F. Reichert is an associate professor of physics at the State University of New York at Buffalo. He is a solid state experimental physicist with a research program in the area of low temperature physics, doing primarily magnetic resonance studies. He received his B.S. in physics and astronomy from Case Institute of Technology, his Ph.D. from Washington University in St. Louis, and did postdoctoral studies at Harvard University. He has been active in undergraduate affairs throughout his academic years, serving as director of undergraduate studies for eight years and receiving two outstanding teaching awards. He has taught all the introductory physics courses for more years that he can recall. In the early 1970s, he began plotting his personal revolt against the "catechism" set out by the "prophets" Sears and Zemansky and Halliday and Resnick. This has gradually led to the development of a new one-semester introductory course in mechanics, beginning with conservation laws and particle physics, which he wants to share with this conference.

Robert Resnick is Professor of Physics and Hamilton Distinguished Professor of Science Education at Rensselaer Polytechnic Institute in Troy, New York. He is a graduate of Johns Hopkins University, where he earned his A.B. (1943) and Ph.D. (1949). He worked for NASA from 1944–46. Professor Resnick is the author or coauthor of seven different textbooks in relativity, quantum physics, and general physics. These books are now in 13 editions and have been translated into no less than 40 languages. He has research publications in aerodynamics, atomic physics, nuclear physics, the history of physics, and physics education.

Professor Resnick's honors include: Oersted Medal of AAPT, the Association's highest honor (1974); Honorary Research Fellow and Visiting Professor at Harvard University (1964–65); Fulbright Professor, Peru (1972); Honorary Visiting Professor, People's Republic of China (1981 and 1985); Exxon Foundation Award for Outstanding Teaching (1954); Distinguished Faculty Award, RPI (1971); Fellow of American Physical Society and of American Association for the Advancement of Science. He was Vice-President of the American Association of Physics Teachers in 1986 and President-Elect in 1987.

James Ruebush is a faculty member at St. Charles High School in St. Charles, Illinois. Born in February 1947, he was raised and educated in Illinois. He graduated with a B.S. in Ed. from Western Illinois University (1969) and an M.S. in Physics Teaching from Eastern Illinois University (1982). He is married and has two children of high school age. He has taught 17 years. In the summers he is the physics coordinator for the afternoon sessions of the Fermilab Summer Institute for Science Teachers. His hobbies include astronomy, photography, and music.

H. Ryan

Barbara Susan Saur received a B.S. degree in physics from the University of California, Los Angeles in 1977, and an M.A. degree in education from Stanford University in 1983. She is presently in her third year of teaching physics and mathematics at Kentwood High School in Kent, Washington. Ms. Saur is very active in promoting science and mathematics education, particularly for young women. Prior to becoming a teacher, she worked as a physicist at Hughes Aircraft Company and TRW in Southern California. She is a member of AAPT, Sigma Pi Sigma, and Phi Beta Kappa.

Donald A. Schaefer has 36 years of teaching experience. He was a teacher of math and science at West Union High School, West Union, Iowa from 1949–1956. Since 1957 he has been teaching at Bettendorf High School, Bettendorf, Iowa. He is presently teaching honors physics (PSSC-5th Edition), honors advanced science (physics and chemistry), and planetarium director. He holds a B.S. degree from Upper Iowa University (math, physics, and philosophy) and an M.S. degree from U. of Wisconsin (science education). He has participated in NSF and other institutes (primarily in physics) at Carnegie Tech, Michigan State, Reed College, Denver Research Institute, and Knox College. He has been a staff member/consultant for institutes at Nebraska Wesleyan University, Knox College, University College of Rhodesia and Nyasaland (Africa), North Bengal University and Agra College (India), University of Iowa, and San Diego State.

Walter P. Schearer was born and raised in eastern Pennsylvania. After receiving a B.S. in Physics from Villanova University in 1964, he spent seven years teaching various physics and chemistry courses in the suburban Philadelphia area. During this time he also completed work on an M.Ed. at West Chester State College and an M.S. at Marquette University. Since 1971 he has taught IPS, PSSC physics, A.P. Physics, and a two-year integrated chemistry/physics course at Glenbard North H.S. in Carol Stream, Illinois. Outside the classroom he has coached track, negotiated for the teachers' union, and served on numerous academic committees. He has participated in NSF,

AEC, and DOE summer programs at Knox College, Notre Dame, University of Illinois, and Argonne National Lab. In 1983 he received an outstanding teacher award from the University of Chicago, and in 1984 he was named an Illinois Master Teacher. Presently he serves as a lab instructor in Fermilab's Summer Institute for Science Teachers, as well as directing Physics West. His wife, Patricia, and his five children live in rural St. Charles.

Robert Sears, Jr. obtained his undergraduate degree from Centre College in Kentucky and his Ph.D. from the University of Colorado. His graduate research was in experimental high energy physics. He has been teaching at Austin Peay State University since 1968. He has been Chairman of the Department of Physics since 1977. He is Section Representative of the Tennessee Section of AAPT and has been Chairman of all the AAPT Section Representatives since 1983.

Robert Sells was born in 1948. He graduated from Ohio Wesleyan in 1970 and has taught at Choate Rosemary Hall from that year on. He earned a graduate degree from the University of Purdue. While at CRH, he has instituted a very successful method of individualized instruction, developed a physics course for freshman and and sophomore students, introduced a term-based program for juniors and seniors, and created a nuclear physics course. He served as science department head of the winter school for one year and currently serves as the science department head for the summer school. In that last capacity, he helped develop the Connecticut Scholars Program. In 1981, he visited the Fermi labs for a symposium on the history of modern physics. In 1983 he visited China and Russia on an AAPT trip. He has four children and a wonderful wife.

Raymond A. Serway is currently professor of physics and Department Head at James Madison University, a position he has held since 1980. He received his Ph.D. from IIT (1967), an M.S. from the University of Colorado (1961), and a B.A. from Utica College (1959). He was a member of the physics faculty at Clarkson University for 13 years (1967–1980). He spent a sabbatical in 1974 with IBM Research Laboratory in Zurich, Switzerland, and was a visiting scientist at Argonne National Laboratory in the summer of 1973. He was also employed as a research physicist at IIT Research Institute, (1963–1967) and as a project engineer at Rome Air Development Center (1961–1963). He has published about 40 research papers in the areas of electron spin resonance and semiconductor physics, and has written three textbooks and three student study guides. The first two texts, now in their second edition, are introductory calculus-based texts written for science and engineering students. The third text, *College Physics,* co-authored with Jerry Faughn, is an algebra-trigonometry-based text published in January 1985. He was the recipient of the Distinguished Teaching Award at Clarkson University in 1977, and the Alumni Achievement Award at Utica College in 1985.

John M. Sherfinski was born and raised in Ashland, Wisconsin. He attended the University of Wisconsin–Madison as a General Motors Scholar, where he earned a B.A. in physics and a Phi Beta Kappa certificate. He spent two years at Caltech and earned an M.S. in physics. Following a year teaching physics at a public high school in Ogdensburg, NY, where he met his wife Janet, he returned west to UCLA and worked on a law degree. Los Angeles then began to wear a bit, so he moved to Hartford, CT, and was a patent attorney for two years. For the past ten years he has been teaching physics and calculus at Kingswood-Oxford School, an independent prep school in West Hartford, and for the past eight years he has been an adjunct faculty member in physics and math at the University of Hartford. At Kingswood he has coached boys' basketball at the ninth- and tenth-grade levels for all ten years, which is a big plus as it gives him the perfect opportunity to play every day all winter! Three years ago his perspective on everything was favorably altered with the birth of a daughter, Kristin, and within two months he and his wife should be parents for the second time.

Kenneth A. Soxman was born in January 1930. He received his undergraduate degree from Missouri Valley College. A graduate of McCormick Theological Seminary, he has earned advanced degrees from Washington University, St. Louis, and New Mexico Highlands University. Married since 1952, his family includes two daughters, two sons-in-law, and three granddaughters. He holds a life teaching certificate in Missouri public schools, having taught science for six years in secondary schools in that state. The opportunity to teach at the college-university level was accepted in 1963. The initial positions were in small, church-related liberal arts colleges; Buena Vista in Iowa, and Drury in Missouri. For the last 17 years he has been part of the physics faulty at Southwest Missouri State University. Best described as a generalist in the field of physics teaching, his deepest interest is in studying the life of physicists, and their ideas, as they sought/seek to understand or describe nature. In an attempt to make the work of physicists available to more students and in palatable portions, he was instrumental in the introduction of a mini-physics curriculum at SMSU. Over the years he has had continued enriching experiences in numerous institutes and workshops sponsored by: NSF, ERDA, Shell Oil and DOE. He continues to teach and study in Springfield, Missouri.

John D. Spangler was born on November 18, 1936 in Lincoln, Nebraska. He was raised in rural central Minnesota, where his father was a veterinarian. After graduating from Atwater Minnesota High School in 1954, he attended Kansas State University, receiving a B.S. in physics in 1958. He received his Ph.D. at Duke University in 1961. His thesis was on the spectroscopy of organic solids at low temperatures under the direction of professor Hertha Sponer. He served as a lieutenant in the U.S. army from 1962-64 when he became an assistant professor of physics at DePauw University. In 1965, he joined the physics faculty at Kansas State University and served as an assistant professor, associate professor, and professor until 1985. In 1985 he accepted a position as professor of physics at St. Norbert College in DePere, Wisconsin. He has been married since 1958 and has four daughters and one son.

Bruce Spatz

Judith Fibkins Tavel was born in Somers, Connecticut, on June 1, 1947. She graduated from The Chaffee School in Windsor, CT, in June 1965. She graduated with honors in physics from Vassar College in Poughkeepsie, New York, in June 1969. While at Vassar College she attended Ohio State University on a Summer NSF Fellowship to study molecular spectroscopy. She attended the State University of New York at Albany as an NDEA Fellow and received an M.S. in physics in January 1971 and Ph.D. in physics in June 1978. Her Ph.D. thesis was a theoretical analysis of the thermodynamics of small metallic particles with irregular surfaces. Since 1972 she has been teaching at Dutchess Community College, Poughkeepsie, NY, where she is currently an Associate Professor. She is responsible for a wide variety of courses, including physics, astronomy, geology, and history of science and technology. She is a member of the Two Year College Committee of the American Association of Physics Teachers and is very interested in problems of teaching at two-year Colleges.

Saul Téllez Minor is a professor of physics in the Escuela Superior de Ingeniería Mecánica y Eléctrica (ESIME) of the Instituto Politecnico Nacional (IPN), in México City. Téllez Minor holds his B.S. in physics and mathematics in the Escuela Superior de Fisica y Matemáticas, also of the IPN. His thesis was about the theory of the pion-deuteron dispersion in the region of the δ (1236). In 1976 he was in the Hebrew University of Jerusalem, where he took a training course in the field of teaching physics. He is currently project director in the field of solar energy in the IPN, and was recently appointed coordinator of a multidisciplinary research group that works in two areas: alternative sources of energy and food. He is interested in solar energy, cosmogony, and mainly in education.

Frederick E. Trinklein teaches physics, astronomy, and religion at Long Island Lutheran High School in Brookville, New York, and is an adjunct professor of astronomy at Nassau Community College in Garden City. He graduated from Concordia College (Illinois) and Northwestern University Graduate School. He is married and the father of five grown children. Trinklein is the co-author of several physics and astronomy texts, including *Modern Physics,* the widely used high school textbook. He has led nine solar eclipse expeditions to various parts of the world, most recently to Indonesia in 1983.

Piotr Trzesniak received his B.S. in physics for teaching and research from the University of São Paulo, where he also received his M.S. He is presently engaged in doctoral work. His area of interest is in condensed matter, particularly luminescence and conductivity. He began teaching mathematics and physical sciences to high school students while he was a university student. In 1974, he began his work at EFEI (Escola Federal de Engenharia de Itajubá). Piotr serves as head of the Department of Applied Sciences, for which he was elected in 1982 and again in 1984.

Larry Weathers was born and raised in the Chicago area. He attended Marquette University, Milwaukee, Wisconsin, (1965–69) where he received a B.S. in Chemistry. He completed an M.S. degree in Chemistry at Michigan State University, E. Lansing, Michigan (1971). Deciding to pursue a career in education, he remained at M.S.U. until 1973, broadening his science background in biology and physics. While teaching introductory courses at the Science and Mathematics Teaching Center, M.S.U. and teaching physical science at a local prison through Montcolm Community College, he worked on obtaining secondary teaching credentials. In 1973 he accepted a secondary chemistry teaching position at Notre Dame Academy, Worcester, Massachusetts, where he became department head in 1975. In 1978 he became coordinator of the laboratory program of the biology department at Holy Cross College, Worcester, MA, where he took additional coursework in Physics. In 1979 he accepted his current position as physics and chemistry teacher at The Bromfield School, Harvard, MA. Several additional professional activities included being elected head of the Division of Finance and Administration of the Board of Trustees of Notre Dame Academy, and currently he is a member of the Peace and International Relations Committee of the Massachusetts Teachers Association.

Joseph C. Wesney was born and reared in Columbus, Ohio. He married Anita Quick in 1962 and they have two sons, Tom (age 16) and Doug (age 19). They now reside in Cos Cob, Connecticut. His educational background includes completion of a B.S. in chemistry at Ohio State in 1964, an M.S.T. in chemistry at Cornell in 1966, an M.S. in physics at Purdue 1969, and his Ph.D. in science education and physics at Cornell in 1977. He has taught chemistry, physics, and earth science in Grove City, Ohio; physics in Worthington, Ohio; physics at Cornell; and physics and astronomy in Greenwich, Connecticut. He is currently Division Chairman of Science at Greenwich High School, where he splits his time between administering a 20-member science faculty, and teaching C-level A.P. Physics and Astronomy. In addition, he serves on the boards of directors for the Connecticut Science Fair and the Science Education Center. He has also been a Stauffer Chemical Fellow in the Project to Increase Mastery of Mathematics and Science (PIMMS) at Wesleyan University for the past two years, and will work with other PIMMS Fellows and Wesleyan faculty on a project for gifted innercity students and

339

teachers in physics and astrophysics this summer. His personal areas of study currently include the conceptual, historical, and philosophical aspects of astrophysics, cosmology, and quantum physics.

Myra Ragle West is an assistant professor of physics at Kent State University in North Canton, Ohio. She received a B.S. in Math and Physics Education from Kent State in 1965 and an M.S.T. (Physics and Math) from Wisconsin State University at Superior in 1970. She has taught high school mathematics and physics for four years and is completing her 15th year at Kent State. She has taught several of the nonprerequisite courses such as "The Seven Ideas That Shook the Universe," "Energy and the Environment," and "Forensic Physics" as well as the general college physics sequence. She has recently been teaching some developmental (remedial) math courses in arithmetic and introductory algebra.

Mary E. Williams-Norton received a B.S. degree, magna cum laude, with a major in physics from Bates College in Lewiston, Maine, and M.S. and Ph.D. degrees from Rutgers University in New Brunswick, New Jersey. After participating in additional research in experimental low energy nuclear physics as a research associate at Florida State University, she joined the faculty of Ripon College in Ripon, Wisconsin, (part-time) in 1975. In addition to participating part-time in research at the University of Wisconsin–Madison, she has developed courses in astronomy at Ripon College and has been leading science activities workshops for area elementary school teachers. She lives in Poynette, Wisconsin, with her husband, Gregory Norton, and their daughters Jeanne and Laura.

Jack Wilson is AAPT Executive Officer and Professor of Physics at the University of Maryland in College Park. He also holds a joint appointment in the Science Teaching Center at the University. His 15-year career in teaching and research includes two years as a Physics Department Chair and one year as Division Director for the Physical Sciences and Engineering at Sam Houston State University. Wilson has served on many evaluation and accredation teams for high schools and universities, founded the annual Sam Houston High School Science Symposium, and operated workshops for high school teachers and students from 7-18. He is a member of the governing board of the American Institute of Physics, the Steering Committee of the Presidential Awards in Science Teaching, the Organizing Committee of the Triangle Coalition (Business, Labor, and Academia) for Science Education, and the Steering Committees for Conferences on "The Education of Physicists," "Teacher Institutes and Workshops," and "Education for Professional Work in Physics." His research and professional interests have included structure of liquid crystals, Mossbauer spectroscopy of sickle cell hemoglobin, computers in the advanced physics laboratories, training of physics teachers, and other topics in chemical physics and physics education.

Douglas Wylie received his Ph.D. in radiation damage in solids from the University of Connecticut in 1961 following a B.Sc. from the University of New Brunswick and M.Sc. from Dalhousie University. He started teaching physics at the University of Maine in 1951 and has devoted his entire career to teaching. All courses at the undergraduate level have been taught at one time or another. During the early 1970s he was heavily involved with the teacher training institutes sponsored by NSF. Included in all summer and in-service institutes were units on modern physics including special relativity, elementary particles, and nuclear physics. Whenever appropriate, modern topics are used in all courses to illustrate the unified nature of physical principles.

Ted Zaleskiewicz was born and raised in Schenectady, New York, and attended Union College (B.S. 1961). He spent some time in New England (Dartmouth, MA, 1973; Brown, MAT, 1966) and then came to Western Pennsylvania as instructor of physics at the University of Pittsburgh, Greensburg campus. He continued graduate work at Pitt (Ph.D., 1974—"Design, Implementation and Evaluation of an Auto-Tutorial Course Sequence for Secondary School Physics Teachers"). His responsiblity is teaching calculus-based engineering physics.

Dean Zollman, professor of physics at Kansas State University, has been working in the field of physics education for about 15 years. During that time he has developed a number of teaching tools. Much of his recent work has involved the use of interactive video to teach introductory physics. He is the co-author of four interactive videodiscs and a large number of computer-based lessons which use interactive video. He has also developed a physics course for elementary education majors. With Jacqueline Spears, he wrote *The Fascination of Physics,* a textbook for nonscience students. At present he is teaching a modern physics course for students who have completed at least one semester of physics and are not majoring in physics or engineering.

Earl Zwicker has taught physics at the physics department, IIT Center, Chicago, Illinois since 1953. He received his B.S. in chemical engineering at the University of Wisconsin, 1948, M.S. in 1953 and Ph.D. in 1959 at IIT. He has done research in physical electronics, effects of radiation on sliding friction, flash photolysis investigation of energy transfer processes in organic molecules, co-discoverer of hydrated electron. Over the last 15 years his interest has turned to the teaching and learning of physics, and working with both high school and university colleagues in this area. Zwicker believes in a phenomenological or "hands-on" approach. He edits a column titled "DOING Physics" which appears in The Physics Teacher, and writes the informal "ISPP Reminder" for a Chicago area physics group conducting monthly meetings.

Julio Gratton was born in Milano, Italy, on August 24, 1940. He is an Argentine citizen, married to Nieda Esther Padilea, and has no children. He studied physics at Cordoba University and Instituto Balseiro; obtained Licenciado degree in 1962 and Doctor degree at the University of Buenos Aires in 1970. He specialized in Plasma Physics and has done theoretical and experimental research at the Plasma Physics Laboratory at Buenos Aires, which he helped to establish. He has taught physics since 1967. He has been a full Professor of the University of Buenos Aires since 1977. Julio Gratton is a researcher of CONICET, presently is Investigator Principal (Senior Researcher); has been director of the Plasma Physics Laboratory at Buenos Aires; Director of the Physics Department; and held many other academic positions. He is the author of 46 scientific papers; has taught Optics, Thermodynamics, Quantum Mechanics, Laboratory Physics and Mechanics, and post-graduate courses on Plasma Physics. He has been in charge of the Introductory Physics course for students of biology and geology since 1984.

YOU CAN WRITE